Methods, Implementation, and Application of Cyber Security Intelligence and Analytics

Jena Om Prakash
Ravenshaw University, India

H.L. Gururaj
Vidyavardhaka College of Engineering, India

M.R. Pooja
Vidyavardhaka College of Engineering, India

S.P. Pavan Kumar
Vidyavardhaka College of Engineering, India

A volume in the Advances in Information Security, Privacy, and Ethics (AISPE) Book Series

IGI Global
PUBLISHER of TIMELY KNOWLEDGE

Published in the United States of America by
IGI Global
Information Science Reference (an imprint of IGI Global)
701 E. Chocolate Avenue
Hershey PA, USA 17033
Tel: 717-533-8845
Fax: 717-533-8661
E-mail: cust@igi-global.com
Web site: http://www.igi-global.com

Library of Congress Cataloging-in-Publication Data

Names: Jena, Om Prakash editor. | Gururaj, H. L., 1988- editor. | Pooja, M. R., 1982- editor. | Kumar, S. P. Pavan, 1989- editor.
Title: Methods, implementation, and application of cyber security intelligence and analytics / Jena Om Prakash, H.L. Gururaj, M.R. Pooja, and S.P. Pavan Kumar, editors.
Description: Hershey, PA : Information Science Reference, an imprint of IGI Global, [2022] | Includes bibliographical references and index. | Summary: "This book looks at the intersection of Artificial Intelligence and Cyber security which has emerged as one of the most promising use cases for the advancement of Cyber security in the technological age"-- Provided by publisher.
Identifiers: LCCN 2022012051 (print) | LCCN 2022012052 (ebook) | ISBN 9781668439913 (h/c) | ISBN 9781668439920 (s/c) | ISBN 9781668439937 (ebook)
Subjects: LCSH: Computer networks--Security measures--Data processing. | Artificial intelligence.
Classification: LCC TK5105.59 .H3525 2022 (print) | LCC TK5105.59 (ebook) | DDC 005.8--dc23/eng/20220524
LC record available at https://lccn.loc.gov/2022012051
LC ebook record available at https://lccn.loc.gov/2022012052

This book is published in the IGI Global book series Advances in Information Security, Privacy, and Ethics (AISPE) (ISSN: 1948-9730; eISSN: 1948-9749)

British Cataloguing in Publication Data
A Cataloguing in Publication record for this book is available from the British Library.

All work contributed to this book is new, previously-unpublished material. The views expressed in this book are those of the authors, but not necessarily of the publisher.

For electronic access to this publication, please contact: eresources@igi-global.com.

Advances in Information Security, Privacy, and Ethics (AISPE) Book Series

Manish Gupta
State University of New York, USA

ISSN:1948-9730
EISSN:1948-9749

Mission

As digital technologies become more pervasive in everyday life and the Internet is utilized in ever increasing ways by both private and public entities, concern over digital threats becomes more prevalent.

The **Advances in Information Security, Privacy, & Ethics (AISPE) Book Series** provides cutting-edge research on the protection and misuse of information and technology across various industries and settings. Comprised of scholarly research on topics such as identity management, cryptography, system security, authentication, and data protection, this book series is ideal for reference by IT professionals, academicians, and upper-level students.

Coverage

- Security Classifications
- Security Information Management
- Global Privacy Concerns
- Computer ethics
- Internet Governance
- IT Risk
- Device Fingerprinting
- Access Control
- Electronic Mail Security
- CIA Triad of Information Security

IGI Global is currently accepting manuscripts for publication within this series. To submit a proposal for a volume in this series, please contact our Acquisition Editors at Acquisitions@igi-global.com or visit: http://www.igi-global.com/publish/.

The Advances in Information Security, Privacy, and Ethics (AISPE) Book Series (ISSN 1948-9730) is published by IGI Global, 701 E. Chocolate Avenue, Hershey, PA 17033-1240, USA, www.igi-global.com. This series is composed of titles available for purchase individually; each title is edited to be contextually exclusive from any other title within the series. For pricing and ordering information please visit http://www.igi-global.com/book-series/advances-information-security-privacy-ethics/37157. Postmaster: Send all address changes to above address.

Titles in this Series

For a list of additional titles in this series, please visit: www.igi-global.com/book-series

Information Security Practices for the Internet of Things, 5G, and Next-Generation Wireless Networks
Biswa Mohan Sahoo (Amity University, India) and Suman Avdhesh Yadav (Amity University, India)
Information Science Reference • © 2022 • 325pp • H/C (ISBN: 9781668439210) • US $250.00

Global Perspectives on Information Security Regulations Compliance, Controls, and Assurance
Guillermo A. Francia III (University of West Florida, USA) and Jeffrey S. Zanzig (Jacksonville State University, USA)
Information Science Reference • © 2022 • 309pp • H/C (ISBN: 9781799883906) • US $240.00

Handbook of Research on Cyber Law, Data Protection, and Privacy
Nisha Dhanraj Dewani (Maharaja Agrasen Institute of Management Studies, Guru Gobind Singh Indraprastha University, India) Zubair Ahmed Khan (University School of Law and Legal Studies, Guru Gobind Singh Indraprastha University, India) Aarushi Agarwal (Maharaja Agrasen Institute of Management Studies, India) Mamta Sharma (Gautam Buddha University, India) and Shaharyar Asaf Khan (Manav Rachna University, India)
Information Science Reference • © 2022 • 390pp • H/C (ISBN: 9781799886419) • US $305.00

Cybersecurity Crisis Management and Lessons Learned From the COVID-19 Pandemic
Ryma Abassi (Sup'Com, University of Carthage, Tunisia) and Aida Ben Chehida Douss (Sup'Com, University of Carthage, Tunisia)
Information Science Reference • © 2022 • 276pp • H/C (ISBN: 9781799891642) • US $240.00

Applications of Machine Learning and Deep Learning for Privacy and Cybersecurity
Anacleto Correia (CINAV, Portuguese Naval Academy, Portugal) and Victor Lobo (Nova-IMS, Naval Academy, Portugal)
Information Science Reference • © 2022 • 315pp • H/C (ISBN: 9781799894308) • US $250.00

Cybersecurity Capabilities in Developing Nations and Its Impact on Global Security
Maurice Dawson (Illinois Institute of Technology, USA) Oteng Tabona (Botswana International University of Science and Technology, Botswana) and Thabiso Maupong (Botswana International University of Science and Technology, Botswana)
Information Science Reference • © 2022 • 282pp • H/C (ISBN: 9781799886938) • US $240.00

Advances in Malware and Data-Driven Network Security
Brij B. Gupta (Asia University, Taiwan)
Information Science Reference • © 2022 • 304pp • H/C (ISBN: 9781799877899) • US $215.00

IGI Global
PUBLISHER of TIMELY KNOWLEDGE

701 East Chocolate Avenue, Hershey, PA 17033, USA
Tel: 717-533-8845 x100 • Fax: 717-533-8661
E-Mail: cust@igi-global.com • www.igi-global.com

Table of Contents

Detailed Table of Contents

attacks also pose a major threat in daily activities. Being able to recognize the cyber security challenges, one must be able to secure themselves at both individual and organization levels. In this chapter, the authors deal with the continuously changing nature of cyber-attacks that has enhanced the difficulty of handling and avoiding emerging threats along with technological solutions that can address the concern. The techniques for demands of privacy and protection will also be discussed.

Digital protection is the act and protocol to protect personal computers, mobile, electronic devices, and data from malicious attacks. Artificial intelligence (AI) and machine learning (ML) are presently essential for our daily existence, and this incorporates network protection. AI/ML can recognize vulnerabilities and lessen occurrence reaction time. Due to this unprecedented test, AI-based instruments for network safety have arisen to help data security groups decrease break chance and further develop their security to act proficiently and successfully. Simulated intelligence and AI developed basic advances in information security, as they can quickly examine a huge quantity of datasets and recognize various kinds of network stranded and breaches – from malware, one can pursue advantage to distinguishing unsafe comportment which might prompt a phishing attack or downloading malicious program. The advancements learned over the system can be used to identify new types of attacks in the future.

In this digitized world, everything is changing from offline to online. Data plays a vital role in this digital network. The theft or loss of USB devices, computers, or mobile devices by an unauthorized person who gains access to your mobile or laptop devices, email account, or network is generally termed as a data breach. Securing data from theft and breaches is a challenging issue. It is very hard to identify data breaches in complex networks. Adding extra intelligence using machine learning (ML) approaches will be efficient in identifying such attackers. In this chapter, various ML techniques to identify data breaches such as malware attack, man in the middle (MIM), spear phishing attack, eavesdropping attack, password attack, cross-site scripting attack will be depicted with suitable case studies.

Cyber threat intelligence (CTI) has emerged as a critical pillar in a well-developed cyber security strategy. When used correctly, threat information may assist security teams in defending against an ever-more sophisticated threat landscape before, during, and after an attack. Groups can design more effective, more delicate, and durable cyber defenses by evaluating attackers and understanding their methods and aims. As a result, the purpose of this chapter is to give an overview of how CTI promotes cyber resilience by utilizing intelligent technologies such as artificial intelligence (AI) and machine learning (ML).

Smart grids use the potential of information technology and advanced analytics to intelligently supply energy via bidirectional communication while also meeting environmental criteria through the facilitation of green technology integration. Communication technology's fundamental vulnerability has exposed the system to a slew of security concerns. Several survey publications have addressed these issues and proposed solutions. The majority of these publications, however, categorized assaults based on confidentiality, integrity, and availability, but they did not include responsibility. Furthermore, present solutions are focused on preventing individual assaults or defending certain components, but there is no comprehensive method to securing the complete system. The authors evaluate the security requirements, describe some serious cyber-attacks, and present a cyber-security approach to identify and counter these assaults in this work.

We all know that every positive integer has a unique Fibonacci representation, but some positive integers have multiple Gopala Hemachandra (GH) representations, or some positive integers haven't any GH representation. Here, the authors found the first k-positive integer k=(3 2^((m-1))-1) for which there is no Zeckendorf's representation for Gopala Hemachandra sequence whose order m. Thus, the authors formulated the first positive integer whose Zeckendorf's representation can't be found in terms of its order. The authors also described the fourth, the fifth, and the sixth order GH representation of positive integers and obtained the fifth and the sixth order GH representations of the first 26 positive integers uniformly according to a certain rule with a table. Finally, the authors used these GH representations in symmetric cryptography, and the authors made some applications with a method which they construct similar to Nalli and Ozyilmaz.

The secured data hiding for future transfer is an application proposed for a secured data transmission in which the application alters the originality of the file to an encrypted form and embeds that file into a video file which will be transferred. The motive of the application is to provide the ease for the user to pass the information with the encryption standards and algorithms then store the information in an unreadable format. In the next phase, the application has a reverse process where the embedded video can be de-embedded, and the encrypted file can be decrypted to its actual data upon proper user request. In the process of encryption and decryption, the application should confirm the authenticity of the user. The application is generated using the tiny encryption algorithm to encrypt the data. This algorithm is a 64-bit block cipher with a variable length key. The tiny encryption algorithm is used because requires less memory, and as a compliment, it uses only simple operations; thus, it is easy to implement.

In our daily lives, we conduct billions of payment transactions, yet each payment method requires the transport of a substance. It is typical for users to have a variety of payment materials on hand, such as cash, credit cards, and even mobile phones. Meanwhile, these goods are easily stolen or misplaced. People suffer enormous trauma as a result of these incidents. This chapter details a biometric payment application created to introduce the concept of hardware less payments. It allows users to pay anytime and anywhere by registering their finger without any hardware. This involves registering user information once, and then all subsequent transactions are confirmed and processed by the user's fingerprint, with the application managing the entire process. This solution creates a new payment option and eliminates the risk of shipping valuables abroad. For the company, this application enables an efficient and secure payment system.

The basic utilities such as electricity, gas, and water are the types of services that can be commercialized and delivered at the doorstep of a user. The user is not concerned about the location of the service provider or the method of the service delivery. Computing is a new utility whose nature has been intensely transformed to a traditional utility where users can utilize the services as per their requirement and irrespective of their location. The significant advancement in information and communication technology needs a type of computing, which is required to meet the everyday demands of the generic computing class of users. There are many computing models such as cluster computing, grid computing, cloud computing, and many more which have been used in solving complicated problems of science and engineering. This chapter discusses the latest computing platforms which can provide fundamental computing services to users for fulfilling their computational needs.

Android OS powers the majority of the market share. Malware acts as stimuli to the vulnerabilities in Android devices as it affects a huge amount of user data. Users' data is at high risk when it comes to attacks through varied types of malware. Also, mutations in malware have brought up newer variants in them. Malware families have been expanding, thereby making analysis and classification diverse. Mainly classified into static, dynamic, and alternative or hybrid analysis, the field of malware analysis is facing many repercussions. The development of malware is endless and hence calls for intelligent and self-learning approaches in this regard. However, more distinct techniques are in need and can be served by integrating intelligent and analytical capabilities. This chapter involves a fourfold approach with major contributions to review existing Android malware analysis techniques, intelligent techniques for Android malware detection, determination of future challenges and need of security in this direction, and finally, analyzing possible defense mechanisms possible in this regard.

A network intrusion detection system (NIDS) has a significant role in an industry or organization to protect their data. NIDS should be more reliable to manage huge traffic over the networks to detect the emerging attacks. In this chapter, novel entropy-based feature selection is proposed to select the important features of intrusion detection system. Feature selection reduces the computational time and improves detection rates. In entropy, within-class entropies and between-class entropies are computed for the various classes of intrusion in the KDD dataset. Based on computed entropy values, features are ranked and selected. Radial basis neural network (RBNN) is employed as a classifier. Performances of the proposed entropy-based feature selection algorithm are evaluated using the 10% dataset for training and two other datasets for testing. The proposed system shows significant improvement in the detection rate, reduces the false positive rate (FPR), and also reduces the computational time.

Online social networks (OSNs) are increasingly influencing the way people communicate with each other. Well known sites such as Facebook, LinkedIn, Twitter, and Google+ have millions of users across the globe. With the wide popularity there are lot of security and privacy threats to the users of online social networks (OSN) such as breach of privacy, viral marketing, structural attacks, malware attacks, and profile cloning. Social networks have permitted people to have their own virtual identities which they use to interact with other online users. It is also completely possible and not uncommon for a user to have more than one online profile or even a completely different anonymous online identity. Entity resolution (ER) is the task of matching two different online profiles potentially from social networks. Solving ER has an identification of fake profiles. The solution compares profiles based on similar attributes. The system was tasked with matching two profiles that were in a pool of extremely similar profiles.

Preface

The AI technologies of recent days would enable discovery of endpoint vulnerabilities, protect corporate data, and uphold compliance, with containment plus identity and patch management that most of the industries are aiming at. AI security solutions would play a major role in safeguarding customers' data and are expected to get stronger and smarter over time. This book would facilitate the development of contextual analytics that would sensitize the security events while protecting your endpoints, users, apps, docs and data from one platform. This is the motivation for the edited book.

Chapter 1 explores the fundamental idea towards cybersecurity and cyber intelligence. These chapters are written for educators, practitioners, professionals who want an overview of cyber security which provides insights of understanding of cyber security and its impact on economy. This chapter walks through foundation of cyber security by highlighting its framework, challenges, other application areas of cyber security which transforms society and so on.

Chapter 2 describes the Emerging Trends in Cyber Security. The increasing sophistication of these attacks has prompted numerous organizations to crack down on cybercriminals, resulting in the back-and-forth that characterizes today's cybersecurity trends. For-profit and charitable businesses alike are losing billions of dollars due to security flaws. The bulk of corporations, institutions, and other working professions are being pushed to migrate to remote work by Covid-19. To keep the job going, many new enterprises, institutions, and IT farms must shift to remote working areas quickly and without sufficient planning. This unanticipated movement allows security measures to be bypassed, increasing risk and susceptibility. It has become fashionable to circumvent security measures to gain speedy access, hence increasing cybercrime.

In Chapter 3 the major role of AI-ML Techniques in Cyber Security is presented. Anomaly detection can be used by AI and ML approaches to detect suspicious behavior such as unidentified devices joining a network, anomalous network traffic, and so on. Furthermore, AI and machine learning may detect host-based anomalies such as high CPU consumption, which can indicate the presence of malware.

In Chapter 4 brief out how Machine Learning (ML) Methods are used to Identify Data Breaches. Although we hear a lot about major cybersecurity breaches in non-insurance organizations – Target, Experian, the IRS, and so on – there have also been breaches in the insurance industry, albeit in a less publicized manner. Regardless of how hard organizations try to secure their data and systems, hackers' methods of breaching continue to evolve. This is why risk management and insurance innovation are so critical. Machine learning can collect and process massive amounts of data, analyze historical cyber-attacks, predict the types that may occur, and set up defenses against them.

In Chapter 5 brief out How Cyber Threat Intelligence (CTI) ensures Cyber Resilience using Artificial Intelligence and Machine Learning. Cyber Intelligence is the ability to prevent or mitigate cyber-attacks by studying threat data and providing information on adversaries. It aids in the identification, preparation, and prevention of attacks by providing information on attackers, their motivations, and their capabilities.

Chapter 6 summarizes the role of Cyber-Security in Smart Grid. The smart grid is a communication and information technology-based system for the generation, delivery, and consumption of energy power. Device protection is the third critical component in the smart grid security supply chain. Many research papers and recommendation reports have been published, all of which contribute to endpoint security assurance. Several security technologies, including host IDS, anti-virus, and host data loss prevention, are recommended.

In Chapter 7 The Fifth and The Sixth Order Gopala Hemachandra Representations and The Use of These Representations in Symmetric Cryptography is presented. GH codes are a variant of the Fibonacci universal code that finds use in cryptography and data compression. The most common application of cryptography in electronic data transmission is to encrypt and decrypt email and other plain-text messages.

Chapter 8 emphasizes on handling secured data transmission with the motive of the application to handle data encryption and hiding in an effective way. In the next phase, the application has a reverse process where the embedded video can be de-embedded, and the encrypted file can be decrypted to its actual data upon proper user request. A discussion on the role of the application in performing a confirmation on the authenticity of the user with proposed file encryption and decryption is done.

Chapter 9 discusses a biometric payment application that paves a way to overcome material-based payment. The application combines all crucial upsides and eliminates most drawbacks discussed in the literature review. The application uses automation to solve the problem of the material dependency. The application makes the users less worried about safeguarding their wallet or carrying the exact change to places. The overall user experience is made comfortable for both the payee and the payer.

Chapter 10 discusses the latest trends in the computing platforms which can provide fundamental computing services to the users for fulfilling their computational needs. Also, it details various limitations or issues related to these computing platforms along with their possible solutions with special focus on complex engineering problems.

Chapter 11 discusses data-driven android malware analysis intelligence approaches in android operating systems leading to invasion in mobile systems wherein personal or organizational or financial information may be at risk. Existing security mechanisms involve static, dynamic and hybrid approaches. However, there are many challenges and issues that need to be addressed in this direction. This offers a detection and mitigation area in this sector. The chapter deals with how machine learning-based approaches serve as the prime solution to be used to track such behavior and create alerts regarding suspicious activity. Intrusion detection systems based on apps are solely concerned with a single application and detect assaults directed at it.

Chapter 12 proposes a Entropy Based Feature Selection for Network Intrusion Detection System. By lowering the number of features, entropy-based feature selection is presented for reducing computational time and enhancing detection rate. The radial basis function has been experimented which has been effectively used in a variety of benchmarks.

Chapter 13 proposes a solution for recognizing user portraits for fraudulent identification on online social networks by deploying Entity Resolution and uses it to compare online user profiles from social networks in order to identify matches. The systems opt to compare the two images and try to distinguish between fake and real entities. The solution compares profiles based on similar attributes. The system was tasked with matching two profiles that were in a pool of extremely similar profiles.

Jena Om Prakash
Ravenshaw University, India

H. L. Gururaj
Vidyavardhaka College of Engineering, India

M. R. Pooja
Vidyavardhaka College of Engineering, India

S. P. Pavan Kumar
Vidyavardhaka College of Engineering, India

Acknowledgment

The world is a better place thanks to people who want to develop and lead others. Writing a book is harder than I thought and more rewarding than I could have ever imagined. None of this would have been possible without the contributors who contributed various chapters across the globe for this book. Without the experiences and support from my co-editors Dr. Pooja M R, Mr. Pavan Kumar S P and Mr. Om Jeena this book would not exist.

H. L. Gururaj

Chapter 1
Introduction to Cyber Security

Sudeep Jadey
https://orcid.org/0000-0001-7427-4555
NIE Institute of Technology, India

Girish S. C.
NIE Institute of Technology, India

Raghavendra K.
NIE Institute of Technology, India

Prasanna Kumar G.
https://orcid.org/0000-0002-8962-8034
NIE Institute of Technology, India

Srinidhi H. R.
https://orcid.org/0000-0002-8727-253X
NIE Institute of Technology, India

Anilkumar K. M.
JSS Science and Technology University, India

ABSTRACT

The rapid growth of the internet and the technological revolution in the information and communication sectors has created a gateway for users to connect online. With digital information being stored, security and privacy has been a major concern around the world to protect vital data. The modern tools and techniques used by cybercriminals perform malicious activities which aim at disrupting, distorting, and stealing sensitive data for self-financial gains. More complexities were added to the cyber world with the emergence of AI and ML algorithms in the smart phones. Cybercriminals take advantage of this barrier and build their network to commit cybercrimes across the globe. This chapter focuses on an audience that includes researchers, professionals, academicians and UG, PG students who are interested to know more about cybersecurity. This chapter aims to provide a brief overview of cybersecurity, cyber laws, cyber-attacks and security tools, objectives of cybersecurity, applications of cybersecurity, and so on. Finally, the future of cybersecurity along with some case studies are discussed.

DOI: 10.4018/978-1-6684-3991-3.ch001

INTRODUCTION TO CYBER SECURITY

The Internet has become one of the most important useful source of information for cyber users. The expansion of cyberspace in the digital world has created numerous opportunities and challenges for cyber users to exchange information across the globe. The advent of computers and communication technologies has open the doors for malicious users to weaken security systems of the society. Today, computing devices like laptops, tablets, and smart phones has become a crime instrument for malicious users. Every computing devices is connected to the networks which can access and transmit the data anywhere, making it more vulnerable to cyber-attacks. More complexities were added to the cyber world with the emergence of Artificial Intelligence & machine learning algorithms. Further, Computers with wireless networking creates space for free malicious access. With digital information being stored, security & privacy has been a major concern around the world to protect vital data. The sophisticated modern tools and techniques used by cybercriminals to perform malicious activities aims at disrupting, distorting & stealing sensitive data for self-financial gains. Cybercriminals take advantage of this barrier & build their network to commit cybercrimes across the globe. These cybercriminals create barriers to innovations, financial, economic growth and free flow of information. Thus there is a need to secure these system. The main goal of cyber security is to protect valuable assets. These assets could be hardware, software, network and data (Priyadarshini, I. 2019).

Imagine that you are travelling in a bus having a smart phone with Bluetooth enabled unknowingly in it. A nondescript guy who is sitting away few distance in the same bus might be fiddling with his android phone which syncs with your device through Bluetooth & load the malware into it. All your phone contacts will get an offensive messages which were not sent by you but are from your phone. This is one classic example of security breach through smart devices.

Today, new opportunities are open up for wireless network devices which can be hacked. The airports, restaurants & hotels, libraries, coffee shops & other public places are the prime target sites for cybercriminals to hack assets through wireless networks. Even though these sites are password protected, still it is dangerous because there is slight control who has gained the password. On a wireless network, anyone can listen in on anyone else on the network. A victim in a public place has got no clue who is nearby with respect to wireless network (Waschke, M. 2017, Herrmann, D., & Pridöhl, H. 2020). So let's take a brief overview about cybersecurity

What is Cybersecurity?

Cybersecurity may be defined as the ability to protect and recuperate from cyberattacks. According to NIST (National Institute of Standards & Technology), it can be defined as the ability to defend cyberspace usage from cyberattacks. Cyberspace could be internet, computer systems, telecom networks, embedded controllers etc. The security of any organization completely relies on three key areas namely confidentiality, availability and integrity.

1. **Confidentiality:** The word confidentiality looks alike privacy. The key idea is to prevent unauthorized users from accessing the sensitive information. Confidentiality makes sure that only authorized users are given permission to access sensitive information. Identity theft, credit card fraud, phishing, wiretapping are some examples of confidentiality attacks

2. **Availability:** Availability refers to the genuine access of resources. The information should be made available to the authorized personal. Denial of service attack, Internet Control Message Protocol flood attacks are some examples of availability attacks.
3. **Integrity:** Trustworthiness, Data consistency and accuracy are all ensured by integrity. The main goal is to prevent data modification who aren't authorised to do so. Man in middle attacks, salami attacks and session hijacking are some examples of Integrity attacks. (Priyadarshini, I. 2019)

What Cybercriminals are looking for?

Dark web has created wide range of opportunities for cybercriminals to explore & invade the computing devices to steal assets. The main intent of cybercriminals is to extract money through digital devices i.e. gain access & transfer money from one accounts to another, steal bank accounts, personal & credit card information. Not all cybercriminals have the intent to make money. Some computer enthusiastic have pleasure snooping into computer networks, while others invade with intelligence motive (Waschke, M. 2017).

For Example: Steal sensitive information from government and private firms, steal information from Research & development innovative products.

How do Cybercriminals Get In?

The rapid growth of internet & communication technologies has opened a lot of opportunities for the cybercriminals to invade the computing devices & obtain information. The ripples of attacks are emerging & spreading like waves. The internet usage has become necessity medium for part of our lives. A small system loophole is enough for cybercriminals to take away everything from victims. There are few sources through which cybercriminals sneak into system which are as follows. (Waschke, M. 2017)

1. Social Engineering
2. Remote Access Tools
3. Man in Middle
4. Trojans
5. Botnet Invasions & Denial of Services
6. Password cracking

Objectives of Cybersecurity

Cybersecurity strives to secure cyberspace and safeguard the infrastructures by recovering from incidents of cyberattacks and take appropriate response & recovery resolutions. A secure cyberspace ensures a legal framework to prevent cyber threat & protect cyberattacks. The following are few objectives listed below.

1. **Prevent Threats:** It is critical to assess attacks and verify the design, implementation, and operation of essential network control protocols to prevent threats. Identification of threat indicators and incident reporting guidelines must be established. Certain dangers are avoided by following best practices & spotting malicious technologies through extensive research.

2. **Identify threats & system hardening:** The prime goal of cybersecurity is to harden the system by identifying threats at initial stages. It involves threat assessment & adopting security measures. Mitigate certain risk associated with security is the prime purpose of system hardening.
3. **Contingencies preparation:** The concept of contingency planning entails being ready for cyberattacks. It contains policies, procedures, best practices and plans for recovery.
4. **Information Allocation:** The allocated information that is disseminated across the system must be effective. Issuing alerts could be used to report vulnerabilities, cyber threats and other incidents. The data can be disseminated successfully across multiple platforms.
5. **Specialized training:** A specialized security training has to be provided to the every personnel. Information services must be available to federal partners in order for the workforce to be ready in the event of cyber disaster.
6. **Fault tolerance of System:** Vulnerability assessment can be used to calculate a system's fault tolerance. A high processing systems may be able to withstand different types of cyberattack.
7. **Vulnerability reduction:** To reduce vulnerabilities, many security practices like usage of strong passwords and firewalls prevents malicious access to systems.
8. **Usability Improvement:** The term usability refers to how simple something is to use. Along with trustworthy technology, usability requirements should build into the systems.
9. **Cyberspace Authentication:** Identity verification of user is an important procedure in cybersecurity. Authentication like one factor, or multifactor to be set up depending on the device.
10. **Automate Security methods:** Automation leads to increased efficiency, better behaviour prediction, & quicker execution. Cyberattacks can be avoided by implementing automation. Automation can correlate data, encourage faster prevention than attack spread & detect infections/anomalies in network.
11. **Establish Security measures:** Cyberattacks can be detected, corrected & prevented by establishing security measures. Use of firewalls, access control, secure remote access, password protection, network segmentation, establish policies, create training programs are some examples of security measures (Priyadarshini, I. 2019,Waschke, M. 2017)

Challenges of Cybersecurity

Today individuals, enterprises and governments are all concerned about cybersecurity. Cybersecurity is one such concept that can assist firms in becoming more competitive. Cybersecurity is all about taking preventive measures before any threat exploits systems. In this era of digitization, protecting privacy from cybercriminals plays an important role as hackers are getting smarter every day (Taddeo, M. 2019). According to (Isha, U. 2020), in terms of cyber-attacks, India ranks 11th globally and has witnessed 2,299,682 incidents. Let's take a look at some of the cybersecurity challenges

1. **Potential misuse of 5G Networks:** Today youths are more curious about the developments of 5G networks. The effective usage & adoption of the gadgets by the current generation makes it more vulnerable to cyberattacks. It may be emotional or physical attacks. The cybercriminals enter into wireless 5G networks through endpoints containing complex architecture & misuse data using these speedy gadgets. The network attack surface can be exploited easily by cybercriminals which has more traffic routing points in 5G dynamic software systems. Thus, there is a need to find the

identities of unauthorized access of these endpoints & securing the privacy of network to these attacks.

2. **Artificial Intelligence controlling cybersecurity:** Artificial Intelligence has changed the world by not acting on the defensive side but also on the offensive side. Artificial Intelligence tools are being adopted by health care industry & supply chain sectors. The fundamental flaw in adopting AI system is that passwords & biometric logins are constantly changed by patients, distributors & other supply chain players. Hackers/cybercriminals may then target the pain points allowing them to control the surveillance of personal information such as address, bank account information & so on. Health care & supply chain businesses are sensing malware threats, with ransomware undermining the incentivize growth, because AI tools can execute with least human input in real time.
3. **Popularity growth of IOT devices:** The use of IOT devices is becoming popular due to their quick response times and lower processing cost which benefits the cyber technology. The popularity of these devices has created opportunities for cybercriminals to acquire the control of expanding business operation within an organization. The IOT devices once deployed in the physical environment are seldom updated which increase chances of being a victim for cyberattacks. (Gupta, A. 2021)
4. **Lack of expertise:** Skills are the cornerstone of successful organization developments. To be successful, businesses must adopt cybersecurity to improve their cybersecurity infrastructure. Organization should have a proactive plan of business to maintain & build its own cybersecurity workforce. Organization should explore nurturing security talent as expertise becomes increasing difficult to find and retain. It is important to employ experienced professionals & provide long term benefits. (Pipikaite, A., Barrchin, M., & Crawford, S. 2021, Tyagi, R. 2020)

DEVELOPMENT HISTORY

Many people are unaware that internet and cybersecurity existed long back ago. Many species evolved at the same time, each attempting to gain advantage over the other. Cyber criminals who attempt to exploit the system flaws for their personal benefit have matured alongside with evolvement of technology. This section throws light about the developments of cybersecurity which took over time

a) **Early 1940's:** In 1943, the first digital computer was created and carrying out cyberattacks was tricky for about decades. The massive electronic devices were only accessible to small number of people and these devices were not networked. Most of these devices were noisy, large & difficult to use. A few experienced professional knew how to use them and the threat was minimal or non-existent. Yet, by the end of the decade, several people have evolved virus theories, but Jon Von Neumann speculated that these computer programs could reproduce and damage the devices. These programs has the ability to replicate itself in the same way that existing virus does. It also has the potential to spread it to new host devices. Later in 1966, he developed theory of "self-reproducing automata" and published a paper on it.

b) **Early 1950's:** Hacking didn't start out to gather data via computers. The origins of computer hacking may be more related to early telephone usage. This was clear in the 1950's, when the practice of "phone phreaking" became popular. People who are phone phreaks have a strong interest in how phones work. These people attempted to exploit the protocols in place that allowed engineers to work remotely on the network. It enabled people to make free calls and pay lower long distance tolls. For some time, this practice continued. Unfortunately for the phone companies, there was no way to stop the phreaks albeit the practice did fade away in 1980.There are reports that apple founders, Steve Jobs and Steve Wozniak were interested in phone phreak community . Apple computers later adopted digital technologies based on similar notions.

c) **Early 1960's:** The earliest mention of malicious hacking occurred in the student newspaper at Massachusetts Institute of Technology. Even the mid of 1960, the majority of the computers were massive mainframes housed in temperature controlled rooms. Because these computing devices were so expensive, even the programmers had the limited access. Early hacking aimed at gaining access to the existing systems and causing trouble so as to improve it. New, faster and more efficient hacking methods emerged over time. In 1967, a group of students was invited inside IBM's office to try out the designed computer. Students worked deeper learning system languages & gained access to various parts of the system. This gave IBM with insight into the system vulnerabilities. As a result, a defensive mindset developed, implying that computers require a security measure to keep cybercriminals out. This could have been the industry first instance of ethical hacking. This was the significant step in the establishment of cybersecurity plans. The size reduction & cost of computers made large organization to invest in technologies for storing and managing of data systems. It didn't seem beneficial or workable to lock the computers in a room. A work access is needed to many employees which resulted in development of computer access & use of passwords.

d) **Early 1970's:** In 1970 the real birth of cybersecurity occurred. A research project called Advanced Research Project Agency Network (ARPANET) was developed to provide network connectivity prior to the internet. This project developed protocols for distant computer networks. Researcher Bob Thomas invented a computer software called **"Creeper"** that wander around this network that leaving a trail whenever it went which read **"I am the creeper, catch me if you can".** Another Researcher, Tomlinson Ray (Email Creator) developed first antivirus software program called **"Reaper"** (self-replicating worm) which chased and removed Creeper. Figure 1 shows the taunting message of creeper program.

Figure 1. Taunting message of Creeper Program
(Chadd, K. 2020, Sentinel, O. 2019)

```
BBN-TENEX 1.25, BBN EXEC 1.30
@FULL
@LOGIN RT
JOB 3 ON TTY12 08-APR-72
YOU HAVE A MESSAGE
@SYSTAT
UP 85:33:19    3 JOBS
LOAD AV    3.87    2.95    2.14
JOB TTY   USER       SUBSYS
1   DET   SYSTEM     NETSER
2   DET   SYSTEM     TIPSER
3   12    RT         EXEC
@
I'M THE CREEPER : CATCH ME IF YOU CAN
```

In this emerging technologies, challenging vulnerabilities becomes more important as most of the organization started to use cellular phone to build isolated networks. Each connected chunk of hardware created a new access point that needs to be secured. With increased computers & growth of network, it became evident for the regimes that security was crucial, & illegal access to data & systems could be disastrous. Early computer security developed by Electronic System Division (ESD) and Advanced Research Project Agency (ARPA) in collaboration with US Airforce & other organization produce a safety kernel for Honeywell Multics (HISlevel68) PC systems. The project at ARPA looked at operating system security, identifying alternative techniques for identifying software weaknesses. In 1979, a 16 year old cybercriminal Kevin Mitnik got arrested for hacking into ARK. It was massive system used for developing operating system which was located at Digital Equipment Corporation (DEC). He was arrested and jailed for making copies of the software. This was the classic example of the first cyberattack

e) **Early 1980's:** The advent of cyber-attacks brought numerous problems to the computer network. In the 1980, high profile attack become more common, including those at Los Alamos National Laboratory, AT&T and National CSS. It wasn't until 1983, that the new terms for these attacks were coined. **"Computer virus"** and **"Trojan Horse"** were one among them. The risk of cyberespionage evolved throughout the cold war. In 1985, The US Defence department created the Trusted Computer System Evaluation Criteria which was called "Orange book". It aims to assess the degree of trust that is put into software that uses any type of sensitive information. In the 1986, a German hacker named Marcus Hess utilized cyberspace gateway in Berkeley, California to connect to the ARPANET. He attempted to sell intelligence to the KGB by hacking 400 defence systems which includes mainframes at the Pentagon. From this point forward, security became more of a priority.
f) **Early 1980's:** Although there are different claims about who invented the first antivirus product, 1987 was the birth year of the commercial antivirus. At the same time, kai Figge and Andreas Luning developed the first Atari ST antivirus solution and released ultimate virus killer. The early version of the NOD antivirus was created by three Czechoslovaks. John McAfee created McAfee software & published virus scan in United States. The same year, a German researcher "Bernd Fix" nullified the infamous Vienna virus (an early example of malware that transmitted & corrupted data). The encrypted virus (cascade) infected the ".COM files" for the 1st time. Cascade created a significant issue in IBM's Belgian office a year later, which sparked the company antivirus development of product. Previously, any antivirus solutions produced by IBM were restricted to internal use. Figure 2 shows the falling text of cascade virus program at the bottom of the screen.

In 1988, several antivirus companies had started around the world which includes "Avast". Simple scanners were used in the early days of antivirus software to find unique viral code by doing context searches. Many of these scanners contained immunizers which tricked viruses to think that computer had already infected & hence not to attack it. Soon, these immunizers became ineffective as the number of virus grew in size

g) **Early 1990's:** The internet's amazing growth and development spanned the entire decade. In 1990, the first polymorphic viruses was created. It contains codes that mutated by keeping the new algorithm intact to avoid detection. When British computer magazine "PC Today" released a free CD edition that contained Disk Killer virus, which infected ten thousand machines. Antivirus was developed which used a signature based, system binaries to a virus signature database. As a result, early antivirals produced a high number of false positives & used lot of processing power that annoyed customers and hampered production. As more antivirus scanners became available, attackers retaliated & the first antivirus was released in 1992. New malware and viruses have exploded, which used innovative techniques and methods to set new challenges for stealing information. Somewhere in the mid of 1990, the first firewall programme was created by a NASA researcher who modelled it after the physical structures that prevent fires from spreading in buildings. The Melissa virus got released in the year 1999. It infiltrated the users machine using a word document, then sent copies of itself to the first 50 email accounts in Microsoft Outlook.

Figure 2. Falling text of Cascade virus (Chadd, K. 2020, Sentinel, O. 2019)

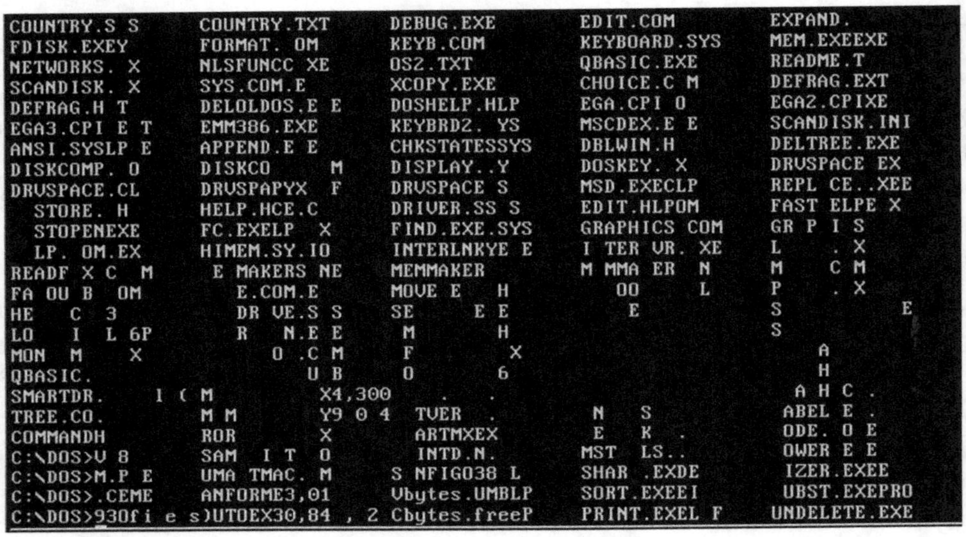

h) **Early 2000's:** The availability of internet in homes and offices across the globe created opportunities for the cybercriminals to exploit more no of devices and software vulnerabilities to target than ever before. There was more to plunder as most of data kept online. In 2001, a new type of infection emerged. Users did not need to download the files. Viewing an infected website was sufficient, as cybercriminals swapped clean webpages with infected ones on legal websites. These types of hidden malwares was damaging which even infiltrated instant message services. The protagonist were hard on their trail as crime organizations began to fund professional cyberattacks. In 2000, the Open Antivirus project releases the first open source antivirus engine. In 2001, ClamAV & Avast launched free open source antivirus software engines that was commercial with full featured security solution. Antivirus software has the drawbacks of slowing down the computer performance. One solution was to move the software to the cloud. In 2013 & 2014, cybercriminals broke into yahoo accounts. Further, 3 billion people's yahoo accounts got hacked, which contained personal information. The cybercriminals used phishing methods that created opportunity through backdoor access. In 2017, a ransomware called "WannaCry" infected 230,000 devices in a single day. A year later, in 2019, the stock market in New Zealand shutdown temporary due to many DDOS attacks. The ongoing digitization of life aspects offered cybercriminals to exploit new opportunities. As cybersecurity evolved to address the growing range of attack types, cybercriminals responded with their own innovation such as multi-vector attacks and social engineering. New solutions for inimitable threats were created which includes use of new technology. (Chadd, K. 2020, Sentinel, O. 2019)

CYBER LAWS, CYBER-ATTACKS AND SECURITY TOOLS

The advent of computer has made human life easier. Computers are now used for various purposes, ranging from individuals to large firms across globe. Computers are utilized for flawed purposes, either for personal or for benefits of others. This led to the development of cybercrime. Technological advancements in cyberspace have led to the opening of doors for cyber laws to prevent malicious activities in cyberspace (Farah Minhaj Aditya Agrawal, U. K. 2021). This section explores on importance of cyber laws in digital space, different types of cyber-attacks & security tools.

Cyber Laws: Cyber law had created to put an end to the digital crimes which occurs in cyberspace. In short, a law that governs computers and internet is called cyber-law. Cyber Law is a branch of law that deals with legal issues that arise from the use of interconnected information technology. The expansion of ecommerce has prompted the need for more active and effective regulatory policies to enhance the legal infrastructure that is important to its success. All these legal structures & governing mechanism comes under cyber law domain. (Farah Minhaj Aditya Agrawal, U. K. 2021).

1. Why Cyber Law is Important?

In this modern era of technology, cyber law is extremely significant. Cyber law is important because all aspects of transactions & activities involving World Wide Web, internet and cyberspace are touched. Every action and reaction in cyberspace has some legal perspectives.

2. Why Cyber Law required in India?

When the internet was invented, the founders of internet had no idea that it could transform itself into all-pervading revolution that might be used for criminal activities which required regulations. Due to anonymous nature of internet, there are many distressing things happening in cyberspace. People with intelligence have been exploiting the facet of internet to perpetuate illicit actions in cyberspace. Thus there is a need for cyber law in India. (Farah Minhaj Aditya Agrawal, U. K. 2021).

3. Awareness programs on Cyber Law?

Transacting over the internet has far reaching legal ramifications since it disrupts the commercial traditional practices. The legal issues of e-transactions must be addressed from the onset in-order to build enduring relationships with online customers. To keep informed about cybercrime, one should have the following knowledge:

a) The Cyber law should be thoroughly read.
b) Understanding of the Internet & its security is essential.
c) Examine the cases of cybercrime. By reviewing those cases, one can learn how to avoid committing similar crimes.
d) Protection of data or sensitive information can be done from a trusted site or trusted application.
e) The impact of technology on crime. (Tripathi, V. 2021).

4. The (IT Act 2000) Information Technology:

The constitution of India has made several averments by introducing the Information Technology Act (IT Act or ITA-2000) at the parliament house on 17th October 2000. The prime goal of this Act is to provide legal recognition to electronic commerce and facilitate to file electronic records with the government. This Act divides into 13 chapters that contains 94 sections. Each chapter contains digital signatures, attributions, e-governance, regulation of certifying authorities, dispatch and acknowledgement of electronic records, security of e-records and digital signature, subscriber duties to digital signature certificates, cyber regulation appellate tribunal, liabilities & offenses of network service providers. There are 4 schedules under this Act that lay down amendments to be made in Indian Penal Code (IPC), Indian Evidence Act, Bankers Books Evidence Act & Reserve Bank of India Act. The various sections & punishments of the Information Technology Act is listed below

a) **Section 65-Tamper Computers Source Documents:** Whoever intentionally destroy, conceal, modifies source code of any computer, or computer programme or computer system or computer network
Punishment: Person involving in such crimes is sentenced up to 3 years imprisonment or with fine of Rupees 2 Lakhs or with both.

b) **Section 66- Alteration of data or Hacking computer system:** Any individual who has the intent or purpose to cause any loss, damage, destroy, delete or alter any information stored on a public or a private computer
Punishment: Person involving in such crimes is sentenced up to 3 years imprisonment or with fine of Rupees 5 Lakhs or with both

c) **Section 66A- Offensive message sending using any communication services:** Any offensive message containing threatening characters sent through communicating services with a motive to annoy, insult, inconvenience, danger, injury, or criminal intimidation .
Punishment: Person found to commit such crimes could be sentenced up to 3 years imprisonment along with fine.

d) **Section 66B-Steal computer resources dishonestly or communication device:** Whoever fraudulently receives or retains computer resources stolen or any communication devices
Punishment: Any individual involved in such crimes could be punished for a term not less than three years of custody or with fine of Rupees one Lakh or both.

e) **Section 66C- Identity Theft:** Anyone who makes fraudulent or dishonest use of another person electronic signature, password or other unique identification feature
Punishment: Persons involved in such activity could be penalized for a term up to 3 years jailed or with fine of Rupees 1 Lakhs or both

f) **Section 66D- Internet fraud:** Whoever tries to deceive someone by impersonating another person using communication devices or computer resources is a committed fraud.
Punishment: Any individual involved in such activities could be penalized for a period up to 3 years or with fine of Rupees 1 Lakhs or both.

g) **Section 66E- Publish private images:** Whoever violates an individual privacy intentionally or with the goal of publishing, disseminating or recording the photos of their private areas or private parts without their consent.
Punishment: Individuals involved in such crimes could be taken into custody for a period of 3 years or with fine not exceeding more than Rupees 2 Lakhs or both.

h) **Section 66F-Acts of Cyberterrorism:** Anybody who threatens the unity, integrity, sovereignty, or security of the people or any group of people with the goal to terrorize them causing injury, death or damage the properties and deny access to computer resources
Punishment: Any individual who is indulged in such types of crimes shall be penalized to life time detention.

i) **Section 67- Electronic form Publishment of pornographic materials:** Anyone who transmits, publishes, or causes obscene contents to publish in electronic form
Punishment: Anyone involved in such types of crimes shall be sentenced for a period up to 5 years with fine not exceeding more than 10 Lakhs rupees

j) **Section 67A- Publish materials containing sexual acts:** Whoever publishes or transmits materials containing sexual explicit contents or acts is guilty of crime.
Punishment: Any individual who is involved in this crime shall be sentenced either for a term extended up to 7 years of imprisonment or with fine of Rupees 10 Lakhs

k) **Section 67B-Publish sexually explicit materials in electronic form:** Anyone who publish or transmit materials that depict sexual explicit act of children in any electronic form.
Punishment: Any individual who is involved in this crime shall be imprisoned for period of 5 years & fine of Rupees 10 Lakhs (for first conviction) & sentenced for 7 years with fine of Rupees 10 Lakhs (on Second conviction)

l) **Section 67C- Record maintenance failure:** Intermediaries must keep and preserve any information that the central government may need for the period & in the format specified by the central government. Any middleman who intentionally violate sub sections provision.
Punishment: Any individuals who compels such activity shall be imprisoned for a period that extends up to 3 years (Sarmah, A., Sarmah, R., & Jyoti Baruah, A. (2017).

Apart from these sections listed above, there are many other sections in the IT Act 2000. The following Table 1 shows the different types of offenses and sections that come under Information Technology Act

Table 1. Offense types & section of IT Act (Animesh Sarmah, Roshmi Sarmah, and Amlan Jyothi Baruah- 2017)

Section Under IT Act 2000	Type of Offenses
Section 43	Mutilation to computer or Computer Systems
Section 69A	Power to impose orders prohibiting public access to any information stored on any computer resources
Section 69B	Power to authorize the collection of traffic data or information, as well as monitoring of any computer resources for cybersecurity purpose
Section 70	Unauthorized entry into password protected system
Section 71	Punishment for misrepresentation
Section 72	Violation privacy & confidentiality
Section 73	Dissemination of fake digital signature certificates
Section 74	Fraudulent Publication with purpose
Section 75	Act for offense or contravention committed outside India
Section 77	Penalties or seizure not to interfere with other punishments
Section 85	Felonies by companies
Section 503 IPC	Criminal Intimidation
Section 499 IPC	Criminal defamation
Section 420 IPC	Cheating and dishonestly inducing delivery of property or cyber frauds
Section 463 IPC	Forgery or email spoofing
Section 383 IPC	Extortion or web jacking
Section 500 IPC	Email Abuse or punishment for defamation
Section 507 IPC	Criminal Intimidation by an anonymous communication
NDPS Act	Online sale of drugs
Arms Act	Online sale of Arms

Cyber Attacks: Every year cybercrime has increased drastically as attackers has become more efficient & sophisticated. Cyberattacks can occur for a variety of reason & in a variety of formats. However, cybercriminals will always look to exploit the vulnerabilities in an organization security policies, methods or technology (Shruthi, M. 2021). So, let's understand the concept of cyberattack.

1. What is Cyber Attack?

A hacker or attacker attempting to gain unauthorized access to an IT system for the purpose of theft, extortion, disruption or other evil purpose is known as Cyber Attack. A hacker or attacker is a person who carries out the cyber-attack. Several negative effects are there for cyberattacks. Any attack can result in data breaches, which leads to data loss or manipulation. Financial loss incurred, customer trust is harmed & reputational damage occurs. To protect ourselves from cyberattacks, cybersecurity got implemented. (Jefferson, B. 2020).

2. Types of Cyber Attacks

In today's world, there are many different types of cyberattacks. Knowing the different types of cyberattacks makes it easier to defend the systems and networks against it. Let's take a close look at different types of cyberattacks.

a) **Malware Attack:** One of the most common type of cyberattacks is malware attack. It is a malicious software that is installed on the system users without their consent. It can be hidden inside a code, applications or on the internet. Malware infects computers by exploiting a security flaw & then download the malicious software to the host computing environment. Worms, spyware, ransomware, adware & Trojans are some examples of malicious software viruses.
b) **Phishing Attack:** The most prominent widespread type of cyberattack is Phishing Attack. It is one type of social engineering attack in which attacker impersonates as trusted contact and sends fake emails to the victims. The victim, unaware of this, opens the email and clicks on the malicious link or open the attachments. As a result, the attackers get access to sensitive data and account credentials. A phishing attack can also use to download the malware. There are different types of phishing attacks such as deceptive phishing, spear phishing, whaling and pharming. Fraudulent emails from banks, tax department, government & private firms and other trusted entity are some common examples of Phishing attack.
c) **Man in Middle Attack:** Eavesdropping attacks are known as Man in Middle Attacks. In this attack, the attacker intercepts between two party communications. The attacker exploits the network security vulnerabilities & hijacks the session between host and the client. I. (2019, May 1). The 2 common entry points of Man in middle attack are
    i) **Unsecure public Wi-Fi-** Attacker inserts between visitor device and the network.
    ii) **Breach device-** Once device gets breached by a malware, attacker can install the software to gain all information of victim.

The following figure 3 represent the man in middle attacks

d) **Denial of Service Attacks:** One major significant threat to most of the organization is the denial of service attacks. In Denial of service attacks, the attacker floods traffic into systems, server, or networks to exhaust resources and bandwidth. As a result, the system is no longer able to process & fulfil legitimate requests. Denial of service attacks overwhelm a system resources in order to stifle service demands. In addition to these, there is Distributed Denial of Service attacks. The main goal of DDoS attacks is to infect multiple host machines by achieving service denial & bringing system offline such that it allows another attack to penetrate the network. TCP SYN flood attack, smurf attack, teardrop attack, ping of death attack and botnets are some of the examples for DoS and DDoS attacks.

Figure 3. Man-In-Middle Attack I
(2019, May 1)

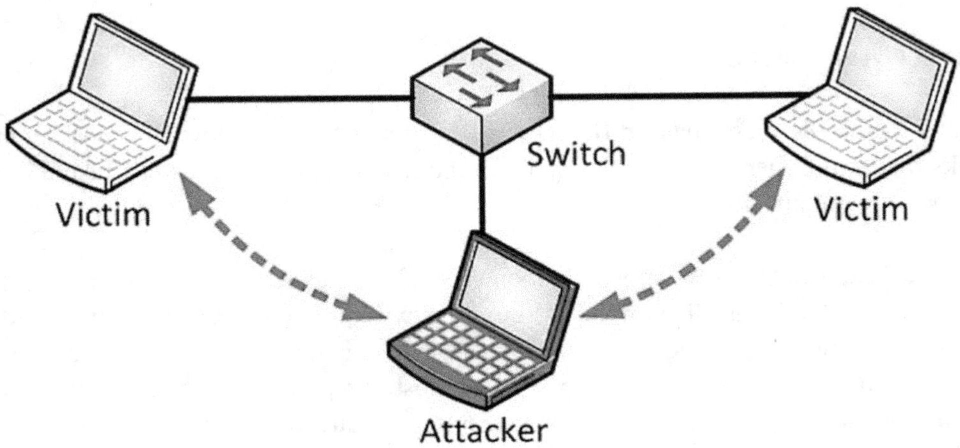

e) **SQL Injections:** An attack which is specific to SQL databases is called SQL injections. This will occur when an attacker inserts a malicious code to manipulate database and access sensitive contents. An SQL database queries this data using SQL commands, which gets executed via an HTML form on a webpage. If the database permissions are not set properly, the attacker exploits HTML forms to execute queries like create, read, modify and delete data stored in the database. Authentication bypass, Information disclosure, remote command execution, Data integrity are some of the examples of SQL injections attacks.

f) **Zero Day exploit Attack:** It is another type of cyberattack in which the attacker learn about the vulnerability used in software application, operating system & targets organization which are using that software to exploit the vulnerability before developer fixes it. Depending on type of vulnerability, developer could take any amount of time to resolve the issues. The disclosed vulnerability is the target of the attackers. Bugs, URL redirects, Buffer overflows, missing data encryptions, broken algorithms and problems with password security are some of the examples of zero day exploit attacks.

g) **Password Attacks:** One of the most common method of authenticating access to a secure information system is often referred as password. As a result, it makes an appealing target for attackers. In this type of attack, the attacker can gain access to confidential or acute data & system by gaining access to user passwords that includes the ability to manipulate & control systems. Many different tools & techniques for cracking user passwords are adopted. Some of password cracking tools are Cain, Aircrack, Abel, Hash cat and John the ripper, Brute force attack, key logger attacks, dictionary attacks, credential stuffing, password spraying and rainbow table attack are some of the examples for different types of password attacks

h) **Crypto Jacking Attack:** In this digital era, users are more inclined towards doing financial transactions & shopping online. With the rise of cryptocurrencies, new threats and risk have evolved which has shifted the focus of attackers to perform crypto jacking. Crypto Jacking is a crypto mining attack in which the attacker gain access into business and personal computers, laptops and mobile devices to install the malicious software. These attackers mine valuable coins & cryptocurrencies like Bitcoins & other digital wallets to steal personal assets. Attackers can access any computer resources & leave unknown victims with increased processor usage, slower computer response and overheating of computer devices. File based crypto jacking, browser based crypto jacking (Coin hive & Authed Mine) and cloud based crypto jacking (Mykings Botnet, Kobe Bryant Wallpaper, Vivin, RedLock) are some examples of crypto Jacking attacks.

i) **Cross Site Scripting Attack:** It is also called XSS attack. It is a variant of SQL injection attack in which the attacker injects malicious code (client-side JavaScript) into victim's websites. It is a client side code injection attack. There are several ways of inserting malicious code. Most popular approach is to either add it to end of the URL or posting directly onto the page that display user generated content. Persistent XSS, reflected XSS, and DOM-Based XSS are some of the common types of XSS attacks.

j) **Birthday Attacks:** It is a type of cryptographic attack that takes the advantage of the mathematics behind the birthday problem in a probability theory. These attacks are used to exploit the communication between two or more parties by launching against hash algorithms that verify message integrity, digital signature of software. This attack is based on fixed permutation number (pigeonholes). (Nageotte, A. 2021, June 25).

Security Tools

As security threats and vulnerabilities continue to evolve over time, it becomes essential for enterprise organizations to protect their data, systems, networks, and applications from different types of cyberattacks. Organization uses cloud based service to operate remotely. Cloud provides anyplace, anytime & no interruption services. Due to this, there is a rise in average screen time and data generation led to spike in the rate of cybercrimes. To preserve the privacy of businesses and individuals, & increased number of threat rates cybersecurity software is required that avoids unauthorized access to data. This section explores the few types of security tools which provides specific functionality.

1. **Wireshark:** It is a free and open source widely used protocol analyser tool which captures packets in real time in a network like internet or a desktop in home or work place. Wireshark helps to analyse the packet deeply inside and see the content of it. The main functions of Wireshark are to capture the packets, visualize the packet data and filter the data packets. Wireshark may be used as learning tool that can be adopted by government agencies, academic establishments, small businesses & several other organizations. (CompTIA, 2021).
2. **Metasploit:** Metasploit can be used as a command line interface penetration testing framework that simplifies the method of hacking for both attackers and ethical hackers. It is an open source framework (BSD licensed) that can be modified to work with different tools. Some of the tools that it supports are Network mapper (Nmap), Simple Network Management Protocol (SNMP) and several other tools. (Porup, J. 2019).
3. **Nessus:** It is a remote protection scanning tools that scans the computers in the connected network and provides alert messages, if it discovers any vulnerabilities that the malicious hackers might take advantage of and get access to any computer in the network. It will do this operations via a method of running over 1200 different checks on a given computer, looking to see if any of these attacks are going to be break into the computer or in the other case, harm it. (Wendlandt, D. 2021)

4. **Aircrack:** It is a software suite that allows user to protect against attacks on Wi-Fi networks. It is not a single tool but a set of tools, each of which plays a specific role. These accommodates a detector, packet sniffer, a WEP-WPA cracker, and so on. The main function of Aircrack is to sniff the packets and test the hashes out of them as to how to crack the password. (Mister, X. 2021, July 5).
5. **Snort:** It is an effective open source tool used for open Intrusion Detection (IDS) & Prevention System (IPS), that monitors the real time traffic that goes in & out of the network and can be deployed by individuals and organizations. It uses a rule based language that combines anomaly protocol and signature inspection methods to find the most likely malicious activity. These tools help network administrators to find denial of services (DoS) attacks & distributed denial of services (DDoS), Common Gateway Interface (CGI) attacks, stealth port scans, & buffer overflows. It makes a series of rules that defines malicious network activity, detects malicious packets & sends user alerts. (Top 11 Most Powerful Cybersecurity Software Tools In 2021. 2021, November 1)

APPLICATIONS OF CYBERSECURITY

The internet has become one of the most vital source of information for users. Securing data, networks and applications has become an essential component for most of the organization as they are more vulnerable to different types of cyber-attacks. The increased demand of Artificial Intelligence, Machine Learning and Blockchain technologies has opened avenues for cybercriminals to exploit sophisticated security tools to breach data, networks and applications. One of the critical issues in the cybersecurity space is the accurate detection of cyber-attacks, cyber threats, intrusions and malicious software programs. Some of important broad areas of cybersecurity applications are listed below

1. Cybersecurity applications for AI & IOT driven consumer devices.
2. Cybersecurity applications in Food and Agriculture sectors.
3. Cybersecurity applications in Financial systems.
4. Applications of Cybersecurity in Electrical power grids & Telecommunication sectors.
5. Application of cybersecurity in automotive sectors.
6. Cybersecurity applications for Healthcare Industry.
7. Application of cybersecurity in Aviation sectors
8. Cybersecurity applications in Education sectors.
9. Cybersecurity solutions for Military Applications.
10. Cybersecurity applications in government and smart cities.
11. Application of cybersecurity in judicial sectors.

FUTURE OF CYBERSECURITY

The rapid growth of technologies has made the business activities & life of people to work in digital mode. As a result, cybersecurity has emerged as a critical concern for the day. Understanding the future of cybersecurity will prepare users, how to make the best use of resources to stay safe. The IT industry is constantly evolving, so does the cybercriminals. Predicting the future is hard due to the shift in cybercriminals behaviour & inculcating new attacks procedures. (Parker, J., & @. 2020, May 19). This section explores on the future trends of cybersecurity.

Artificial Intelligence and Machine Learning

AI &ML are two dominated technologies that are being used to provide a high level of sophistication to cybersecurity business. Speed and automation are the two factors that distinguish between humans and machine Intelligence. Analysing & calculating the data needs lot of time for individuals. On the other hand, it requires fractional time for machines. An organization gets the opportunity to use the machine learning algorithms that automate threat detection & eliminate it. AI & ML simplifies the complex process by incorporating robust model for cybersecurity. The complexity of cyberattacks have grown over time. Attackers might use AI & Ml techniques in non-ethical way by breaking system using automated attacks that are very hard to defend. Attackers use AI & ML approaches to analyse stolen data (password credentials & network traffic) to identify the vulnerabilities. In future, cybersecurity developers adopt the AI & ML methods to detect security issues & address vulnerabilities. **Example-** Cybersecurity developers embed AI methods to warn user about the risky websites (Jyotsna 2019, August 21).

Cloud is the Future Trend of Cybersecurity

The exponential growth of ransomware attacks had created loopholes in the legacy security. Cybersecurity must be migrated to cloud in order to predict & defeat the cyberattacks in real time. Cloud infrastructures employ big data analytics to instantaneously identify & predict the known threats that overwhelm security. To build a global threat monitoring system, cloud security must develop a collaborative strategy that analyses normal & abnormal activity. Since the same cloud environment is used by many different users, cloud security is used to build an environment that predicts threats instantaneously through threat monitoring system that is shared among all cloud users. Cyber attackers continue to disrupt lives by stealing data & seeding malwares. Cybersecurity must actively work to disrupt attackers by leveraging big data analytics thrive within cloud. (O'Neill, E. (2020, May 17).

Future Blockchain Use Case for Cybersecurity

Blockchain begun as a technology underlying bitcoins, but has now evolved as a viable cybersecurity mitigation technology. Blockchain are cryptocurrency system that maintains a secure decentralized record of transactions. Most of the data breaches are caused due to human errors. Blockchain automates data storage by minimizing human factor in data storage systems. Any kind of digital asset can insert into Blockchain, making it more transparent. Blockchain provides no privacy or confidentiality for any transactions. Below are some of the future beneficial Blockchain use case to strengthen cybersecurity which are as follows

a) Secure private message in social media.
b) Secure IOT devices in decentralized system.
c) Secure DNS & DDoS
d) Decentralize medium storage
e) Computer software provenance (Legrand, J. 2020, September 4)

CASE STUDIES OF CYBERSECURITY

The prime goal of cybersecurity is to keep the data, network, and devices protected against cyber threats & ensure risk-free secure environment. In order to achieve these goals, it is very important to identify and classify the assets based on priority by identifying potential risks. It is essential to choose appropriate security strategy for each identified threat & monitor the breach activity. Finally, it becomes imperative to respond to a security issue through iterative maintenance by updating security policies to handle risk. This section explores few case studies of cybersecurity which are related to Electric energy sector, Food & beverages automation sector. So let's take a brief look at each of these case studies in detail.

Cybersecurity for Smart Charging Electric Vehicles: A Grid Perspectives

The automobile industry is migrating away from traditional gasoline fuelled automobiles with the introduction of electric vehicles (EV). The demand for charging these electric vehicles has grown significantly, which has created electric vehicle charging stations (EVCS) deployed at various commercial outlets. Over the last few decades, electric vehicles (EV) have emerged as one important technology that assists society in meeting clean energy targets. The growth of Electric vehicles has drastically increased annually on an average of 60% in the year 2014 to 2019. China & US being the leader of production of Electric vehicles. This will even continue to grow in the mere future due to several factor which are as follows.

1. Incentives for clean fuel vehicles & carbon reduction efforts
2. Overcome anxiety range in electric vehicle drivers due to advancement in battery charge technologies
3. Continuous Electric vehicle charging experience

The Figure-4 illustrates about the outline of smart electric power grid technology in Electric vehicles

There are two different layers in the smart electric power grid. The top layer is the cyber layer & the bottom layer is the physical layer. The physical layer contains generation facility units or power generators which generates the electricity or power. This flow of power is transmitted using transmission facility that contains transmission lines and transmission substation. The distribution facility contains distribution feeders, distribution substation that is responsible for transmitting the power energy to DER. There is a centralized controlled & monitoring system called as Supervisory Control and Data Acquisition (SCADA) that are used in power grids. SCADA contains the central master terminal units with subsystem controllers that are networked through Local Area network (LAN) such as TCP/IP or UDP. SCADA & Corporate LAN used by operators are disconnected with Virtual Private Network (VPN), firewall & Intrusion detection systems. The four main components of SCADA are

a) A central terminal master unit supported by other subsystems such as energy management system, a Distributed Energy Resource (DER) Management system, Geographic information system & Demand Response (DR) automation.
b) To assist system operators to manage SCADA, a Human Machine interface is used to support the functionality.
c) Field units such as Remote Terminal Unit (RTU) and Programmable Logic controller (PLC) are used.
d) Communication channels for transmission and information flow.

Figure 4. Cyber physical view of smart power grid
(Acharya, S., Dvorkin, Y., Pandzic, H., & Karri, R. 2020, November 27)

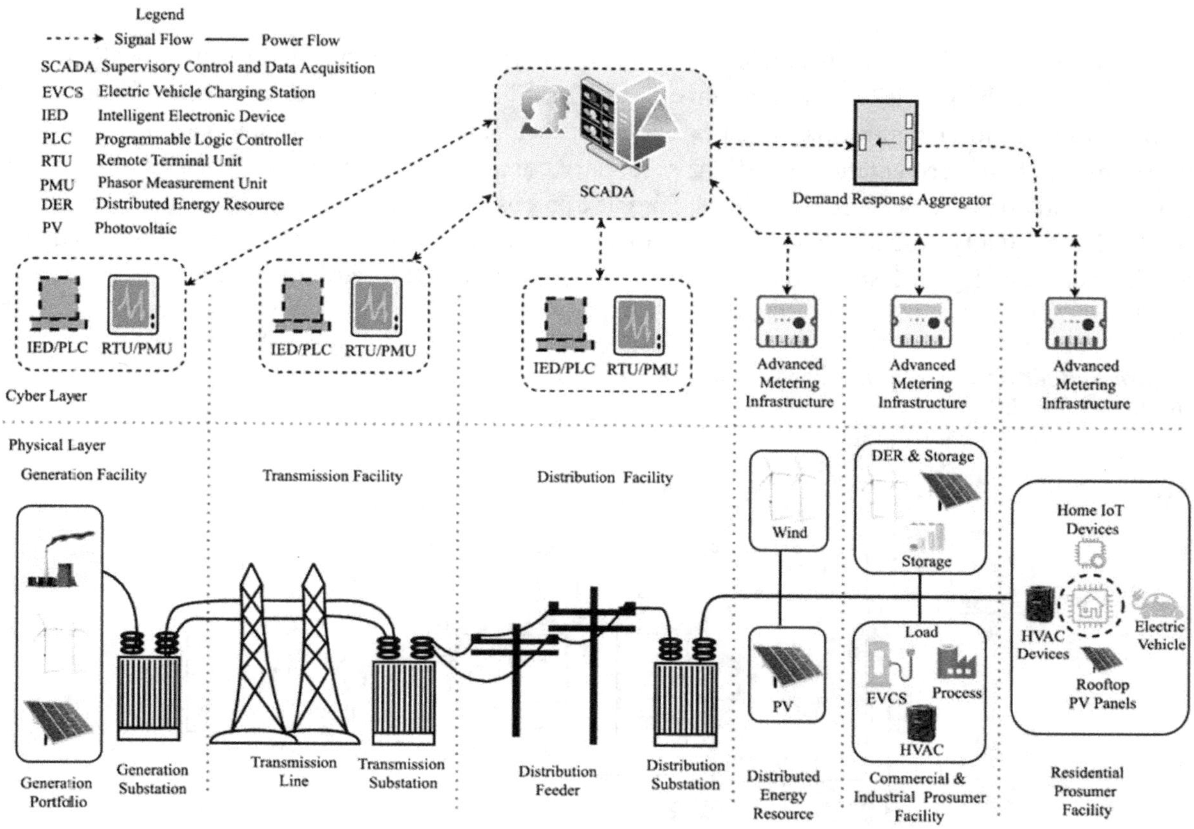

The SCADA field units comprises of Intelligent Electronic Devices (IED), PLC, RTU and PMU. These IED contains microprocessor devices, such as relay, sensors, and circuit breakers. The job of Remote Terminal Unit (RTU) is to transmit measurements to PLC and monitor IED. The RTU acts an intermediary between SCADA & IED which sends the control signals between them. Phasor Measurement Units (PMU) are power grid monitoring devices that record measurements in microseconds. Every power grid installs Advanced Metering Infrastructure (AMI) like Smart Meters (SM) that enables communication between consumer and Distributed Energy Resource (DER). The DER operators are required to safeguard their SM with authenticated Virtual Private Network (VPN). All Residential consumers use the IOT enabled devices such as smartphones to get connect to the network which is similar to that of Smart Meter. All field units communicate each other using field bus protocols. These protocols are more vulnerable to cyberattacks. The attackers can manipulate the data from each field units & force computer operators to take fallacious decisions. Some kind of cyber security measures needs to be employed at different grid layers of Electric Vehicle & its Charging Stations protecting it from vulnerable attacks. (Acharya, S., Dvorkin, Y., Pandzic, H., & Karri, R. 2020, November 27).

Smart Beer Production: An Automatic Control in Industrial Process (Possibility of Cyberattack)

Beer is defined as a beverage generated by the alcoholic fermentation of microorganism called "saccharomyces carlsbergensis" & "saccharomyces cerevisiae" mixed with malt & aromatized lupine. Beer belongs to a group of light drink which is rich in extract, saturated with carbonic gas. The toadstool cause the alcoholic fermentation as well as secondary compounds that affect the features of the product. Fermentation is considered as the most complex process. The usage of enzymes leads to increased production of quality products. The fermented food has gained significant attention from both academia and industry. The figure-5 shows the beer production process & the hazard main points.

Figure 5. Beer production process with hazards points
(Belulil, V. M. 2019)

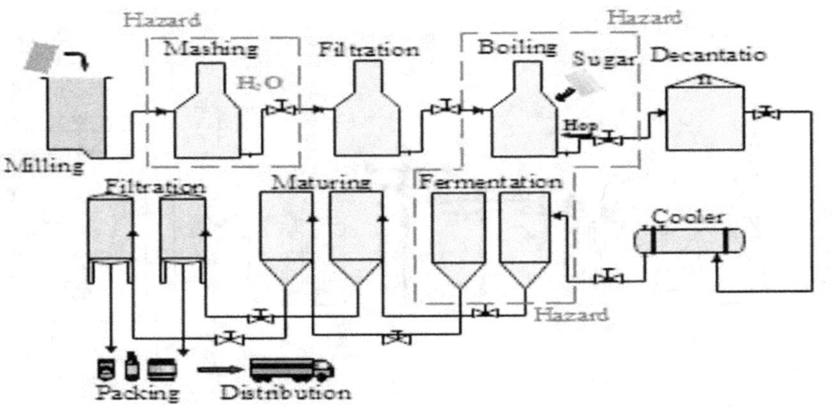

Today, the advancement of technology has created a lot of opportunities and appropriate conditions for the food product management to create intelligent process monitoring system. The technical complication in beer industry or brewing industry is comparatively less than any other industries. It is very easier to create a technical complication in beer industry through cyberattacks in automatic process control. The human experts are completely replaced by computer aided systems in bioengineered process industry. Due to non-uniformity and invariability of raw materials, food process are difficult to automate & there is lack of sensors to monitor the key processes. The rapid growth of computer hardware & software technology had created rising interest in industrial automation.

This case study explores about the professional risk involved in beer industry or brewing industry which uses bioreactors and fermenters. The bioreactors provide technical support for understanding the physical, chemical and biological effects of control systems. There could be a possibility of cyberattacks in these bioreactors that can degrade the industrial process by causing explosion which cannot be controlled or managed. The main points of the cyberattack in the beer process are in bioreactors & fermenters. The main goal for exploiting these two strong points are due to high pressure contained during the work process, that causes outburst and release heat up to 270 degree Celsius from fermenters and biomass reactors. The Figure-6 shows how the operators controls the automatic beer process

Figure 6. Operators control automatic beer process
(Belulil V. M. 2019)

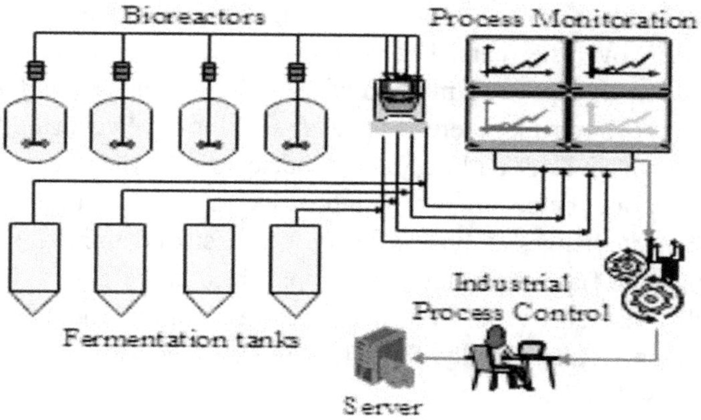

These bioreactors creates large biochemical reactions containing enzymes that produce alcohol. Attention has to be paid towards the physical condition such as temperatures that is used for boiling biomass. During the process of boiling, any change in temperature can result in damaging the process of alcohol production. Cyber attackers identify two major vulnerabilities in industrial production of beer manufacturing.

a) In the boiling sector, high speed accident can occur inside or outside the industry due to intervention of cyberattacks by increasing the temperature unobtrusively. This requires the attacker to have a complete knowledge about the current system dynamics and the type of communication protocol to be used.
b) Another possibility of outburst within brewing industry can happen at the fermentation sector. In this process, cyber attacker fetch the CO2 excess release valve in the tank that creates high gas pressure which creates explosion leading to loss of human life

The scientific study reveals the main vulnerabilities of cyberattack is the computer used to capture & monitor the beer process parameters controlled by operators of beer industry.

One way to control these cyber-attacks is by adhering the operating procedures by following the safety guidelines in the process industry. To ensure the safety, productivity, quality and security, the industrial operations must follow proper procedure to all the stake holders involved in the beer industry system. The deployment of additional sensor technology to a disaster controlling system overcomes traditional standalone monitoring. Finally, any industry must have a secret room for mainframe system installed with upgraded internet security solutions that can easily detect the cyberattack instance (detection of virus or ransomware or any malware) of automated production operation and send the notification alarm to the control Center to take further actions. (Belulil, V. M. 2019).

CONCLUSION

The cybersecurity has evolved as one of the emerging technologies that has eye catched many users, developers, and researchers over the years. The rapid use of information and communication technologies (ICT) have pervaded in all aspects of modern life enriches, improves, and interacts the world. Due to tremendous increase in cybercrime, there is always need for cybersecurity. In conclusion, we have made a fair attempt to provide a brief outline about cybersecurity at the fairly deep level. At the outset, we introduced the concept of cybersecurity, highlighted its importance at the cyberspace. We have addressed the objectives and challenges of cybersecurity. We discussed about the historical developments of cybersecurity highlighting flaws in system. We emphasized about cybercrime & its laws, different cyber-attacks types and cybersecurity tools. Furthermore, we discussed about the broad application areas of cybersecurity in various sectors. Finally, the future of cybersecurity, along with some case studies with solutions, are discussed in detail.

REFERENCES

Acharya, S., Dvorkin, Y., Pandzic, H., & Karri, R. (2020, November 27). *Cybersecurity of Smart Electric Vehicle Charging: A Power Grid Perspective.* https://ieeexplore.ieee.org/stamp/stamp.jsp?arnumber=9272723

Beluli1, V. M. (2019). *Smart Beer Production as a Possibility for Cyber-attack Within the Industrial Process In Automatic Control.* . doi:10.1016/j.procs.2019.09.043

Chadd, K. (2020, November 24). *The History of Cybersecurity.* Avast. https://blog.avast.com/history-of-cybersecurity-avast

CompTIA. (2021). *What Is Wireshark And How To Use It.* Default. https://www.comptia.org/content/articles/what-is-wireshark-and-how-to-use-it

Farah Minhaj Aditya Agrawal, U. K. (2021, October 26). *An Overview on the Emerging Issues in Cyberlaw and Cyber Security.* Design Engineering. http://thedesignengineering.com/index.php/DE/article/view/5697

Gupta, A. (2021, June 15). *Top 10 Cybersecurity Challenges In 2021.* GeeksforGeeks. https://www.geeksforgeeks.org/top-10-cybersecurity-challenges-in-2021/

Herrmann, D., & Pridöhl, H. (2020, February 11). *Basic Concepts and Models of Cybersecurity.* Springer Link. https://link.springer.com/chapter/10.1007%2F978-3-030-29053-5_2

I. (2019, May 1). *Cybersecurity 101: Intro To the Top 10 Common Types Of Cybersecurity Attacks.* Infocyte. https://www.infocyte.com/blog/2019/05/01/cybersecurity-101-intro-to-the-top-10-common-types-of-cyber-security-attacks/

Isha, U. (2020, August 28). *Top 10 Challenges Of Cyber Security Faced In 2021.* Jigsaw Academy. https://www.jigsawacademy.com/blogs/cyber-security/challenges-of-cyber-security/

Jefferson, B. (2020, August 20). *The 15 Most Common Types of Cyber Attacks*. Lepide Blog: A Guide to IT Security, Compliance and IT Operations. https://www.lepide.com/blog/the-15-most-common-types-of-cyber-attacks/

Jyotsna (2019, August 21). *How Will AI Impact Cyber Security In the Future.* Jigsaw Academy. https://www.jigsawacademy.com/how-will-ai-impact-cyber-security-in-the-future/

Legrand, J. (2020, September 4). *The Future Use Cases of Blockchain for Cybersecurity*. https://www.cm-alliance.com/cybersecurity-blog/the-future-use-cases-of-blockchain-for-cybersecurity

Mister, X. (2021, July 5). *Aircrack-ng. Aircrack-ng*. https://www.aircrack-ng.org/

Nageotte, A. (2021, June 25). *Top 10 of the Different Types of Cyber Attacks.* Oodrive. https://www.oodrive.com/blog/security/top-10-different-types-cyber-attacks/

O'Neill, E. (2020, May 17). *The Future Of Cybersecurity Is In the Cloud.* https://technologymagazine.com/cloud-and-cybersecurity/future-cybersecurity-cloud

Parker, J.. (2020, May 19). *What Is the Future Of Cybersecurity?* TechRadar India. https://www.techradar.com/in/news/what-is-the-future-of-cybersecurity

Pipikaite, A., Barrchin, M., & Crawford, S. (2021, January 21). *These Are the Top Cybersecurity Challenges of 2021*. World Economic Forum. https://www.weforum.org/agenda/2021/01/top-cybersecurity-challenges-of-2021/

Porup, J. (2019). *What Is Metasploit? And How To Use This Popular Hacking Tool.* CSO Online. https://www.csoonline.com/article/3379117/what-is-metasploit-and-how-to-use-this-popular-hacking-tool.html

Priyadarshini, I. (2019, March 25). *Introduction on Cybersecurity - Cyber Security in Parallel and Distributed Computing*. Wiley Online Library. https://onlinelibrary.wiley.com/doi/abs/10.1002/9781119488330.ch1

Sarmah, A., Sarmah, R., & Jyoti Baruah, A. (2017). A brief study on Cyber Crime and Cyber Law's of India. International Research Journal of Engineering and Technology.

Sentinel, O. (2019, February 10). *A Brief History of Cybersecurity*. SentinelOne. https://www.sentinelone.com/blog/history-of-cyber-security/

Shruthi, M. (2021). *10 Types Of Cyber Attacks You Should Be Aware In 2021.* https://www.simplilearn.com/tutorials/cyber-security-tutorial/types-of-cyber-attacks

Taddeo, M. (2019, June 3). *Three Ethical Challenges of Applications of Artificial Intelligence in Cybersecurity.* Minds and Machines. https://link.springer.com/article/10.1007%2Fs11023-019-09504-8

Top 11 Most Powerful Cybersecurity Software Tools In 2021. (2021, November 1). https://www.softwaretestinghelp.com/cybersecurity-software-tools/

Tripathi, V. (2021). *Cyber Law in India.* http://www.cyberlawsindia.net/cyber-india.html

Tyagi, R. (2020, May 27). *Cybersecurity Challenges in 2020 and How to Tackle Them? Managed IT Services and Cyber Security Services Company*. https://www.teceze.com/cybersecurity-challenges-in-2020-and-how-to-tackle-them

Waschke, M. (2017). *Personal Cybersecurity: How to Avoid and Recover from Cybercrime*. Apress Media. doi:10.1007/978-1-4842-2430-4

WendlandtD. (2021). *Nessus*. https://www.cs.cmu.edu/~dwendlan/personal/nessus.html

Chapter 2
Emerging Challenges in Cyber Security

Pooja M. R.
Vidyavardhaka College of Engineering, India

Gururaj H. L.
https://orcid.org/0000-0003-2514-4812
Vidyavardhaka College of Engineering, India

Pavan S. P. Kumar
Vidyavardhaka College of Engineering, India

ABSTRACT

In today's world of digitization where individuals and organizations are connected by the internet, cyber-security has become a major concern. Some of the most common challenges in cyber security that have evolved to a greater extent include cloud attacks, attacks from IoT, and hardware and software attacks along with the huge increase in phishing and ransomware attacks. With most of the organizations and companies using technological advancements for financial management, cryptocurrency and blockchain attacks also pose a major threat in daily activities. Being able to recognize the cyber security challenges, one must be able to secure themselves at both individual and organization levels. In this chapter, the authors deal with the continuously changing nature of cyber-attacks that has enhanced the difficulty of handling and avoiding emerging threats along with technological solutions that can address the concern. The techniques for demands of privacy and protection will also be discussed.

INTRODUCTION

The swift degree of transformation in cyber technologies, growing legal constraints governed by policies, and quick data movement is generating an atmosphere in which economic, technical, and policy issues are powerfully dictating what is which science could be doing, how the science could best be done, and essentialisms that are required for those who can do the right science. A significant task has been control

DOI: 10.4018/978-1-6684-3991-3.ch002

of data . Data is detained by and skillfully controlled by a limited number of workers who can limit the access, how, what and when it could be retrieved, as well as how, or if, the data organizations will be at danger since social security attacks concerning cyber space (Meza et. al, 2009). Malwares embedded in tweeted image, phishing, Phony Facebook up-dates, etc., produce administrative uncertainties extending from product operation to negotiating personnel to acquire admittance to intelligence to damage of data or machines themselves from social media delivering malware(Thakur et. Al, 2019, Thuraisingham et. Al,, 2020). Some of the prevalent applications of cyber security involved with handling traffic data analysis comprise of intrusion detection, botnet detection and malware analysis.(Miao et. al, 2018, Maglaras et.al, 2018.

While on the legal side, guidelines along with regulations are out-of-sync with the brand novel technologies, significantly, the charge of alternate withinside the era is such that upcoming styles of unlawful pursuit are rising at an exceptional charge. Many decision/policy and regulation makers have negligible know-how of the era and so layout coverage and regulation which might be regularly unenforceable, inappropriate or so preventive that they save you the technology from being completed that could inhibit or stumble on initial social cyberattacks(Sarker et.al.,2020)

Owing to heterogeneous and/or interconnected characteristic of smart cities, conditions get up wherein all the problems of every aspect location are present (Pal et al,. 2009) We would be untaken with the poorest of all worlds because the interrelated characteristic of smart cities method that each unstable and non-unstable, open-supply and registered structures are concerned in transactions and facts flows. Therefore, virtual forensics for embedded, Cloud and IoT gadgets is extreme and exciting. This is because now, no longer all IoT gadgets have the equal community and alertness architecture (Lu et. al, 2018). Digital forensics of gadgets hosted withinside the Cloud surroundings is made hard through the absence of third party settlement with the client, that could permit forensics investigators to get entry to the facts saved on the Cloud. Data saved in special nations brings in addition assignment to virtual forensic detectives because of judiciary disparities and records laws. The threats along with the demanding situations to the facts saved on Cloud infrastructure will even practice to the IoT gadgets questioned during forensic investigation. Figure 1 shows some of the common Vulnerabilities and the frequent approaches for adopting defense strategies against such vulnerabilities.

Figure 1. Vulnerabilities and approaches for defense strategies

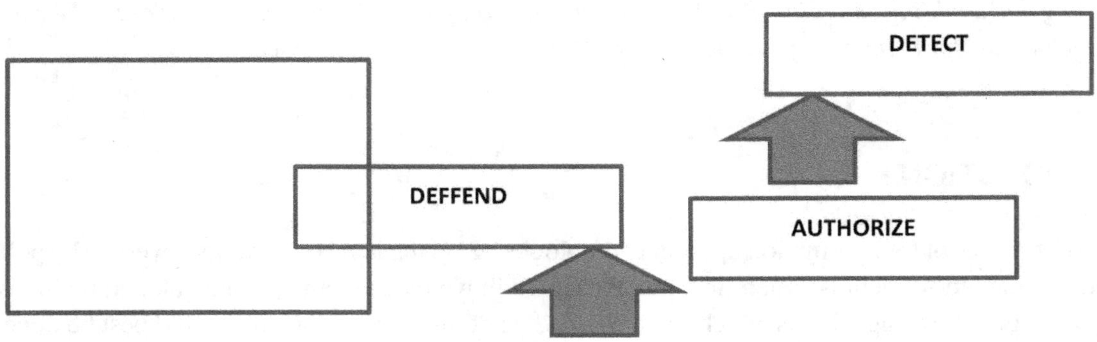

Cyber security is highly important as almost every organization including private or public both, profitable and non-profitable, military and corporate bodies access their data online and retain their data, programs/process, files on clouds and other devices Gupta et. al, 2018. The data would be very important and carries information that may be of personal importance, intellectual property, financial information, unauthorized data etc. which might be accessed by those concerned to cybercrimes and that is the reason cyber security is very important and everyone must know all safety measures which need to be taken while working on networks (Rajasekharaiah et.al, 2020, Benson et. al, 2017) The internet community easily trusts any one online and shares these details. Most of them share all personal detail on the wall of their social media which is usually not recommended. Cloud storage also pose to be risky where people share their identity information like Social security number, bank credentials and so on which would be very unsafe.

CYBER SECURITY CHALLENGES

Some of the common cyber security challenges are discussed below (Conti et. Al, 2018)

1. Network Security - Unwanted interventions and intruders can possibly put the network prone to unwanted risks.
2. Application Security - It involves adding functionality or features to software in order to stop a range of diverse threats including repudiation of service attacks and/or cyberattacks, data breaches including data theft circumstances.
3. Data Security - incorporating techniques that aim at protecting the data from unauthorized access, theft and data corruption.
4. Cloud Security - involves guarding the data deposited on cloud computing platforms from data leakage, theft and obliteration. Firewalls, penetration testing, tokenization, Cloud security involves the defense of data deposited on online cloud computing platforms from theft, deletion and leakage. Approaches of providing cloud security include obfuscation, firewalls, tokenization, virtual private networks (VPN),penetration testing,, and evading internet connection via public connections can work towards ensuring cloud security.
5. Mobile Security - methods intended to defend delicate information deposited on and communicated by laptops, tablets, smartphones, wearable, and other portable devices. At the source of mobile device security is the aim of having illegal users away from accessing the enterprise network.

Connecting Challenges in Cyber Security and Digital Forensics

1. The elementary ladders of digital forensics included gathering of the mode of communication in the scene, maintenance of the content in the mode of communication, confirmation, analysis, interpretation, documentation, exhibition of the result in court (Baig et.al,2017). However, the Cloud forensics is far complicated compared to traditional digital forensics. The main challenges originate from data acquisition, static and/or elastic and live forensics, and segregation of evidence, virtualized environments, modeling multi-jurisdictional laws, staffing, dependency chain of protection and lastly service level agreements(Masys et.al,2014) . The verification along with preservation of the possible evidence, and legal issues must be of high concern. Cloud forensics therefore would

be more compound, expensive and time intensive. Analyzing Cloud artifacts as those acquired from client devices assists one to depict the state of the system, however showing to be interesting as cloud data does not collectively accept standardized representations. A sophisticated approach involving mutual authentication at system level allowing the hardware to authenticate the firmware, then the firmware to confirm the identity of the hardware, by means of two safe protocols TIDS and TIDP was proposed(Choo et. al, 2017).

The technique of protecting systems, networks, handheld platform and data against hostile intrusions is known as cyber security. The various categories of security are operational, network, application and information(K. Ly et.al, 2016).

Over latest years, ransomware has risen in popularity, and it has become one of India's most serious cyberattacks (Yaqoob et. al, 2017). These have infected 83 percent of Indian organisations, according to cyber security company Sophos. The goal of this assault is to get accessibility to an identity data and prevent them from utilising it until a deal is completed (Sobb et. al, 2020) Ransomware certainly bad for individual users, but they're even more damaging for businesses since they won't be able to obtain information they need to carry out their daily operations.

IoT applications are intellectual, physical and technical devices that can independently communicate data over the web. As and when the popularity of Iot systems rises at a breakneck pace, so do cyber security concerns. When IoT devices are hacked, personal users ' data may be exposed. Securing Iot systems are among the most complex problems in Data Security, as gaining access to these devices may lead to more destructive attacks.

Phishing is a form of social engineering that is commonly used to gain sensitive data from consumers, like login credentials and account information. Unlike ransomware attacks, the hacker does not encrypt sensitive user data once it has been obtained. Instead, they exploit it to benefit themselves, mostly for internet retail and bank fraud. Hackers employ them frequently because they may exploit the victim's details until the person realizes. Phishing attacks are still a major Cyber Security worry in India, owing to the population's lack of familiarity with handling sensitive data.

While the terms blockchain and bitcoin may be unfamiliar to the ordinary internet user, they are extremely important to businesses. Since a result, assaults on these frameworks represent significant issues for enterprises in terms of Cyber Security, as they might endanger client data and corporate processes. These inventions have progressed beyond their formative stages, but they have not yet reached a stable advanced state (Zubair et. al, 2017). As a result, various assaults have occurred, including DDOS, Sybil, and Eclipse, to mention a few. Organizations must be aware of the security risks associated with these technologies and ensure that no security gaps exist for hackers to exploit.

The majority of individuals currently utilise cloud services for both individual and commercial objectives. Furthermore, one of the concerns with Network Security for businesses is the infiltration of cloud-platforms in order to steal client data (Ahanger et. al, 2019). The catastrophic iCloud breach, that exposed private photographs of celebirity, is well-known. If an attack on business data is successfully out, it might pose a major threat to the company and perhaps result in its destruction.

While AI and ML technologies have proven to be incredibly beneficial in a multitude of industries, they are not without limitations. Unauthorized individuals can utilise this technology to carry out assaults and pose a danger to businesses. These techniques could be used to locate highly ambitious targets in large datasets (Tonge et. al, 2013). Threats against AI and ML are also a serious concern in India. A sophisticated attack may be too difficult to handle due to our nation's lack of Cyber Security understanding.

Apart from these, Supply chain 4.0 systems will be a next combat for data protection for military and defence systems. The relevance of understanding defence supply chain vulnerability is proven by the changing landscape of technology and the military's involvement, as well as the interconnectedness within commercial and global systems and the growing reliance on this connectivity (Gade et. al, 2014). Military supply networks have distinct mission and targets than commercial supply chains, with military supply chain management having the capacity to impact global defense and human lives.

Machine Learning Methods to Identify Data Breaches

The worldwide cyber attack is rapidly evolving, with an increasing amount of data thefts every year. According to a survey released by RiskBased Security, privacy violations exposed 7.9 billion identities in the very first 9 months of 2019. Machine learning breakthroughs can help prevent four different sorts of data breaches[10].

1. Spear Phishing

Every day, firm employees get emails in their workplace inboxes. Some of them, which may come from unknown sources, may contain dangerous links. Language patterns, such as email subject lines, links, body content/communication patterns, phrases, and even punctuation patterns, may now be identified and classified using machine learning algorithms. If the system is set up appropriately, anomalies may be identified and security experts can investigate, potentially detecting emails before they are opened.

2. Ransomware

This security vulnerability is well-known to most people. The files of the individuals are locked. Customers must then pay a fee in order to obtain an unlock code that will allow them to access their data. These files frequently include sensitive customer information, company secrets that are required for corporate operations. The other sort of ransomware assault simply locks a user's computer and refuses to let them access it until the ransom is paid.

3. Watering Hole

Employees, particularly insurance agents who work on the field, may have preferred coffee shops or lunch places. Let's say a group of coworkers has a favourite restaurant where they routinely order food for delivery or takeaway. There is significantly less protection and an excellent area for hackers to enter a user's credentials via that gateway, whether they are using the Wi-Fi in that watering hole or browsing that company's site to buy online.

4. Webshell

A Webshell is just a short bit of code. It is placed into a site so a hacker may gain access to the server directory and make modifications. The hacker then has access to the database of the system. Hackers frequently target customers' financial and credit card details, and this sort of attack is especially common on e-commerce websites. Medical practises and insurance firms, on the other hand, are absolutely vulnerable since they store a lot of personal information. The activity becomes much more appealing to these hackers when the insured set up automated payments from their bank accounts. Payments are simply redirected to a different location.

Will the need for in-house or hired cybersecurity professionals eventually be eliminated by machines? It's really improbable. At this moment, robots are unable to conduct the in- depth investigations that analysts conduct after people are aware of suspected breaches or abnormal actions. However, machine learning should undoubtedly be included in risk management and insurance innovation. Humans are just incapable of gathering and analysing data at the same rate as computer algorithms. It's only natural to include machine learning as a key component of cybersecurity (Gade et. al, 2014).

Some of the common emerging Cyber Security Threats would be Email Spamming and Phishing (Parkinson et. Al, 2017)

- Malware and Spyware
- Botnet o Social Engineering
- Key loggers
- Denial of Service (DOS)
- Virus
- Worm

Assumptions in Addressing Challenges in Cyber Security

Existing methods are neither characteristic nor definite, as they blend up dissimilar layers of analysis (Carley et. Al, 2018). Further, few of the prevailing frameworks mark conventions about the relation amongst the offensive party and the protective party, while observed evidence hints at the incidence of all kinds of possible combinations, including for instance state- sponsored attacks on foreign enterprises or extremist groups. SAM based framework, which is relies on three dimensions Stakeholder, Activities and Motives, permits a classification of cyber threats that is both distinguishing and convincing (Javaid et.al,2013). Stakeholder dimension would decide, who is executing, mandating and who is pretentious. Goings-on dimension decides what happenings have been carried out and the consequences in terms of defects. The Motive dimension addresses why the activities have been carried out, the fundamental inspirations and purposes.

Organizational approach to implement cyber security Following are some of the approaches that can be taken as initiatives to handle cyber security at organizations.

Originality Protection

Understanding the factual charges and influence of cybersecurity programs and thus distinguishing the need for a new position for cyber- security within a networked world (Mthunzi et. al, 2017). Of course all systems and data would not be treated equal; hence defending critical digital resources is of utmost importance. Equally important is the threat from cyberrisk and hence use active defense" as a Cyberrisk measurement, thus reaching a holistic cybersecurity approach

Engage All-Inclusive Approach for Decision-Making

Implementing a context for improving discussions within organizations and comprehending the decision board's role in managing cybersecurity risks while identifying the factual questions to define government's role in cybersecurity.

Using commercial strategy thereby creating value

Recognizing how decision makers can tackle the task of cybersecurity in the era of the Internet of Things, while familiarizing infrastructure to repel cyber threats(Kremer et al., 2015)

Ensuring Cyber Security in Social Media

1. Users of social media ought to be cautious before any involvement with new workers in order to ensure privacy and security. Most of the social media websites lets the wokers establish dissimilar privacy settings. They could choose with whom to part their information.
2. The users also necessarily must have the following deliberation as effective steps to safeguard digital communication, mainly in social media
    - Restrict individual information
    - Ensure legal and safe download.
    - Following careful and safe shopping.
    - Handle mails thoughtfully.
    - Chat safely and sensibly.
    - Control the accesses with respect to camera.
    - Share what is atmost necessary
    - Follow correct social networking protocol and observe safety precautions.

Patterns in Cybersecurity

A cybersecurity data theft is characterized by event and response. Few of the typical patterns and responses are debated here. A dissatisfied employee installs malware in commercial systems and transfers files to and from servers to USB drive. Few of the hints at this would be where an Investigation with senior directors about temporary files being created and deleted is carried out. Also, Help-desk coupon will be sent to IT security lead. As a reply to this, the IT team in charge of security does not understand that data is being vulnerable and after the data are penetrated, the team tries to control finest way to inform. An Insider possibly gives or trades employee data to a cybercriminal. Cybercriminal uses timeworn nonetheless valid credentials to contact company servers and download. Employee records could containpersonally identifiable information (PII).Few of the clues at this would be loss of data alerts directed to the security principal in the IT organization. However with this, team emphasizes on the concept of forensics of the alert, but being not able to attach it to earlier announcements.

CONCLUSION

Simple safeguards, such as using the most up-to-date gear and software for your digital needs, can help protect your systems and sensors against cyber threats. To establish an added level of security, we need to adopt advanced security measures like deploying a firewall. One of the most difficult aspects of Cyber Security is keeping our data protected. Cyber security issues come in a variety of shapes and sizes. Phishing assaults, Ransomware, Malware attacks and other cyber security threats were few of them which were dealt here. We discussed some of the top most challenges in cyber security and common ways to deal with the same.

REFERENCES

Ahanger, T., & Aljumah, A. (2020). Cyber Security Threats, Challenges and Defense mechanisms in Cloud Computing. *IET Communications*, *14*. Advance online publication. doi:10.1049/iet-com.2019.0040

Baig, Z. A., Szewczyk, P., Valli, C., Rabadia, P., Hannay, P., Chernyshev, M., ... Peacock, M. (2017). Future challenges for smart cities: Cyber-security and digital forensics. *Digital Investigation*, *22*, 3–13.

Benson, V. (2017). *The state of global cyber security: Highlights and key findings*. . doi:10.13140/RG.2.2.22825.49761

Brown, S., Gommers, J., & Serrano, O. (2015, October). From cyber security information sharing to threat management. In *Proceedings of the 2nd ACM workshop on information sharing and collaborative security* (pp. 43-49). ACM.

Carley, K. M., Cervone, G., Agarwal, N., & Liu, H. (2018, July). Social cyber-security. In *International conference on social computing, behavioral-cultural modeling and prediction and behavior representation in modeling and simulation* (pp. 389-394). Springer.

Choo, K. K. R., Kermani, M. M., Azarderakhsh, R., & Govindarasu, M. (2017). Emerging embedded and cyber physical system security challenges and innovations. *IEEE Transactions on Dependable and Secure Computing*, *14*(3), 235–236.

Conti, M., Dargahi, T., & Dehghantanha, A. (2018). Cyber threat intelligence: challenges and opportunities. In *Cyber Threat Intelligence* (pp. 1–6). Springer.

Gade, N. R., & Reddy, U. (2014). *A Study Of Cyber Security Challenges And Its Emerging Trends On Latest Technologies*. Academic Press.

Gade, N. R., & Reddy, U. (2014). *A Study Of Cyber Security Challenges And Its Emerging Trends On Latest Technologies*. Academic Press.

Gupta, B. B. (Ed.). (2018). *Computer and cyber security: principles, algorithm, applications, and perspectives*. CRC Press.

Javaid, A. (2013). *Cyber security: Challenges ahead*. Available at SSRN 3281086.

Kremer, J. F., & Müller, B. (2012, June). *Cyber security: developing a framework to understand the emerging challenges to states in an interconnected world*. In *British International Studies Association and International Studies Association Joint International Conference*, Edinburgh, UK.

Lu, Y., & Da Xu, L. (2018). Internet of Things (IoT) cybersecurity research: A review of current research topics. *IEEE Internet of Things Journal*, *6*(2), 2103–2115.

Ly, K., Sun, W., & Jin, Y. (2016). Emerging challenges in cyber-physical systems: A balance of performance, correctness, and security. *2016 IEEE Conference on Computer Communications Workshops (INFOCOM WKSHPS)*, 498-502. doi: 10.1109/INFCOMW.2016.7562128

Maglaras, L. A., Kim, K. H., Janicke, H., Ferrag, M. A., Rallis, S., Fragkou, P., ... Cruz, T. J. (2018). Cyber security of critical infrastructures. *Ict Express*, *4*(1), 42–45.

Masys, A. J. (Ed.). (2016). *Exploring the Security Landscape: non-traditional security challenges*. Springer International Publishing.

Meza, J., Campbell, S., & Bailey, D. (2009). *Mathematical and statistical opportunities in cyber security*. arXiv preprint arXiv:0904.1616.

Miao, Y., Ruan, Z., Pan, L., Wang, Y., Zhang, J., & Xiang, Y. (2018). *Automated big traffic analytics for cyber security*. arXiv preprint arXiv:1804.09023.

Mthunzi, S. N., Benkhelifa, E., Bosakowski, T., & Hariri, S. (2019). A bio-inspired approach to cyber security. In *Machine Learning for Computer and Cyber Security* (pp. 75–104). CRC Press.

Pal, P., Schantz, R., Rohloff, K., & Loyall, J. (2009, July). Cyber physical systems security challenges and research ideas. In *Workshop on Future Directions in Cyber-physical Systems Security* (pp. 1-5). Academic Press.

Parkinson, S., Ward, P., Wilson, K., & Miller, J. (2017). Cyber threats facing autonomous and connected vehicles: Future challenges. *IEEE Transactions on Intelligent Transportation Systems*, *18*(11), 2898–2915.

Rajasekharaiah, K., Dule, C., & Sudarshan, E. (2020). Cyber Security Challenges and its Emerging Trends on Latest Technologies. *IOP Conference Series. Materials Science and Engineering*, *981*, 022062. doi:10.1088/1757-899X/981/2/022062

Rajasekharaiah, K. M., Dule, C. S., & Sudarshan, E. (2020, December). Cyber Security Challenges and its Emerging Trends on Latest Technologies. *IOP Conference Series. Materials Science and Engineering*, *981*(2), 022062.

Sarker, I. H., Kayes, A. S. M., Badsha, S., Alqahtani, H., Watters, P., & Ng, A. (2020). Cybersecurity data science: An overview from machine learning perspective. *Journal of Big Data*, *7*(1), 1–29. doi:10.118640537-020-00318-5

Sobb, T., Turnbull, B., & Moustafa, N. (2020). Supply Chain 4.0: A Survey of Cyber Security Challenges, Solutions and Future Directions. *Electronics (Basel)*, *9*(11), 1864. https://doi.org/10.3390/electronics9111864

Thakur, K., Hayajneh, T., & Tseng, J. (2019). Cyber security in social media: Challenges and the way forward. *IT Professional*, *21*(2), 41–49.

Thuraisingham, B. (2020, May). The Role of Artificial Intelligence and Cyber Security for Social Media. In *2020 IEEE International Parallel and Distributed Processing Symposium Workshops (IPDPSW)* (pp. 1-3). IEEE.

Tonge, A. (2013). Cyber security: Challenges for society- literature review. *IOSR Journal of Computer Engineering.*, *12*, 67–75. doi:10.9790/0661-1226775

Yaqoob, I.,, & Ahmed, E., Habib ur Rehman, M., Ahmed, A. I. A., Al-Garadi, M., Imran, M., & Guizani, M. (2017). The rise of ransomware and emerging security challenges in the Internet of Things. *Computer Networks, 129.* Advance online publication. doi:10.1016/j.comnet.2017.09.003

Zubair, Szewczyk, Valli, & Rabadia. (2017). Future challenges for smart cities: Cyber-security and digital forensics. *Digital Investigation, 22*, 3-13. doi:10.1016/j.diin.2017.06.015

Chapter 3
The Role of AI-ML Techniques in Cyber Security

Sima Das
Maulana Abul Kalam Azad University of Technology, Kolkata, India

Ajay Kumar Balmiki
Maulana Abul Kalam Azad University of Technology, Kolkata, India

Kaushik Mazumdar
Indian Institute of Technology (Indian School of Mines), Dhanbad, India

ABSTRACT

Digital protection is the act and protocol to protect personal computers, mobile, electronic devices, and data from malicious attacks. Artificial intelligence (AI) and machine learning (ML) are presently essential for our daily existence, and this incorporates network protection. AI/ML can recognize vulnerabilities and lessen occurrence reaction time. Due to this unprecedented test, AI-based instruments for network safety have arisen to help data security groups decrease break chance and further develop their security to act proficiently and successfully. Simulated intelligence and AI developed basic advances in information security, as they can quickly examine a huge quantity of datasets and recognize various kinds of network stranded and breaches – from malware, one can pursue advantage to distinguishing unsafe comportment which might prompt a phishing attack or downloading malicious program. The advancements learned over the system can be used to identify new types of attacks in the future.

INTRODUCTION

A cyber-physical system (CPS) or smart device is a system which can be controlled as well as monitored the given task through computer primarily related algorithms. Though in cyber physical system, applications of physical and software additives are thoroughly intertwined, able to feature on remarkable spatial and temporal scales, show off multiple and notable behavioral modalities, and have interaction with each different in approaches that exchange with context. CPS involves transdisciplinary methods,

DOI: 10.4018/978-1-6684-3991-3.ch003

merging principles of cybernetics, mechatronics, layout and technique science. The task manager is regularly called embedded structures. In embedded systems, the emphasis has a bent to be greater at the executive factors, and much less on a severe hyperlink among the computational and physical factors. CPS is likewise similar to the Internet of Things (IoT), sharing the same simple structure; though, cyber physical system represents a greatest sum up and synchronization among physical and execution factors.

Introduction to Cyber Physical System

Cyber Physical System (CPS) is generally not defined as a computer or some high-level device or apparatus like sensor and embedded system. Though the system is having good ability, scalability, efficiency as compared to the present system or existing system. This helps the system to interlinked more with the machine or system for better connection and hence for natural connection with physical elements. The assumption regarding cyber physical System (CPS) is that it will dynamically evolve and it will fabricate the new advanced system or new advanced technology that can be the amalgamation of the cyber part and physic part. This new system will be ready to change life in new ways that will enhance the way of living with advanced technology and high-tech features and configuration. This will be the more advanced and more reliable embedded system that we ever used before. Previously, the factors such as protection, quality standard, optimization, monitoring and controlling related fields like healthcare and agriculture, flexibility, advanced systems in vehicles like car and aeroplan, also the main reason for automation in systems is security. These all are related points or important factors which was the need at that time and now the advanced system like Cyber physical system (CPS) has been introduced to us to resolve all these problems. This all summed up in such a way to exclude the problem and hence making the system more reliable, affectable, more secure, advanced mechanism using advanced algorithms, advanced systems with fully automated and any more. The important factor is speed by which all digital systems and industries are dependent. If one can do the comparison between today and yesterday then the result will be astonishing that within such a short time of span the development is done in an advanced system and techniques. Lots of transformation and changes also arise in apparatus, devices, sensors, and many more. They became advanced and their services also became way faster now. When we talk about the information or data transfer then this advanced technology and advanced algorithmic structure and hence computational execution is taking place in each and every single area and sector. Day after day this computational algorithmic structure is becoming more complex when it comes to data or information related operations and hence the development of system and structure in sector or area is also going on increasing. The system or technology is morphing from simple system to advanced system or from simple technology to advanced technology. When it comes to technology and digitalization then an important point also arises that is "Internet". The Internet has become an important part of our daily routine, and it makes our life very easy in each and every sector which is related to advanced systems and technology, such as Schools, Institutes, Office sector, Industrial area, Science and research field and many more places. It just transforms the way of living in human beings from normal to advanced. With the help of the internet, we moved into an advanced communication system and hence this makes the world more social than before. It gives us numerous platforms for communication or the place where we can easily come up with something without any barrier for oneself. The Internet removed many obstacles but it also generated questions about security and protection. But these days with the help of high-level technology and advanced algorithms, we can also manage these things easily, yet there are more to be discovered in this security related field. We can operate this platform only with the help of the internet, that's why the internet performs a very important role for us. The reason is very short and simple: the internet is very important for technology or systems and also for personal computers. The evolution and revolution in the world of technology by the help of the internet regarding systems makes many or almost all systems automated. That means we can fully monitor and control the system operation; we do not have to do it manually. That means this is the generation of wireless systems. In Wireless systems the operation is done fully automatically that means monitoring and controlling is done remotely of course we need the help of the internet and data details or information regarding this operation. Wireless technology is getting famous nowadays and though this is the sector where everyone is focusing, that's why new-new opportunities, evolution and development of technology is taking place in a huge number. Now we are controlling this system and technology but we are not able to understand the fact that only we are the people who are getting surrounded by this advanced and high-level technology. The system which is connected or inter linked with the internet and hence if this device can fully operate automatically and also monitored with information or data storage then this system is generally known as Cyber physical system (CPS). Not only this but it will also play an important part in security and protection as well. If the evolution is done in the system somewhere then the security will also become the main factor which will be concerned. But the fact that if digitalization and advancement will take place then also there will be the evolution and upgradation in security as well. So that the system will be more reliable and secure in terms of other systems.

Introduction to Machine Learning

Recently pattern and structure recognition are very important and famous but to process this operation a method has been discovered and coined as Machine learning also called ML. This method has proved itself that it is very useful and effective and hence used for security purposes also. In Machine Learning there are several algorithms which are used to enhance the system and process. Though it is very beneficial also in the case of data analytics and data prediction. The methods and algorithms help the system to read the pattern or recognize the pattern. There are some steps and processes that we can follow to perform such operations such as selecting the important characteristic, we have to train some of the data set which is responsible for the system and the train data set is generally used for prediction techniques just like weather forecast. Weather prediction is also done by the help of Machine Learning, at first, we train the data set, it is nothing but the cluster of previous reports and information, then we feed it to the system, there are also some powerful algorithms that are used for operation, and then the system predict the output for us. Though it is very tough to anticipate the exact result, the anticipation or prediction is almost the same as the real result. This can be done in the case of spam or fraud messages and real messages. We can train the data set of fraud and genuine message then we can feed data or information to the system, by the help of algorithms which are used in Machine learning it will help the system to do the operation and meaningful task of anticipate or prediction of the desired outcome and by the help of prediction method it will help to recognize that which messages are fraud and which are genuine. It is also very helpful in anticipation of those task and work which are having less amount of information and details, but we already have the number of fabrications, this numerous numbers of fabrication helps system to predict the desire output, though the information we have or gathered is less in amount, then also it will operate the task and provide us the desire output. Lots of other techniques are also there where machine learning is very effective, for example genetic algorithms, these methods currently are in the growing phase. Currently this technique is used by everyone in the lab for forensic research and study, and will give more advance and beneficiary results in future. Classification of machine learning are as discussed in this section. Lots of sectors and areas are there where machine learning is used in the system. These are further divided into some classification such as Unsupervised learning and Supervised learning. These two methods are based on the type and characteristic of the data or information and hence we use this method as we want or as per our need. Within these methods there are further divisions and methods are also present. In Unsupervised learning, we have to train the data set and we do not have the prior data set or information regarding the system, so we have to train the system by feeding the data. The example or the method we use in unsupervised learning is clustering. This method is very different from supervised learning. In this method we do not use manual type of data as an input, in this machine learning method we have to collect the data details or set of data information to fetch the process in the system. But there are some problems that arise in this machine learning method that is related to output, for the pattern recognition the output is not so specific and reasonable. Though supervised there is no guarantee calculation regarding result and output but we have the calculation regarding the features and factors. Only one thing here not count as a benefit that in this model we have to trained or ready the data no pre clue or hints are present, but the fact is different- different model have their own pros and cons or its own feature so it's totally depend upon our methodology and process that which method will be more useful and can give benefit to us if we are working in totally new project then this model will be used. In supervised learning, we have to train some of the data set because in supervised we already have the set of data and information regarding the operation which we want to perform as per our need. It is a learning technique by which one can learn the data set, within this data set a group of functions is present, which are going to be used. In supervised learning the training data set contains some values like a set of data with corresponding output, judging the new output from the given data set which contains data as well as output. The job of supervised learning method is to anticipate the function variable in the input data set, this data set contains the group of input information and output set, which is going to give an output. In supervised learning of machine learning algorithms, we generally use the algorithm first to represent the data set which contains input and output set to receive or to gain the desired output. This is very much differed from unsupervised learning and hence we have one benefit also that we already have the respective data set or trained data set by which the work become more easier and it do not require lots of permutation and combination as compare to the other machine learning technique that is Unsupervised learning technique. This method is also having its own characteristic and feature and totally depend upon the project or system work, if the project or system work is more familiar and hence, we only want to add some new feature or characteristic or we want to anticipate the output but we have enough data set or record then we can go for this model. So, selecting which model is best to work. Every model is designed to enhance the work of the system. Reinforcement learning is the algorithm which is used to make appropriate decisions in the environment and surroundings. Reinforcement algorithms are also very helpful for many situations and conditions. Reinforce learning algorithm is deployed by numbers of software and system to find the possible path as an output. Though the Reinforcement algorithm is very much different from the supervised learning, in supervised learning there is a trained model data set and in supervised learning the output is already calculated before or we get the anticipated result or output. But in the Reinforcement algorithm the system itself decides and takes the action on the given task. Every day the world becomes more advanced and digitalization is also making the technology more and more advanced though this also points against an important factor that is security. Nowadays security has become the main concern worldwide and new-new features are evolving day after day. On one hand there is revolution and development in its peak, and on other hand there is also some concern about security. The security threat is increasing day after day and every single center is busy trying to find the exact method and way. But the fact is till now also no can guarantee full proved security regarding system and machine. But this machine learning is easier because in this model few powerful algorithms are present, and by the help of this algorithm we can achieve security. This algorithm is designed in such a way that they can perform many functions or it can be executed in other places. That means it is used for security purposes as well as other purposes too.

Introduction to Artificial Intelligence

The field of technology and science are evolving day by day and this evolving is emerging very rapidly and vigorously, in this developing phase Artificial Intelligence has also evolved in this sector of technology. Artificial Intelligence makes possible the way of interaction or inter communication between human and machine. This interaction is between the human mind or human thinking with machines. This field has grown so far that it holds many advanced systems and features such as real time interaction gaming, implementation of robotics, theory representation and calculation proving, automatic advance system, other problem-solving strategy, implementation in healthcare also in precision agriculture and other sectors. Lots of techniques and algorithms which are embedded in Artificial Intelligence are based on complex features and methods or we can also say that based on this complexity the techniques and algorithm is designed or invented. Artificial intelligence makes the software as well as hardware to work faster and hence it improves the speed of interaction between the systems. The systems which are having network technologies and properties, or the software which are embedded for network are designed to work in Artificial Intelligence, and by the help of Artificial Intelligence this operation of execution as well as implementation and interaction becomes faster. This system becomes more advanced and able to achieve the solving technique for complex problems and issues. This high-tech level of Artificial Intelligence system is on demand and having a popular state in the technology market too. Almost all industries are concerned in the Artificial Intelligence market. A huge amount of investment is also done in this sector and field to gain the more advanced system. And the sector of Artificial Intelligence is growing rapidly due to the evolution and revolution in technology. Lots of new fields are also taking interest in it. Therefore, not only the business sector and industries but also schools, colleges, institutions and other sectors related to science and technology. Days are passing, the requirement of new and advanced technology is also increasing. The researchers are using this technology to improve laboratory performance such as bioinformatics.

Bioinformatic is deliberately adapting the technique of Artificial Intelligence, the vast change has occurred and rapid growth is observed after including the Artificial Intelligence technology and methodology in Bioinformatics. Researchers are doing research in the field of genetic information and genetic mapping with the help of Artificial Intelligence, and this became easier than previous days. The systems, equipment, and technology has become more advanced by the help of Artificial Intelligence. face detection and fingerprint are easy to implement. This advanced technology using Artificial Intelligence makes our work so easy that the researcher can easily find the guilty person or criminal by using gene mapping or genetic information. Generally, for photoshop department the Artificial Intelligence technique emerges as a miracle, in photoshop world previously lots of things was next to impossible, but now it's just like a piece of cake such as previously there was no live and 3D photos was available also there was no tools available to purify the old image or photo, but now the things have been changed and therefore lots of tools and technology is been discovered in Artificial Intelligence to tackle this obstacle. Nowadays the photoshop world and photoshop related technology are rocking in the present world and also the demand for this technology and requirement is also increasing with the new emerging of new technologies.

Healthcare systems are also using Artificial Intelligence due to its advanced technology and also security and protection. Doctors are using this technology to secure the patient history and record against data infringement. Actually, lots of cases have been seen previously of data infringement, the data of patients getting leaked by the organization or by the intruder itself. So, this has become a serious issue and the fact that everyone is afraid of security threats, to secure and protect this channel and system Artificial Intelligence plays an important and vital role in the healthcare system. For example, the doctor patient portal, and the robots which are used in hospitals, this was designed for the patient and to ease the work for doctors, because doctors can not present everywhere at the same time, so this robot will perform the task which will be given by doctor such as medication of patient in time. With the help of Artificial Intelligence only this can possibly be done and implemented. And also, if we talk about the security in healthcare again Artificial Intelligence plays an important role it removes the vulnerability and makes the system more robust.

Security is also the main factor which plays a very important role in it. These days security threats are increasing very rapidly and hence new-new technologies and methods are also emerging. One of the technologies and models is Artificial Intelligence which contains powerful algorithms and methodology to protect against security threats.

Artificial Intelligence has lots of models and algorithms such as Support vector Machine (SVM). In Support vector machines we execute classification techniques, this algorithm is used for pattern recognition and prediction of output such as fault, optimization for treatment and diagnosis. The next is Artificial Neural Network (ANN) that is used to detect the inter connection of neurons, this interconnection is a type of physical connection. This algorithm is utilized to execute further many issues and problem statement like mathematical model which are used to anticipate the action and reaction of brain perception, perception in now a day is growing fastest as its can, lots of perception is been discovered and yet to discover, some of the perception is hand perception, brain perception, game perception and many more. Normally this is a methodology to interlink the connection between human and machine. Here the main work is done based on the brain signal, just the same as gesture recognition or 6th sense projects. Perception is the vast and broad field in Artificial Intelligence and the important part in perception is mathematical model and calculation, only this can help the output to match with the desired output. Natural language processing is also one of the best inventions in the field of AI, normally it reads the human language not only read but it is also able to recognize and understand the human language. Reason Problem solving is also one of the creations by Artificial Intelligence. This helps humans to execute and solve problems like mathematical puzzles, this was done step-by -step with the help of algorithms and also some deduction part was also done. Previously, this was not sufficient for broad area problems or larger and huge mathematical problems, but now the conditions have been changed and it is nearly possible with the help of Artificial intelligence. Motion detection is also a part of Artificial Intelligence, generally it is used in healthcare in robotics for the location and findings, it maps the location to us and indicates the path. Deep learning is also a technique which is also a part of Artificial Intelligence and deals with the neural network. It is also in demand and deep learning enhances the ability and performance of many sectors and fields which are related to Artificial Intelligence such as recognition method for speech, classification method for image and computer vision. This is also emerging very fast and the set of algorithms we are using and also the set of models and techniques we are using that are very much useful to everyone and worldwide. Many sectors and fields have got new life with the help of Artificial Intelligence and through linking with Artificial Intelligence and Deep learning lots of problems have come to an end. This gave the world a new place to browse and one can easily run its technology idea with the help of this field and sector and the Intelligence is making our base strong.

Rest of the chapters are as follows: background study, advantages, limitations and applications of AI based cyber security, and lastly in conclusion part whole chapter is summarized.

SCOPE OF THE STUDY

The main purpose of this study is to analyze and determine the workflow of Artificial Intelligence (AI) and Machine Learning (ML) in the Cyber Security sector. The key points and factors which are related to cyber security, the powerful algorithms, and techniques of AI and ML which is very essential and important for cyber security to emphasize itself and manage its roots for the purpose of security as well. This study also signifies the advantages, limitations as well as application sectors associated with AI and ML-based security. The agenda of this research is to provide efficient knowledge of AI and ML-based security based on the study and also enhance the AI and ML sector as well.

CASE STUDIES

This section synthesizes the literature and normally identifies the gaps in know-how that the empirical look at addresses; it may also provide a theoretical basis for taking a look at, presence of the studies problem as shown in *TABLE I*, and justify the research as one which contributes something new to validate the techniques.

ADVANTAGES, LIMITATION AND APPLICATION AREA

Advantages, limitations and applications of AI-ML based security are discussed in this section.

Advantages of AI-ML Based Security

Maintaining information and community security isn't clean in today's enterprise surroundings. We will take a decisive step toward being safer by adopting AI to bolster your security infrastructure. There are numerous advantages of using ai for enterprise safety and we expect that very soon synthetic intelligence turns into an imperative part of enterprise cybersecurity. The advantages of AI-ML based security are shown in Figure.1.

- AI learns over time

AI technology is smarter, and it makes use of its potential to improve community safety over time. It uses machine getting to know and deep getting to know to research a commercial enterprise network's conduct over the years.

Table 1. Literature survey on AI-ML based cyber-physical system security

Ref No.	Objective	Summary	Technology used
(Yeboah-Ofori et al., 2021)	Cyber threat predictive analytics for improving cyber supply chain protection	The Cyber Supply Chain system is complicated which includes special sub-structures acting on numerous obligations. Safety in the supply chain is due to the susceptibilities and threats of all portions of the device which may be broken at all points in the delivery chain. The source of simple interruption happening is the general enterprise endurance. Consequently, it's principal is critical to recognize and detect the threats in order which group can accept important manipulation measures for the supply chain protection. The paper aims to analyze and predict threats to enhance cyber delivery chain protection. We have implemented CTI with ML methods to examine and are expecting the threats are built totally at the CTI residences. That permits to perceive the characteristic of CSC vulnerabilities therefore the suitable management moves can be take on for the complete cybersecurity development. To reveal the applicability of our method, CTI statistics is amassed and some of ML procedures, that is LG, SVM, RF, and DT, are used to advance analytical analytics using the Microsoft Malware Prediction dataset. The test reflects attack and TTP as input constraints and vulnerabilities and IoC as output parameters. The outcomes referring to the detect and monitoring that Spyware/Ransomware and phishing are the maximum expectable threats in CSC. They take additionally recommended applicable controls to tackle those threats. We recommend the usage of CTI statistics for the ML based detection system for the general CSC cyber protection development.	Indicator of Compromise (IoC), Cyber Threat Intelligence (CTI), Logistic Regression (LR), Support Vector Machine (SVM), Random Forest (RF), and Decision Tree (DT), Machine Learning (ML)
(Lee et al., 2020)	Impersonation assault detection thru part computing the use of deep autoencoder and characteristic abstraction	An increased range of computation gadgets interrelated via wi-fi systems compressed within the cyber-physical-social structures and a full-size amount of touchy community information transmitted amongst them have raised protection and privacy worries. Intrusion detection is known as a powerful defense mechanism and these days ML methods are used for its development. But, IoT devices regularly have restricted computational assets consisting of confined energy source, computational power and reminiscence, as a consequence, conventional ml-created total ids that require huge computational sources aren't appropriate for working on such devices. This look at for that reason is to design and expand a lightweight ml-based ids tailor-made for the resource-restricted gadgets. Specifically, the look at proposes a light-weight ml-based totally ids model namely effect (impersonation attack detection the use of deep car-encoder and feature abstraction). That is based totally on deep feature mastering with gradient-based totally linear SVM to install and run on resource-confined gadgets through decreasing the number of functions thru function extraction and selection the use of a SAE, MI. The impact is trained on AWID to takeoff attack. Numerical effects display that the proposed impact finished 98. 22% accuracy with ninety-seven. Sixty-four% detection price and 1. 20% fake alarm rate and outperformed existing brand-new benchmark models. Some other key contribution of this look is the research of the functions in AWID dataset for its serviceability for similar improvement of ids.	Internet of Things (IoT), Support Vector Machine (SVM), Stacked Autoencoder (SAE), Mutual Information (MI), Aegean Wi-Fi Intrusion Dataset (AWID)
(Funchal et al., 2020)	ML enabled Security for a multi-agent cyber-physical conveyor	One most important substance of industry 4. 0 which is connectivity of gadgets and structures, the use of IoT, in which CPS performs like spine organization constructed on disbursed and distributed systems. The method gives vast advantages, particularly progressed concert, awareness and maintainability, however additionally carries about complications in phrases of security, because the strategies and systems turn out to be susceptible to cyberattacks. This paper defines the enactment of numerous algorithms to boost the safety in a self-prepared CPS, based on MAS and building up with special separate segmental and sensible transporter segments. IDS is formed to supported by using ML strategies that analyzes the communique among agents, permitting to screen and examine the occasions that occur inside the device, removing signs of intrusions, collectively they help to decreases cyberattacks.	Internet of Things (IoT), Cyber-Physical System (CPS), Multi Agent Systems (MAS), Intrusion Detection System (IDS)

continues on following page

Table 1. Continued

Ref No.	Objective	Summary	Technology used
(Coulter et al., 2020)	Facts-driven cyber safety in angle smart visitor's evaluation	Social and internet site visitor's examination is essential in noticing and protecting cyber-attacks. Conventional methods resorting to physically described rules are regularly exchanged by means of automatic methods authorized with the aid of device learning. The revolution is multiplied by massive datasets that guide system monitoring models with awesome overall performance. Within the framework of a records-pushed paradigm, this article evaluations current analytical studies on cyber traffic concluded societal webs and the cyberspace via the use of a hard and fast of common concepts of similarity, correlation, and collective indication, and through sharing safety desires for classifying community host or packages and users or tweets. The potential to achieve this is not determined in separation, nonetheless as a substitute drawn for a huge use of many unique communities or social flows. Moreover, the flows showcase many traits, along with constant size and a couple of messages among supply and terminus. This text establishes a new studies technique of DDCS and its used in social and net visitor's evaluation. The background of the DDCS procedure includes three additives, this is, cyber safety statistics processing, cyber features safety engineering, and cyber safety system. Demanding situations and destiny instructions on this area are also mentioned.	Data Driven Cyber Security (DDCS)
(Farooq & Otaibi, 2018)	Threat prediction using ML enable optimal solution	Through the scramble in cyber threats, agencies are now determined for higher information removal strategies in order to research security logs acquired from their infrastructures to make certain powerful and automatic cyber risk recognition. ML based totally analytics for security gadget records is the following emerging fashion in cyber protection, geared toward mining safety statistics to discover superior targeted cyber threats actors and diminishing the functioning expenses of retaining static association policies. Though, collection of the most reliable machine studying procedure for safety log analytics nevertheless remains an impeding aspect against the success of statistics science in cyber protection because of the hazard of large variety of fake-high quality detections, especially in the case of huge-scale or worldwide SOC environments. This fact brings a dire need for an efficient machine studying based cyber chance recognition, able to minimize the fake detection charges. In this paper, we're proposing a gold standard system gaining knowledge of procedures with their application framework based on logical and experiential evaluations of accrued results, even as the use of diverse prediction, type and forecasting algorithms.	Security Operations Center (SOC)
(Bakhshi & Shahid, 2019)	Profiling of bio-cyber interfaces using ML	The IoBNT is a developing model on the transportations of IoT, e-medical and artificial natural science. Iob nt is based on modernizing molecular communication among AI nanodevices to help in medicine shipping, far flung specialist care and health care control. The net. IoBNT consequently, needs strong safety primitives to properly discourse patient secrecy apprehensions, prosperous health practitioner self-belief in addition to limit (any) malicious operation ensuing in bio-terrorism. The present work centered on securing the bio-cyber interfacing of IoBNT skills and considered 3 widely wide-spread bioelectric transduction techniques along with bioluminescence, redox modality and BioFETs. The use of the ultra-modern ML algorithms, the proposed protection framework hired parameters profiling the awesome working capabilities of bio-cyber interfaces and aimed to become aware of anomalous operation. At some stage in validation, a finest summarizing precision between 88-91% range was verified.	Internet of Bio-Nano Things (IoBNT), Internet of Things (IoT), bio-electrical transduction techniques, bioluminescence, redox modality and biological field effect transistors (BioFETs).

- Artificial intelligence identifies unknown threats

A person may not be able to become aware of all the threats a corporation faces. Every 12 months, hackers release masses of thousands and thousands of assaults with exceptional motives. Unknown threats can cause large damage to a network. AI has validated to be one of the exceptional technologies which is mapping and preventing unknown threats and save systems.

- Discover hidden threats over traffic

AI can deal with quite a lot of information and plenty of pastime occurs on an organization's network. An average mid-sized agency itself has massive site visitors. Meaning there's loads of facts transferred among clients and the business day by day. This statistic wishes safety from malicious people and software programs. However then, cybersecurity personnel can't take a look at all of the traffic for possible threats. AI is the satisfactory solution that will help you discover any threats masked as ordinary pastime. Its computerized nature permits it to skim through huge chunks of records and site visitors. Technology that uses AI, which includes a residential proxy, permits you to switch data. It may also stumble on and discover any threats hidden within the sea of disordered traffic.

- Higher vulnerability control

vulnerability management is fundamental to prevent a network. As stated in advance, a typical concern deals with many threats daily. It desires to hit upon, perceive and save you to be secure. Evaluating and calculating the present safety features over AI investigation can help in susceptibility control. AI allows us to verify structures faster than cybersecurity employees, in that way growing our trouble-fixing capability. It detects susceptible factors in commercial enterprise networks and facilitates businesses attention on crucial safety responsibilities. It is possible to manipulate susceptibility and secure devices within a period.

- Better threat security

the threats that business networks face change now and again. Hackers trade their techniques each day. That makes it tough to prioritize safety obligations at a corporation. You could cope through a phishing attack through ransomware at a cross. The solution right here is to install AI to discover all sorts of attacks and prioritize and save our devices.

- Reducing duplicative procedures

Primary safety satisfactory practices continue to be the same every day. In case you lease someone to deal with those responsibilities, they'll get bored along the way. Or they may experience worn-out and complacent and miss a critical safety challenge and reveal your community. AI, at the same time as mimicking human characteristics and fulfilling the shortcomings, protects duplicative cybersecurity tactics that could drag our cybersecurity personnel. It enables tests for basic security threats and saves devices from them on an ordinary basis. It additionally analyzes your network extensively to see if there are protection holes that would be destructive in your network.

- Speeds up detection and reaction instances

Threat detection is the start of protecting your enterprise's community. It might be pleasant in case you detected things like untrusted facts quickly. It will save you from the irreversible harm on your network. The satisfactory manner to stumble on and respond to threats in time is through integrating AI with cybersecurity. AI examinations our whole system and tests for any possible threats. AI will identify threats and easily protect devices.

- Securing authentication

Maximum websites have a user account function wherein unity logs in to become right of entry to offerings or purchase merchandise. Some have interaction with administration that visitor's requirement to fill with intuitive information. As a business enterprise, you want an additional security layer to run this kind of website as it involves non-public statistics and sensitive records. The extra protection layer will make sure that your site visitors are secure at the same time as browsing your network. Ai secures authentication whenever a consumer wants to log into their account. AI makes use of various equipment including facial reputation, captcha, and fingerprint scanners amongst others for identity. The data accrued with the aid of these features can assist to discover if a log-in try is actual or now not.

Figure 1. Advantages of AI-ML based cyber security

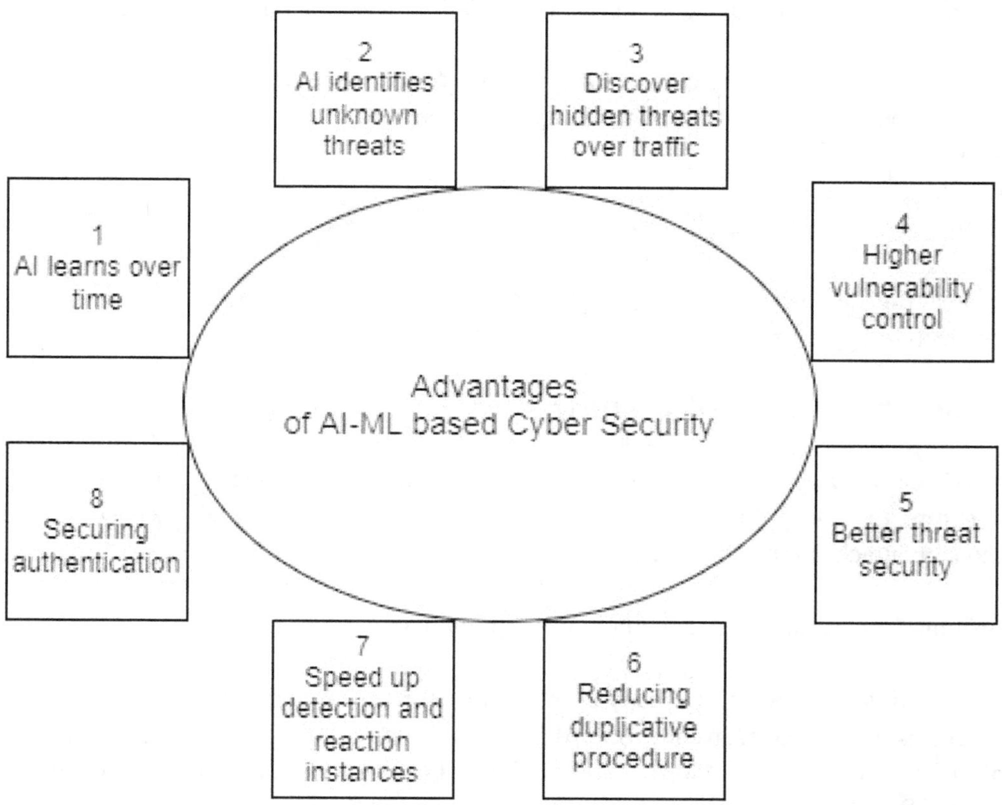

Limitation of Artificial Intelligence Enable Security System

Prediction of AI techniques is totally based on availability of records, this is also a barrier for AI security. To overcome this, we should have a clean strategy from the outset for sourcing the statistics that your AI will require. Some other key barriers to AI adoption are the abilities shortage and the provision of technical teams of workers with the revel in and schooling essential to efficiently deploy and function AI solutions. Investigate indicates skilled statistics scientists are in short supply as are other specialized statistics professionals skilled in device learning, schooling accurate fashions, etc. Value is any other key consideration with shopping ai technology. Companies that lack in-house abilities or are strange with ai frequently must outsource, that's in which demanding situations of fee and renovation come in. Due to their complex nature, clever technology can be steeply-priced and you could incur in addition costs for restore and ongoing maintenance. The computational fee for training statistics fashions etc. also can be an additional fee. Software program packages want ordinary upgrading to adapt to the converting business surroundings and, in case of breakdown, give a hazard of dropping code or important data. Restoring that is frequently time-consuming and expensive. But this hazard is not any extra with AI (Yarali, 2022a, 2022b) than with different software program improvement. Supposed that the system is designed nicely and that those buying AI apprehend their necessities and options, those risks can be mitigated. The limitations of AI-ML based security are shown in Figure.2.

Other AI barriers are as following:

- Delay Operational execution time
  Operation execution times, which may be extended on what you are attempting to enforce.
- Need professional
  Integration demanding situations and lack of knowledge of the latest systems.
- Limitation on usability and interoperability.
- Trust issues.
- lack of transparency.
- Technological complexity.
- Task losses
- The capability of automation technology to give upward thrust to task losses.
- Smart technology increases overall cost of the system

Application of AI-ML Based Security

The usage of system gaining knowledge of to hit upon malicious pastime and detect attacks

device getting to know procedures determine that assistance industries to quickly perceive malicious attacks and stop attacks earlier than they begin.

- To examine, detect and protect cell phone from malicious software

Machine gaining knowledge is already going mainstream on cellular devices, however to this point maximum of this hobby has been for riding progressed voice-based experiences on the likes of google now, apple's Siri, and amazon's Alexa (Lit et al., 2021; Lee et al., 2020; Yıldırım et al., 2019; Krueger & McKeown, 2020; Edu et al., 2021; Haesler et al., 2018). Yet there's software for safety too. As noted above, google is the use of gadget learning to analyze threats in opposition to cellular endpoints, at the same time as employers are seeing an opportunity to protect the growing range of bring-your-own and select-your-own cell devices. Each device uses its own ML algorithm that gets to know a set of rules to stumble on potential threats.

Figure 2. Limitations of AI-ML based cyber security

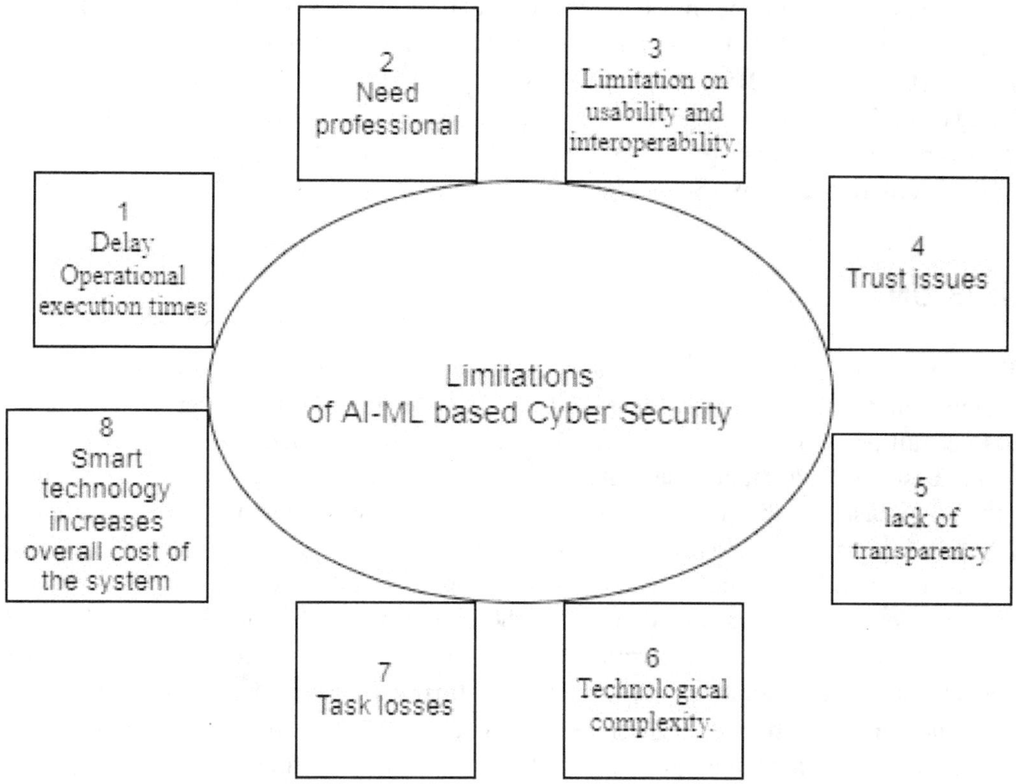

- The usage of machine learning to enhance human analysis

On the coronary heart of machine learning (Rueden et al.,2021; Reznik et al., 2022; Wang et al., 2022; Das & Bhattacharya, 2021; Das et al., 2020) in protection, there's the assumption that it benefits human analysts with all elements of the work, together with predicting malicious assaults, evaluating the community, termination safety and susceptibility valuation. There's arguably most excitement although round danger intelligence. Reviewing hundreds of thousands of logins each day, the device turned into able to filter records and skip it onto the humanoid predictor, decreasing indicators right down to round one hundred in step with day.

- Using system understanding to automate repetitive safety

The actual advantage of system getting to know is that it is able to automate repetitive duties, allowing personnel to recognize greater important work.

- Other applications areas:

Other applications are of cyber physical systems are as follows: Smart workshop (Leng et al., 2019), cyber security (Lee et al., 2019), safe human robot collaboration (Nikolakis et al., 2019), manufacturing technology(Leng et al., 2019), production planning(Ma et al.,2019), Health security((S. Das et al., 2022; Zhang et al., 2017), adaptive shop floor (Mourtzis & Vlachou, 2018; Zhang et al., 2017), industry 4.0(O'Donovan et al.,2018), cost efficient resource management(Gu et al., 2017), optimisation (Liang et al., 2018), anomaly detection1(Schneider et al., 2018).

CONCLUSION

This paper provides us with an idea or logic of Artificial Intelligence important for cyber security, not only Artificial Intelligence but also Machine Learning is beneficial for cyber security. The fact that we cannot get full proof and guarantee regarding security, but digitalization is on peak nowadays, and giving us the best service related to security as much as it can. But if we step in digitalization and talk about it then we must also discuss the vital points and facts such as its pros and cons, limitations and others. The cyber physical system is also introduced to us for our own benefit and need, though nowadays everyone is using the internet and hence by the help of the internet we can operate the system very easily. Machine learning has its own demand in the market which has already been discussed in this paper. And this demand in machine learning led this sector and area to more and more growth of new and advanced technology. People are ready to invest in this sector as much as they can because investing lets their business in a growth phase as well as this is making the world more digitized and advanced. By the help of machine learning techniques and models and a few powerful algorithms we can also achieve protection in the system and also Artificial Intelligence helps with its complex algorithm too in security system and protection. This paper also tells us the broad spectrum of Artificial Intelligence we are using right now and also the application which can be used or going to be discovered in future as well. We must also identify the sector where this is applicable with proper deep studies and research, this paper contains all the information which are discussed in above. Future scope of this field is growing rapidly due to modernization and digitalization and hence the evolution in technology is taking the world into a different and advanced state.

REFERENCES

Bakhshi, T., & Shahid, S. (2019). Securing Internet of Bio-Nano Things: ML-Enabled Parameter Profiling of Bio-Cyber Interfaces. *2019 22nd International Multitopic Conference (INMIC)*, 1-8. 10.1109/INMIC48123.2019.9022753

Coulter, R., Han, Q.-L., Pan, L., Zhang, J., & Xiang, Y. (2020). Data-Driven Cyber Security in Perspective—Intelligent Traffic Analysis. *IEEE Transactions on Cybernetics*, *50*(7), 3081–3093. doi:10.1109/TCYB.2019.2940940 PMID:31634146

Das, S., & Bhattacharya, A. (2021). ECG Assess Heartbeat rate, Classifying using BPNN while Watching Movie and send Movie Rating through Telegram. In J. M. R. S. Tavares, S. Chakrabarti, A. Bhattacharya, & S. Ghatak (Eds.), *Emerging Technologies in Data Mining and Information Security. Lecture Notes in Networks and Systems* (Vol. 164). Springer. doi:10.1007/978-981-15-9774-9_43

Das, S., Das, J., Modak, S., & Mazumdar, K. (2022). *Internet of Things with Machine Learning based smart Cardiovascular disease classifier for Healthcare in Secure platform*. Academic Press.

Das, S., Ghosh, L., & Saha, S. (2020). *Analyzing Gaming Effects on Cognitive Load Using Artificial Intelligent Tools*. Advance online publication. doi:10.1109/CONECCT50063.2020.9198662

Edu, J., Ferrer Aran, X., Such, J., & Suarez-Tangil, G. (2021). SkillVet: Automated Traceability Analysis of Amazon Alexa Skills. IEEE Transactions on Dependable and Secure Computing. doi:10.1109/TDSC.2021.3129116

Farooq, H. M., & Otaibi, N. M. (2018). Optimal Machine Learning Algorithms for Cyber Threat Detection. *2018 UKSim-AMSS 20th International Conference on Computer Modelling and Simulation (UKSim)*, 32-37. 10.1109/UKSim.2018.00018

Funchal, G., Pedrosa, T., Vallim, M., & Leitao, P. (2020). Security for a Multi-Agent Cyber-Physical Conveyor System using Machine Learning. *2020 IEEE 18th International Conference on Industrial Informatics (INDIN)*, 47-52. 10.1109/INDIN45582.2020.9478915

Gu, L., Zeng, D., Guo, S., Barnawi, A., & Xiang, Y. (2017). Cost Efficient Resource Management in Fog Computing Supported Medical Cyber-Physical System. *IEEE Transactions on Emerging Topics in Computing*, *5*(1), 108–119. doi:10.1109/TETC.2015.2508382

Haesler, S., Kim, K., Bruder, G., & Welch, G. (2018). Seeing is Believing: Improving the Perceived Trust in Visually Embodied Alexa in Augmented Reality. *2018 IEEE International Symposium on Mixed and Augmented Reality Adjunct (ISMAR-Adjunct)*, 204-205. 10.1109/ISMAR-Adjunct.2018.00067

Krueger, C., & McKeown, S. (2020). Using Amazon Alexa APIs as a Source of Digital Evidence. *2020 International Conference on Cyber Security and Protection of Digital Services (Cyber Security)*, 1-8. 10.1109/CyberSecurity49315.2020.9138849

Lee, E., Vesonder, G., & Wendel, E. (2020). Eldercare Robotics – Alexa. *2020 11th IEEE Annual Ubiquitous Computing, Electronics & Mobile Communication Conference (UEMCON)*, 820-825. . doi:10.1109/UEMCON51285.2020.9298147

Lee, J., Azamfar, M., & Singh, J. (2019). A blockchain enabled Cyber-Physical System architecture for Industry 4.0 manufacturing systems. *Manufacturing Letters*, *20*, 34–39. doi:10.1016/j.mfglet.2019.05.003

Lee, S. J. (2020). IMPACT: Impersonation Attack Detection via Edge Computing Using Deep Autoencoder and Feature Abstraction. IEEE Access, 8, 65520-65529. doi:10.1109/ACCESS.2020.2985089

Leng, J., Zhang, H., Yan, D., Liu, Q., Chen, X., & Zhang, D. (2019). Digital twin-driven manufacturing cyber-physical system for parallel controlling of smart workshop. *Journal of Ambient Intelligence and Humanized Computing*, *10*(3), 1155–1166. doi:10.100712652-018-0881-5

Liang, Y., Lu, X., Li, W., & Wang, S. (2018). Cyber Physical System and Big Data enabled energy efficient machining optimisation. *Journal of Cleaner Production*, *187*, 46–62. doi:10.1016/j.jclepro.2018.03.149

Lit, Y., Kim, S., & Sy, E. (2021). *A Survey on Amazon Alexa Attack Surfaces. 2021 IEEE 18th Annual Consumer Communications & Networking Conference*. doi:10.1109/CCNC49032.2021.9369553

Ma, S., Zhang, Y., Lv, J., Yang, H., & Wu, J. (2019). Energy-cyber-physical system enabled management for energy-intensive manufacturing industries. *Journal of Cleaner Production*.

Mourtzis, D., & Vlachou, E. (2018). A cloud-based cyber-physical system for adaptive shop-floor scheduling and condition-based maintenance. *Journal of Manufacturing Systems*, *47*, 179–198. doi:10.1016/j.jmsy.2018.05.008

Nikolakis, N., Maratos, V., & Makris, S. (2019). A cyber physical system (CPS) approach for safe human-robot collaboration in a shared workplace. *Robotics and Computer-integrated Manufacturing*, *56*, 233–243. doi:10.1016/j.rcim.2018.10.003

O'Donovan, P., Gallagher, C. V., Bruton, K., & O'Sullivan, D. (2018). A fog computing industrial cyber-physical system for embedded low-latency machine learning Industry 4.0 applications. *Manufacturing Letters*, *15*, 139–142. doi:10.1016/j.mfglet.2018.01.005

Reznik, L. (2022). Hackers versus Normal Users. In *Intelligent Security Systems: How Artificial Intelligence, Machine Learning and Data Science Work For and Against Computer Security* (pp. 247–313). IEEE. doi:10.1002/9781119771579.ch5

Schneider, P., & Böttinger, K. (2018). High-Performance Unsupervised Anomaly Detection for Cyber-Physical System Networks. *Proceedings of the 2018 Workshop on Cyber-Physical Systems Security and PrivaCy*. 10.1145/3264888.3264890

von Rueden, L. (2021). Informed Machine Learning - A Taxonomy and Survey of Integrating Prior Knowledge into Learning Systems. IEEE Transactions on Knowledge and Data Engineering. doi:10.1109/TKDE.2021.3079836

Wang, J., Suo, J., & Chortos, A. (2022). Design of Fully Controllable and Continuous Programmable Surface Based on Machine Learning. *IEEE Robotics and Automation Letters*, *7*(1), 549–556. doi:10.1109/LRA.2021.3129542

Yarali, A. (2022a). Artificial Intelligence Technology. In *Intelligent Connectivity: AI, IoT, and 5G* (pp. 95–115). IEEE. doi:10.1002/9781119685265.ch5

Yarali, A. (2022b). Applications of Artificial Intelligence, ML, and DL. In *Intelligent Connectivity: AI, IoT, and 5G* (pp. 279–297). IEEE. doi:10.1002/9781119685265.ch16

Yeboah-Ofori, A. (2021). Cyber Threat Predictive Analytics for Improving Cyber Supply Chain Security. IEEE Access, 9, 94318-94337. doi:10.1109/ACCESS.2021.3087109

Yıldırım, İ., Bostancı, E., & Güzel, M. S. (2019). Forensic Analysis with Anti-Forensic Case Studies on Amazon Alexa and Google Assistant Build-In Smart Home Speakers. *2019 4th International Conference on Computer Science and Engineering (UBMK)*, 1-3. 10.1109/UBMK.2019.8907007

Zhang, Y., Qian, C., Lv, J., & Liu, Y. (2017). Agent and Cyber-Physical System Based Self-Organizing and Self-Adaptive Intelligent Shopfloor. *IEEE Transactions on Industrial Informatics*, *13*(2), 737–747. doi:10.1109/TII.2016.2618892

Zhang, Y., Qiu, M., Tsai, C., Hassan, M. M., & Alamri, A. (2017). Health-CPS: Healthcare Cyber-Physical System Assisted by Cloud and Big Data. *IEEE Systems Journal*, *11*(1), 88–95. doi:10.1109/JSYST.2015.2460747

Chapter 4
Machine Learning (ML) Methods to Identify Data Breaches

Gururaj H. L.
https://orcid.org/0000-0003-2514-4812
Vidyavardhaka College of Engineering, India

Pooja M. R.
Vidyavardhaka College of Engineering, India

Pavan S. P. Kumar
Vidyavardhaka College of Engineering, India

ABSTRACT

In this digitized world, everything is changing from offline to online. Data plays a vital role in this digital network. The theft or loss of USB devices, computers, or mobile devices by an unauthorized person who gains access to your mobile or laptop devices, email account, or network is generally termed as a data breach. Securing data from theft and breaches is a challenging issue. It is very hard to identify data breaches in complex networks. Adding extra intelligence using machine learning (ML) approaches will be efficient in identifying such attackers. In this chapter, various ML techniques to identify data breaches such as malware attack, man in the middle (MIM), spear phishing attack, eavesdropping attack, password attack, cross-site scripting attack will be depicted with suitable case studies.

INTRODUCTION

The Indian scenario of communication completely changed from the recent past (E. Guven et al., 2016). Nowadays the importance of data is at its height. The users are trying very hard to secure the data in one or another way. Cyber Security is the protection of information, modification of data, data breaches from an unauthorized person (D. C. Le et al., 2019). A crime is conducted by criminals. A crime conducted in which the computer is directly or directly instrumental. The statistical survey according to Reliance on AI in response to cyber-attacks is depicted in Figure 1.

DOI: 10.4018/978-1-6684-3991-3.ch004

'Cyber' is a network that is vulnerable to the outside world. Cybercrime can be defined as any financial dishonesty that takes place in a computer environment or any threats to the computer itself, such as theft of the hardware or software for ransom.

Figure 1. Reliance on AI in response to cyber attacks

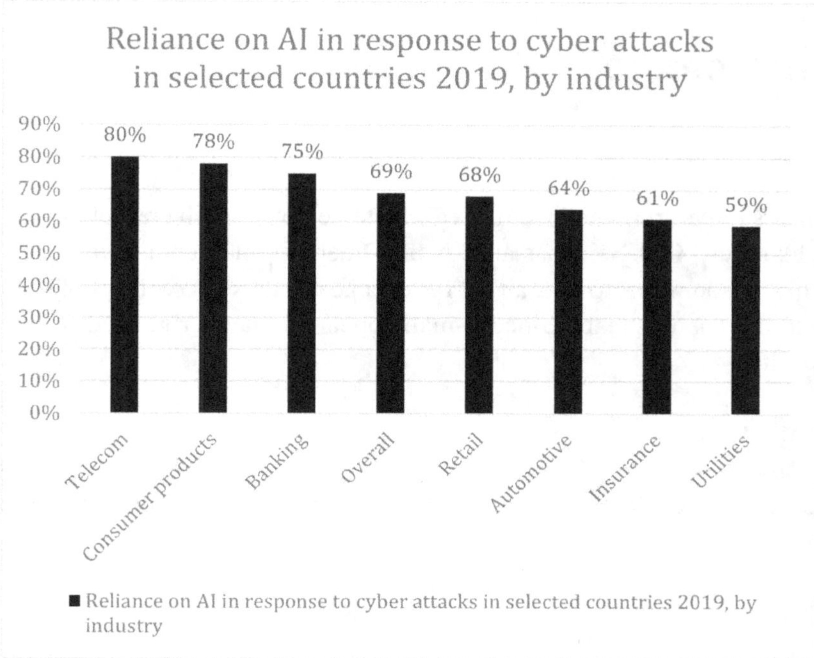

CYBER CRIMES

In this subsection, various cyber-attacks were introduced, and their details are explained. There are two kinds of attacks Techno-crime & Techno-vandalism. Techno-crime is an act against a system, with the intent of copying, steal or modifying the data. This type of attack is possible when the system is connected to the internet for 24x7 (Zincir-Heywood et al., 2019). Techno-Vandalism is a brainless defacement of the websites, such as publicizing someone else information. There are three types of Cybercriminals.

Cybercriminals – hungry for recognition, Cybercriminals – not interested in recognition and Cybercriminals – the insiders.

Some various types of cybercrimes are:

- IT Professionals
- Hobby hacker
- Politically motivators
- Terrorist

- Psychologically perverts
- Financially motivators
- State sponsored hacker
- Organized criminals
- Formal employees seeking revenge
- Economic advantage to damage.

TYPES OF CYBERCRIMES

E-Mail Spoofing

Spoofed mail appears to be originated from a legitimate person (J. Polverari et al., 2018).

For example, let us say, Girl Neha has an E-mail ID neha@gmail.com. But one of her close friends becomes her enemy he/she wants to take an act of revenge on her so he will spoof her E-mail and sends vulgar messages to her friend's team. Since E-mail appears to have originated from Neha, her friends can take offense to her.

Cyber Defamation

Cyberdefamation takes place when someone publishes defamatory information about someone on a pornographic site.

Internet Time Theft

Accessing someone else Internet by stealing a Wi-Fi password without the knowledge of the owner.

Salami Attack

These attacks are used for committing financial crimes (Rolleston et al., 2015). For example, let us say, one of the bank employees inserts a program into a bank server that deducts a small amount of money from the account holder. But no-account holder will notice this kind of deduction. But the employee will make a sizable amount each month.

Web Jacking

Web jacking occurs when someone forcefully takes control of someone else website.

Identity Theft

Identity theft is a fraud involving another person's identity for an illicit purpose (T. Rashid et al., 2016). This occurs when a criminal uses someone else's identity for his/her own illegal purpose.

E-Mail Bombing

E-Mail bombing refers to sending many E-Mails to the victim to crash Victim's E-Mail account.

Password Sniffing

Sniffers are program that monitor and record the credentials such as, username and password of the network users.

The various functions and categories of cyber security (B. Bose et al., 2017) such as Recover, Respond, Detect, Protect and Identify, along with their unique identifiers are shown in Figure 2.

Figure 2. Cyber security model

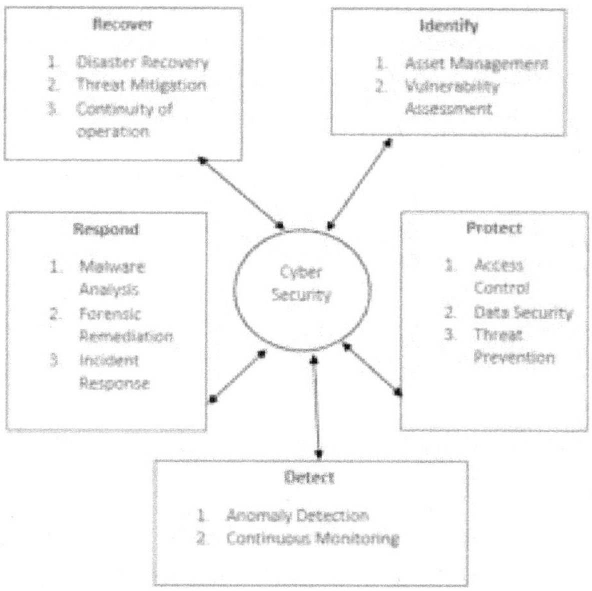

The chapter has been structured as follows: section 2 elaborates the detailed perspective of various authors on data breaches, section -3 and section-4 exemplifies the case studies and at last section-5 draws the conclusion.

DATA BREACHES

Data security is the subbranch of information security which in turn comes under network security. Data theft & data loss is very common in this current situation (W. Wu et al., 2015). Various data breaches and their mitigation techniques at the early stage are elaborated in this section.

E-mail application is one of the most common ways for data breach using spam emails. Nowadays, machine learning models like naïve-Bayes and Decision tree will be used to classify E-mails. Figure 3 depicts how to classify the E-mails whether it is spam or not spam.

Figure 3. E-mail classification

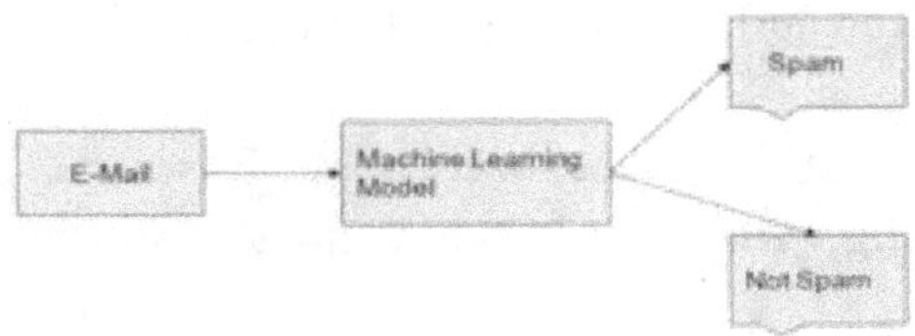

The Behavior Analysis, which targets social media and online advertisements to a carefully defined audience, also has a place in cybersecurity. When combined with machine learning methods, behavioral analytics helps security pros detect harmful patterns of system or network behavior and respond to cyberthreats in real-time. Figure 4 shows how the behavior analysis defends between customer and enterprise against cybercrime.

Figure 4. Behavior analysis defends against cybercrime

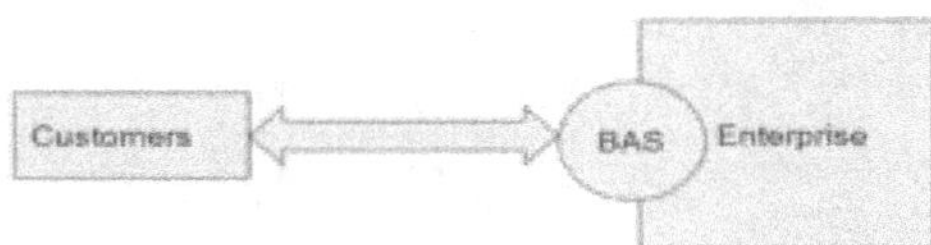

The objective of the Machine Learning algorithms is that allow a machine, rather than human intervention, to view actions learned from previous experience (Abawajy J et al., 2014). Figure 5 explores machine learning applications in cybersecurity and how machine learning algorithms will help combat cyber-attacks. Four different key cyber security initiatives were depicted in Figure 6.

Figure 5. Applications of ML algorithms in cybersecurity

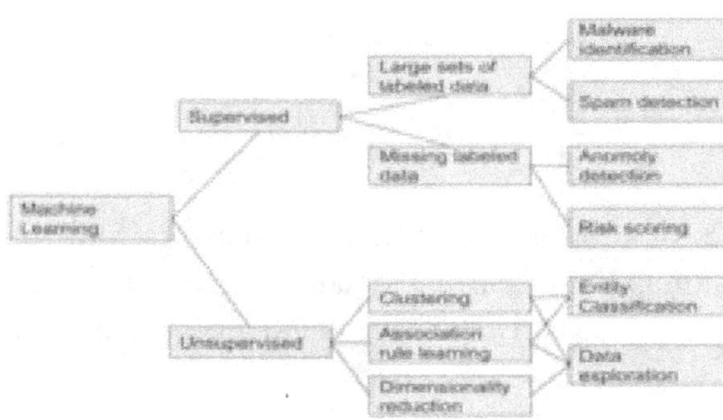

Figure 6. Cybersecurity business initiatives

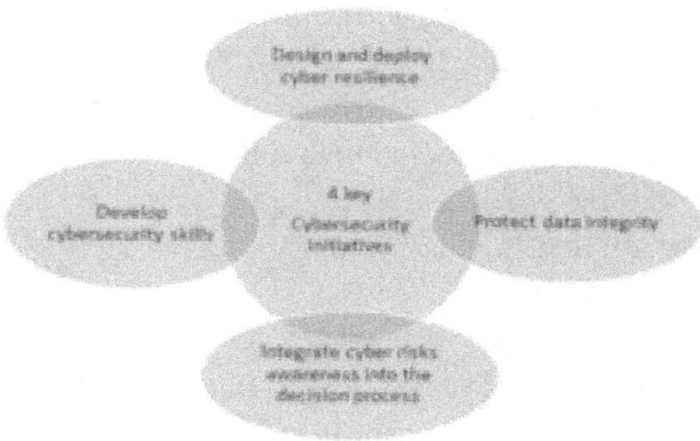

APPLICATIONS

Nowadays, With the ever-increasing advances in technology, cyber-attacks had been on the upward push in latest years. A huge estimation is that cyberattacks value the US trillions of greenbacks each yr. Machine mastering is one method to combat cybercrimes. With such a lot of cybercrimes getting dedicated each day, there are too many for people to forestall on their very own (Kshetri N et al., 2016). Even massive cybersecurity groups can most effective work so fast to stop hackers. Machine studying includes computers helping with the workload that human beings can't preserve up with.

Machine gaining knowledge makes finding safety threats less difficult to do and it makes compiling big quantities of statistics simpler. The system cannot analyze without human beings coaching the device the whole lot it desires to understand (Klimburg A et al., 2012). Machines can procedure information quicker than the human brain, but they cannot apprehend something they may be not taught to recognize. Computer systems are processing facts to identify potential trends. They are slowly however clearly getting better at spotting scams which include phishing emails, online financial institution fraud, and detecting malware.

Thieves and laptop hackers are continuously making changes inside the programs they use and the procedures they get right of entry to facts that companies imply to be non-public. Machine reading permits the pc structures to no longer simplest to understand what they have got seen earlier than, but they could recognize matters they have got now not visible earlier than, and their human contrary numbers can educate the gadget what they should do inside the event that they see the same data sample seem in the destiny. Machines are quick students who in no way get worn out, by no means lose their enthusiasm, and preserve everything that they are taught.

Detecting Network Threats

If there's a first-rate deal of community visitors daily, it can be difficult to perceive nefarious site visitors. Security gear that uses device mastering can examine the IP addresses and strategies of hackers from the beyond and move-references those identifiers to the modern-day situation (Badsha, S. et al., 2020). If they notice any similarities, the computer will send out an alert, indicating that there are signs of a recognized attacker on the network. This alert will even go off if an unauthorized person is trying to hack right into a categorized account.

Unfortunately, there are not only threats from outside assets to take care of. With the rise of jobs in cybersecurity, companies will make a new hire and later discover that the person changed into attempting to hack into the network from the inside. To keep away from this, an awesome idea is to run a public data look at on any new candidate which you are seriously thinking about. This will bring up their crook record file, and you may see if they committed comparable crimes within the past.

Protecting Against New Viruses

There are hundreds of thousands of recent viruses that release into society each year (Jhanjhi, N. et al., 2020). Some of these malware traces perform without binary files, ensuing in them being very tough to discover. This makes it simpler for the malware to infiltrate gadgets and wreak havoc on employer structures. Machine mastering can discover hidden malware that conventional anti-virus software program might leave out. AI algorithms can examine the supply code of malware to decide its legitimacy and block it if required.

Machine Learning techniques for cyber security

Regression - Regression is straightforward. The knowledge about the present statistics is applied to have a concept of the new information. Take an example of residence prices prediction. In cybersecurity, it could be implemented to fraud detection (Mohamed Amine Ferrag et al., 2019). The features determine the probability of fraudulent moves.

As for technical factors of regression, all strategies can be divided into big classes: machine gaining knowledge of and deep getting to know.

Classification - Classification is also truthful. Imagine you have two piles of photographs classified with the aid of type. In phrases of cybersecurity, spam filters out keeping apart spams from other messages can serve as an example. Spam filters are probably the primary ML method applied to Cybersecurity duties. The supervised getting to know technique is normally used for type (Priyanka Dixit et al., 2020).

Clustering - Clustering is much like a class with the handiest however essential distinction. The records approximately the lessons of the facts are unknown. There is not any concept of whether these records can be classified. This is unsupervised getting to know. Supposedly, the first-rate undertaking for clustering is forensic analysis. The motives, direction, and results of an incident are obscure. It's required to categorize all activities to locate anomalies.

Decision Tree - In its simplest form, a determination corner is a classification of flowchart that demonstrates a crystallise course of action to a decision (Moneer Alshaikh et al., 2020). In designations of collections analytics, it is a classification of rule that incorporates counterfactual 'control' declarations to categorise data. determination trees are extraordinarily utilitarian for collections analytics and organization eruditeness thanks to they prison-breaking fine-tune heterogeneous collections into bounteous administrable parts. They're recurrently euphemistic pre-owned in these comedians for prognostication analysis, collections classification, and regression.

- Root Node - This is usually the first node within the path. It is the node from which all other choices, chance, and cease nodes are in the end branch as proven in Figure 7.
- Internal Node - Between the foundation node and the leaf nodes, we will have a range of internal nodes. These can consist of selections and threat nodes.

- Splitting - These sub-nodes may be any other inner node, or they are able to cause an final results.

Figure 7. Decision tree

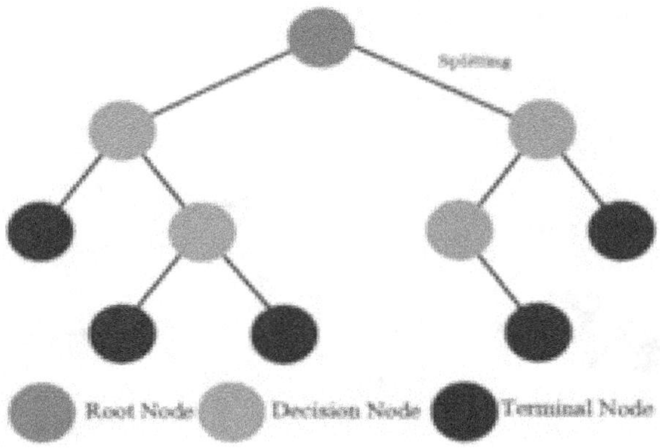

K-Nearest Neighbour - K-Nearest Neighbor is one of the handiest Machine Learning algorithms based mostly on the Supervised Learning approach (Dawn Branley-Bell et al., 2020). K-NN set of guidelines assumes the similarity a number of the new case/statistics and available instances and positioned the new case into the class that is most similar to the available instructions as shown in Figure 8.

K-NN set of guidelines shops all the to be had records and classifies a state-of-the-art information thing based totally on the similarity. This manner when new facts appears then it could be without problems classified right into a well suite elegance using K- NN algorithm.

Figure 8. K-Nearest Neighbour

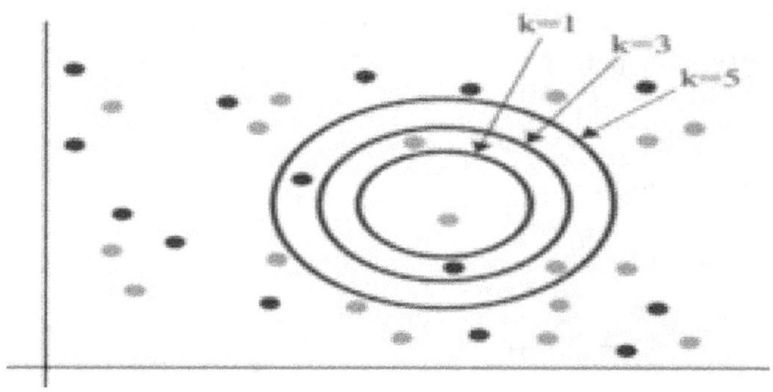

K-NN is a non-parametric set of rules, this means that it does not make any assumption on underlying statistics.

It is also referred to as a lazy learner algorithm because it does not study from the schooling set immediately alternatively it stores the dataset and on the time of type, it plays a movement on the dataset.

Suppose we have a picture of a creature that looks similar to cat and canine, however, we want to apprehend whether it's for a cat or canine. So, for this identity, we will use the KNN algorithm, as it works on a similar diploma. KNN version will locate the identical functions of the brand-new information set to the cats and puppies' pix and primarily based totally on the maximum similar abilities it's going to locate it in both cat or dog class as tested in Figure 9.

Figure 9. KNN classifier

Support Vector Machine - SVM is appropriated into considerateness thanks to the oftentimes euphemistic used and achiever disposition of ML for cyber safekeeping duties, exceptionally for IDS. SVM classifies and breaks apart the two writes down directions supported on the symbol to the perimeter on either expression of the hyperplane. digital audiotape five proffers the pictographic systematization of SVM (Farzana Quayyum et al., 2021). The faithfulness in classifying a write-down factor buoy be maximized via ontogenesis the perimeter and intervals between hyperplanes. The collections specks that complete distortion of the facts at the delimitation of the hyperplane is titled American man agent points. SVM is classed into cardinal virtually far-reaching categories. Its hawthorn is one-dimensional and non-linear particularly supported on the core group feature. It furthermore buoys be one-magnificence and multi-class supported on detection kind. SVM press for diversification of memory for processing and continuance for schooling. SVM be in want of teaching at contradistinctive continuance intervals for more appropriate after-effects to contemplate the enterprising user's behaviour. Figure 10 shows the maximum margin and optimal hyperplane between the support vectors.

Figure 10. Support vector machine

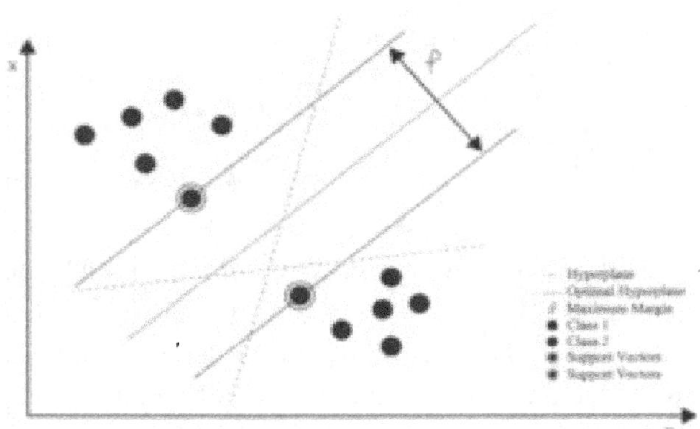

Reinforcement Learning - Reinforcement learning is one of the subfields of machine learning. The machine learning model can gain abilities to make decisions and explore in an unsupervised and complex environment by reinforcement learning (Khan et al., 2020). Reinforcement learning models use rewards for their actions to reach their goal/mission/task for what they are used to. So, reinforcement learning is different from supervised and unsupervised learning models.

The algorithm starts with trials and learns to make decisions by itself to gain maximum rewards. Reinforcement learning models can gain experience and feedbacks (rewards) from their actions which help them to improve their results. This machine learning approach can be best explained with computer games.

Figure 11 shows simple flow for the agent–environment interaction in a Markov decision process.

There are five key elements of reinforcement learning models (B. C. Soundarya et al., 2021):

- Agent - The algorithm/function in the model that performs the requested task.
- Environments - The world in which the agent carries out its actions (Moti Zwilling et al., 2021). It uses current states and actions of the agent as input, rewards and next states of the agents as output.
- States - It refers to the situation of the agent in an environment. There are current and future/next states.
- Actions - The moves are chosen and performed by the agent to gain rewards.
- Rewards - Reward means desired behaviors that are expected from the agent. Rewards are also called feedback for the agent's actions in a given state and are described as results, outputs, or prizes in the model.

Figure 11. Reinforcement learning

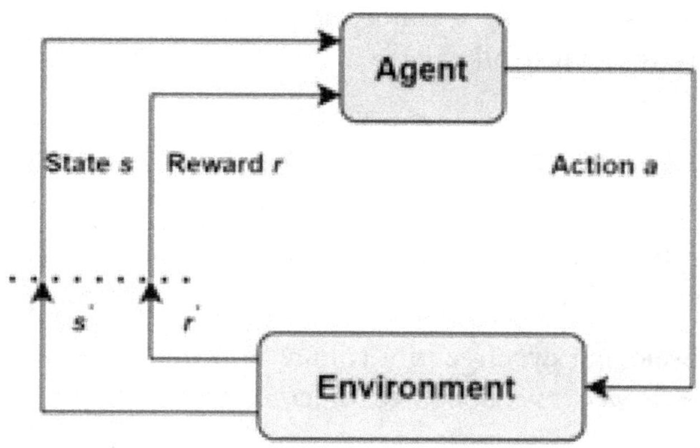

CONCLUSION

Data protection is one of the most distinguished worries companies have today: the common fee of a facts breach for a U.S. enterprise is $4.24 million. Machine Learning can assist in forestalling a breach through detecting malicious activity, examining cellular endpoints, and automating repetitive safety

Table 1. Comparative Analysis of Frequently used ML techniques

Model	Time Complexity	Description	Limitation
SVM	$O(n^2)$	SVM can be used for classification and regression (Axelrod CW et al., 2017).	Unable to address huge or noisier datasets effectively.
Naïve Bayes	$O(mn)$	A probabilistic classifier that takes less computational time (Priyanka Dixit et al., 2020).	Assigns 0 risk if a few classes within the test facts set is not gift inside the schooling data set.
ANN	$O(emnk)$	Adaptive and composed of Interconnected Artificial Neurons.	Assigns zero probability if some class in the take a look at records set isn't always gift in the schooling records set.
Decision Tree	$O(mn^2)$	Continue the manner until the anticipated elegance is acquired.	Difficult to alternate the records without affecting the overall structure.
K-mean	$O(kmni)$	Starts from random centroids refine centroids in iterations till the final cluster analysis.	Difficult to exchange the records with out affecting the general shape. Complex, highly-priced and time - eating.

tasks. A data breach may additionally encompass incidents such as theft or loss of digital media such as pc tapes, tough drives, or laptop computer systems with unencrypted information, posting such statistics on the world broad internet except for suited facts protection precautions, switch of such statistics to a machine which is now not definitely open however is no longer correctly or formally approved for security, such as unencrypted e-mail, or switch of such statistics to the statistics structures of a maybe opposed agency, such as a competing employer or an overseas nation, the place it may additionally be uncovered to extra intensive decryption techniques. Identifying the data breach can be useful to solve many problems including fraud detection, medical diagnosis, etc. Machine learning methods allow to automate anomaly detection and make it more effective, especially when large datasets are involved. In this chapter, we discussed various Machine Learning algorithms used to identify data breaches. In addition to those algorithms, some of the common ML methods used to identify data breaches include LOF, autoencoders, and Bayesian networks.

REFERENCES

Abawajy, J. (2014). User preference of cyber security awareness delivery methods. *Behaviour & Information Technology*, *33*(3), 237–248.

Alshaikh, M. (2020). Developing cybersecurity culture to influence employee behavior: A practice perspective. *Computers & Security, 98*. doi:10.1016/j.cose.2020.102003

Axelrod, C. W. (2017). Cybersecurity in the age of autonomous vehicles. intelligent traffic controls and pervasive transportation networks. In Systems, applications and technology conference (LISAT) (pp. 1-6). Academic Press.

Bose, B., Avasarala, B., Tirthapura, S., Chung, Y. Y., & Steiner, D. (2017). Detecting insider threats using radish: A system for real-time anomaly detection in heterogeneous data streams. *IEEE Systems Journal*.

Branley-Bell, D., Gómez, Y., Coventry, L., Vila, J., & Briggs, P. (2021). Developing and Validating a Behavioural Model of Cyberinsurance Adoption. *Sustainability*. doi:10.3390/su13179528

Buczak, A. L., & Guven, E. (2016). A Survey of Data Mining and Machine Learning Methods for Cyber Security Intrusion Detection. *IEEE Communications Surveys & Tutorials, 18*(2), 1153-1176. Advance online publication. doi:10.1109/COMST.2015.2494502

Choo, Gai, Chiaraviglio, & Yang. (2021). A multidisciplinary approach to Internet of Things (IoT) cybersecurity and risk management. *Computers & Security, 102*.

Dixit, P., & Silakari, S. (2020). Deep Learning Algorithms for Cybersecurity Applications: A Technological and Status Review. *Computer Science Review, 39*. doi:10.1016/j.cosrev.2020.100317

Ferrag, M. A., Maglaras, L., Moschoyiannis, S., & Janicke, H. (2019). Deep learning for cyber security intrusion detection: Approaches, datasets, and comparative study. *Journal of Information Security and Applications, 50*. doi:10.1016/j.jisa.2019.102419

Ferreira, P., Le, D. C., & Zincir-Heywood, N. (2019). Exploring Feature Normalization and Temporal Information for Machine Learning Based Insider Threat Detection. *2019 15th International Conference on Network and Service Management (CNSM),* 1-7. 10.23919/CNSM46954.2019.9012708

Gavai, G., Sricharan, K., Gunning, D., Hanley, J., Singhal, M., & Rolleston, R. (2015). Supervised and unsupervised methods to detect insider threat from enterprise social and online activity data. *Journal of Wireless Mobile Networks, Ubiquitous Computing and Dependable Applications*, *6*(4), 2015.

Ghiasi, M., Dehghani, M., Niknam, T., Kavousi-Fard, A., Siano, P., & Alhelou, H. H. (2021). Cyber-Attack Detection and Cyber-Security Enhancement in Smart DC-Microgrid Based on Blockchain Technology and Hilbert Huang Transform. *IEEE Access: Practical Innovations, Open Solutions*, *9*, 29429–29440. https://doi.org/10.1109/ACCESS.2021.3059042

Humayun, M., Niazi, M., & Jhanjhi, N. (2020). Cyber Security Threats and Vulnerabilities: A Systematic Mapping Study. *Arabian Journal for Science and Engineering*, *45*, 3171–3189. https://doi.org/10.1007/s13369-019-04319-2

Kavya Rani, S. R., Soundarya, B. C., Gururaj, H. L., & Janhavi, V. (2021). Comprehensive Analysis of Various Cyber Attacks. *2021 IEEE Mysore Sub Section International Conference (MysuruCon),* 255-262. doi:10.1109/MysuruCon52639.2021.9641089

Khan, N., Brohi, S., & Zaman, N. (2020). *Ten Deadly Cyber Security Threats Amid COVID-19 Pandemic*. doi:10.36227/techrxiv.12278792

Klimburg A, (2012). *National cyber security framework manual*. NATO Cooperative Cyber Defense Center of Excellence.

Kshetri, N. (2016). Cybersecurity and development. *Markets Globalization Dev Rev., 1*(2). doi:10.23860/MGDR-2016-01-02-03

Lallie, Shepherd, Nurse, Erola, Epiphaniou, Maple, & Bellekens. (2021). Cyber security in the age of COVID-19: A timeline and analysis of cyber-crime and cyber-attacks during the pandemic. *Computers & Security, 105*. doi:10.1016/j.cose.2021.102248

Le, D. C., & Nur Zincir-Heywood, A. (2019). Machine learning based Insider Threat Modelling and Detection. *2019 IFIP/IEEE Symposium on Integrated Network and Service Management (IM)*, 1-6.

Lee, I. (2020). Internet of Things (IoT) Cybersecurity: Literature Review and IoT Cyber Risk Management. *Future Internet, 12*(9), 157. doi:10.3390/fi12090157

Li, Y., & Liu, Q. (2021). A comprehensive review study of cyber-attacks and cyber security. Emerging trends and recent developments. *Energy Reports, 7*. doi:10.1016/j.egyr.2021.08.126

Nifakos, S., Chandramouli, K., Nikolaou, C. K., Papachristou, P., Koch, S., Panaousis, E., & Bonacina, S. (2021, July 28). Influence of Human Factors on Cyber Security within Healthcare Organisations: A Systematic Review. *Sensors (Basel), 21*(15), 5119. https://doi.org/10.3390/s21155119

Polverari, J. (2018). *Why less is more when it comes to cybersecurity 2018.* Available: https://www.forbes.com/sites/forbestechcouncil/2018/06/01/whyless-is-more-when-it-comes-to-cybersecurity/

Quayyum, F., Cruzes, D. S., & Jaccheri, L. (2021). Cybersecurity awareness for children: A systematic literature review. *International Journal of Child-Computer Interaction, 30*. doi:10.1016/j.ijcci.2021.100343

Rashid, T., Agrafiotis, I., & Nurse, J. R. (2016). A new take on detecting insider threats. *Intl. Workshop on Managing Insider Security Threats.*

Sarker, I. H., Kayes, A. S. M., & Badsha, S. (2020). Cybersecurity data science: An overview from a machine learning perspective. *Journal of Big Data, 7*, 41. https://doi.org/10.1186/s40537-020-00318-5

Wu, W., Kang, R., & Li, Z. (2015). Risk assessment method for cyber security of cyber physical systems. *1st International Conference on Reliability Systems Engineering*, Beijing, China.

Zou, B., Choobchian, P., & Rozenberg, J. (2021). (2021). Cyber resilience of autonomous mobility systems: Cyber-attacks and resilience-enhancing strategies. *Journal of Transportation Security, 14*, 137–155. https://doi.org/10.1007/s12198-021-00230-w

Zwilling, M., Klien, G., Lesjak, D., Wiechetek, Ł., Cetin, F., & Basim, H. N. (2022). Cyber Security Awareness, Knowledge and Behavior: A Comparative Study. *Journal of Computer Information Systems, 62*(1), 82–97. doi:10.1080/08874417.2020.1712269

Chapter 5
How Cyber Threat Intelligence (CTI) Ensures Cyber Resilience Using Artificial Intelligence and Machine Learning

Neelima Kant
Sharda University, India

Amrita
Sharda University, India

ABSTRACT

Cyber threat intelligence (CTI) has emerged as a critical pillar in a well-developed cyber security strategy. When used correctly, threat information may assist security teams in defending against an ever-more sophisticated threat landscape before, during, and after an attack. Groups can design more effective, more delicate, and durable cyber defenses by evaluating attackers and understanding their methods and aims. As a result, the purpose of this chapter is to give an overview of how CTI promotes cyber resilience by utilizing intelligent technologies such as artificial intelligence (AI) and machine learning (ML).

INTRODUCTION

Cyber Threat Intelligence is data that helps a company better understand the risks that have been, will be, or are currently being directed at it. This information is used to anticipate, prevent, and identify cyber-threats attempting to exploit valuable resources. Six reasons why CTI is so important:

- Lowering Costs - Because enhanced defenses help to limit an enterprise's risk, CTI can lower overall costs and save commercial firm funds.

DOI: 10.4018/978-1-6684-3991-3.ch005

- Lowering Risks - CTI gives the correct visibility into emerging security threats, reducing the risk of data loss, minimizing or preventing disruption in company operations, and ensuring regulatory compliance.
- Prevent data loss - CTI works as a watchdog when questionable IP addresses or domains seek to communicate with the community in order to capture vital information.
- Maximizing staffing - CTI increases an employer's safety crew's productivity by connecting threat intelligence with anomalies identified by network technologies.
- In-depth Threat Analysis – CTI allows a business to study a cybercriminal's various approaches. The employer can determine whether the safety protection systems can block such attacks by examining such cyber risks.
- Threat Intelligence Sharing - Sharing critical cyber security information, such as how hackers plot a security breach, could help others avoid a similar attack. The more the company is able to thwart these attempts, the less likely the hackers are to carry out such heinous attacks.

BACKGROUND

Researchers have been using various artificial intelligence (AI) and machine learning (ML) techniques to find cyber-attack indicators, malware evaluation, and anomaly detection techniques in recent years. With the advancement of IoT technology, the number of IoT devices/sensors has significantly expanded. Large-scale sensor-based structures are expected to become more prevalent in our societies, necessitating the development of creative approaches for designing and operating these new structures. The Cloud is migrating to the edge of the network, where resources such as routers, switches, and gateways are being virtualized, to help with the computational call of real-time delay-sensitive packages in large part dispersed IoT devices/sensors.

There are 6 major databases (Google scholar, IEEE Explore, ACM Digital Library, Science Direct, Web of Science, Scopus) used in (Xiong & Lagerstrom, 2019; Mckinnel et al., 2019) for providing the systematic literature review on Threat Modelling and AI in Vulnerability analysis and Penetration Testing. The study provides a lot of scope and challenges in cyber security using AI and ML techniques. Viz.

- Direct applications of vulnerability analysis and penetration testing using AI and ML
- Used real data for analysis of false positive rates using different approach and techniques in different research papers.
- It also covers partial applications of AI and ML, used by different researchers.
- Used peered review conferences and journal papers for meta-analysis and literature survey.

Author states that the most of threat modelling work has been done manually, so assuring on validation of results is quite a difficult job. For this literature survey, almost 176 articles published between 2004 to 2021 have been covered, out of which 54 are identified for future prospects for research.

On time recovery, it is mandatory to gather the real time cyber data to block the attacks. For most of the organizations, the technicality behind the Threat Intelligence is actually required. As per (Tounsi & Rais, 2018), the new generation attacks are more complex to handle than earlier attacks. It is necessary to create a boundary for cyber defense for new threats.

As Industry is already moving towards Cloudification, there is also a variation in Indicator of Compromise (IoC). In terms of Limitation author states that

- For protecting Zero-day attacks, its required to get the information on time,
- For dealing with targeted attacks, only solution of fast dispersing of IoC is not adequate.

For this, it is required to refine the threat data with their internal vulnerabilities and weaknesses.

In the context of landscape of CTI, (Mavroeidis & Bromander, 2017) planned CTI model that allows cyber protectors to examine their threat intelligence potentials and judge their place against ever dynamical landscape of CTI.

(Preuveneers & Joosen, 2021) proposed a solution based on the top of MISP, TheHive, and Cortex—three state-of-the-art open source CTI sharing and incident response platforms—to incrementally improve the accuracy of those ML models, i.e., reduce false positives and false negatives with shared counter-evidence, and verify the Models' robustness against ML attacks.

(Ramsdale et al.,2020) looked into the landscape of codecs and languages, as well as publicly available threat feeds, how they operate, and their suitability for providing rich cyber-threat intelligence.

In the context of IoT security, (Mahbub, 2020; Mohanta et al., 2020) focused on the security and privacy challenges in the area of IoT. The major challenge is how to maintain CIA of data and identifies the layer wise issues. Also, identify the solutions to address the security issues using AI, ML and Blockchain technologies. If these technologies get merge with IoT give more security.

In today's world where Cyber-attacks are increasing day by day, there is a requirement of Cyber Security and Forensic expert to perceive, examine and preserve it from real time cyber threats. For this timely recognition of threats and weaknesses with deep analytical skills are required for betterment in the CTI area.

DETAILED LITERATURE SURVEY

In today's world where Cyber-attacks are increasing day by day, there is a requirement of Cyber Security and Forensic specialist to detect, analyze and protect it from real time cyber threats. For this timely recognition of threats and weaknesses with deep analytical skills are required for betterment in the Cyber Threat Intelligence area.

For this Intelligence, we require smart tools using AI (Artificial Intelligence), ML (Machine Learning) and Data mining techniques for analysis of cyber-attacks footprints.

According to the Author (Conti *et al.*, 2018), the main challenges in this area are:

- Attack Vector Reconnaissance: Recognize the point of initiation of attacks and vulnerabilities that can be exploited by cyber attackers/hackers.
- Attack Indicator Reconnaissance: Because of increasing cyber-attacks, attackers/hackers use anti forensic and evasion solutions in their malicious codes to reduce or make the CVSS (Common Vulnerability Scoring System) score less efficient.

This paper (Mavroeidis & Bromander, 2017) suggested a CTI (Cyber Threat Intelligence) model that allows cyber security professionals to assess their threat intelligence capabilities and assess their position in the ever-changing CTI ecosystem. This methodology is also used by the authors to analyse sharing standards, ontologies, and taxonomies. According to the results, the entire field of threat intelligence is not providing the correct ontologies (Figure 1).

Figure 1.

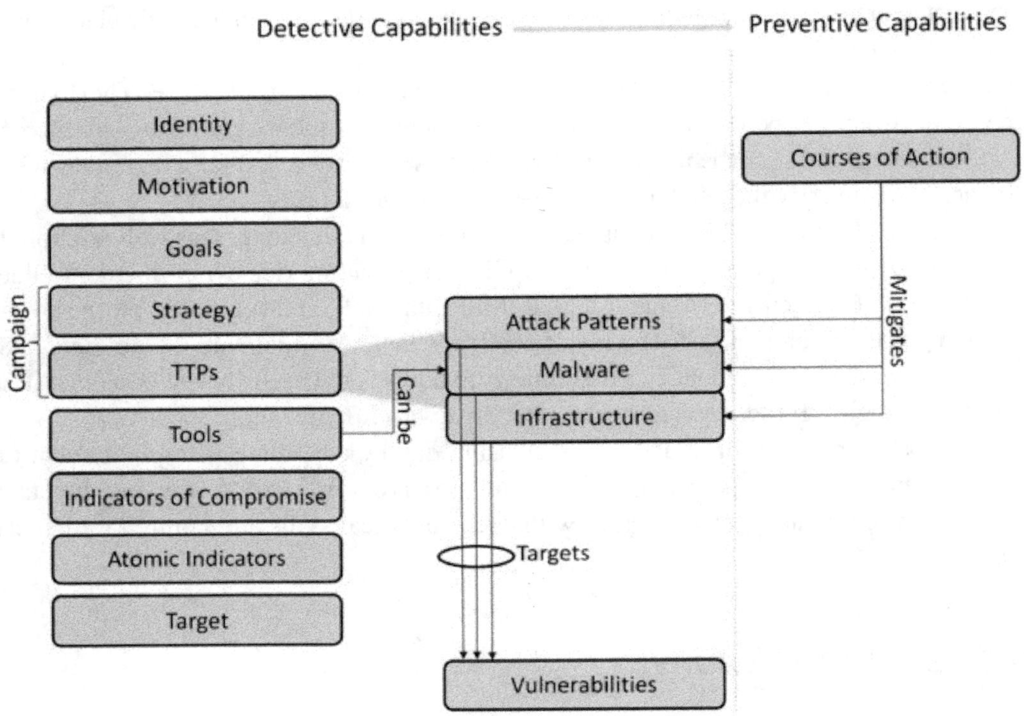

The CTI Model is a representation of the types of data required for advanced threat intelligence and attribution of possible attacks. It improves the organization's security posture by implementing detective and preventive procedures.

According to the author, the following are major issues with CTI:

- Indefinite terminology produces unneeded doubts among specialists and additional labor to develop ontologies
- There is no standard representation of essential information as a result
- Layers of abstractions are not described adequately

The DML (Detection Maturity Level Model) in Figure 2 is used to describe an organization's maturity in terms of its ability to consume threat information and take action on it. It focuses on enhancing the level of abstractions used to detect cyber-attacks, with the assumption that the incident response team's maturity level is low.

Figure 2.

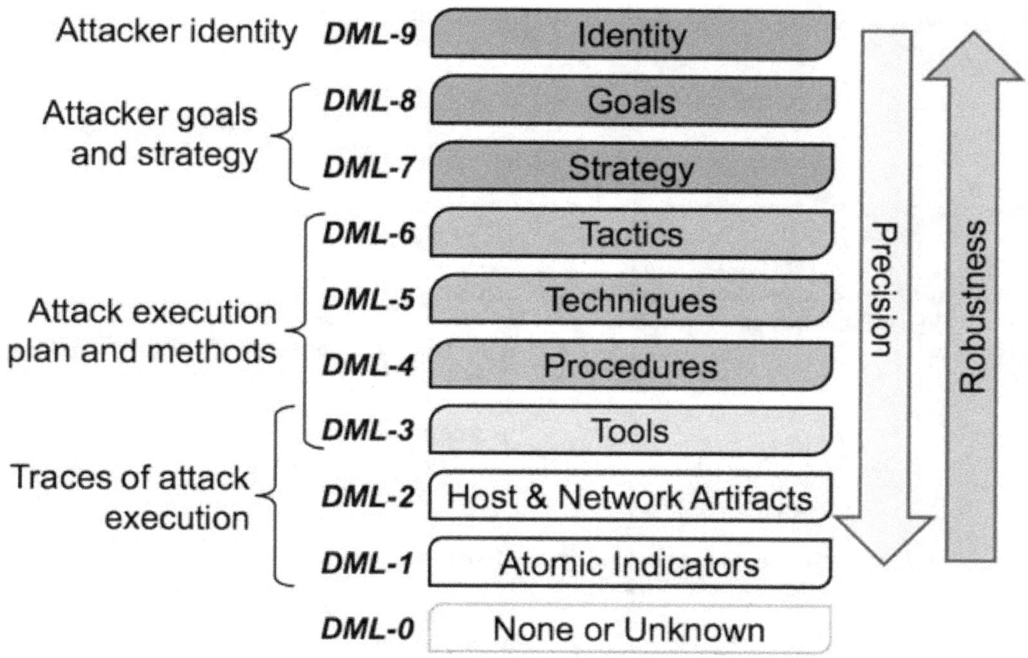

Author concludes that there is no ontology available to use within CTI. The main drawback is the lack of result explanations because of poor development and abstraction layers does not contain the relevant data and information in terms of productive Cyber Threat Intelligence.

This paper (Powell, 2016), includes the study on identifying the intelligence, Intelligence Cycle, intelligence analysis as well as proactive and reactive cyber security Intelligence measures. This paper also explains:

- Why CTI (Cyber Threat Intelligence) is required?
- What function does CTI play in cyber-security and operations?
- What are the major challenges are there which needs attention?

A Literature survey is performed by the author, which defines the traditional Intelligence processes with the new and advanced CTI products and services. These products and services constitute intelligence data feeds, reports analytical platforms, Indicators of compromise etc.

In this paper (Bellaby, 2016), author states that on growing cyber threats in cyber space creates pressure on the Cyber Intelligence community to create new technologies to assess the attack information. By collecting all the digital Information moving in cyberspace, it is not only detailing that what end user have done or are currently doing but also analyze what will be their next step might be. The lot of unrolled discussions stopped between the ethical uses of CTI to prove its potential for preventing attacks with undue harm. The author focuses on explaining the detailed cyber intelligence practices. The paper also produces the combination of Just Cyber Intelligence principles, designed based on Just war tradition for giving the justification of CTI.

The main agenda of this article is to establish the Intelligence way of collection, which cause limited harm, but in many situations to prevent those harms, which requires political responsibility displayed in Table 1.

Table 1.

Threat level	Case example	Response action
Very High	Case 1: Those who make and contribute to a website that plans an attack	a) Detain in real life b) Access email / voice-call content c) In-depth check on real-world electronic footprint d) In-depth check on other online activity e) In-depth check on personal databases and computer searches, including: a. bank and money b. wider electronic information c. medical records
High	Case 2: Makers of websites promoting or advocating violent action or harm to others	a) Detained in real life b) Access email / voice-call content c) Check on real-world electronic footprint d) In-depth check on other online activity e) In-depth check on personal databases and computer searches, including: a. bank and money b. wider electronic information c. medical records
Medium	Case 2: Those who visit a website promoting or advocating violent action or harm to others	a) Identify the individual – keep a record b) Superficial check on web activity (before the first /) c) Check other security databases for red flags
Low	Case 3: Those who make websites with potentially harmful information but with no ill intent	a) Identify individual – no record b) Check other security databases for red flags c) Keep a check on who visits the website
Very Low	Case 3: Those who visit websites with potentially harmful information but with no ill intent	a) Identify individual – if unthreatening, destroy record b) Check other security databases for red flags
Sub-Low	Case 4: Web activity that indicates no threat	No action

The purpose of this paper (Shea, 2016), is giving the explanation about the analysis and actions taken for Intelligence failures. It also explains the:

- Causes of Intelligence Failures
- Action applicable to Intelligence failures
- Past Learning and remedies used to prevent Intelligence failures
- Discussed various case studies
- Explaining the perspective at policy level

Author also states that due to rise of Digital age, it gives new threats and create new area of cyber war and cyberspace. It is also necessary to take care of any wrong application of CTI will cause potential failures. The proper analysis and learning help to apply techniques in CTI to get accurate results shown in Table 2.

Table 2.

Threat level	Case example	Response action
Very High	Case 1: Those who make and contribute to a website that plans an attack	a) Detain in real life b) Access email / voice-call content c) In-depth check on real-world electronic footprint d) In-depth check on other online activity e) In-depth check on personal databases and computer searches, including: a. bank and money b. wider electronic information c. medical records
High	Case 2: Makers of websites promoting or advocating violent action or harm to others	a) Detained in real life b) Access email / voice-call content c) Check on real-world electronic footprint d) In-depth check on other online activity e) In-depth check on personal databases and computer searches, including: a. bank and money b. wider electronic information c. medical records
Medium	Case 2: Those who visit a website promoting or advocating violent action or harm to others	a) Identify the individual – keep a record b) Superficial check on web activity (before the first /) c) Check other security databases for red flags
Low	Case 3: Those who make websites with potentially harmful information but with no ill intent	a) Identify individual – no record b) Check other security databases for red flags c) Keep a check on who visits the website
Very Low	Case 3: Those who visit websites with potentially harmful information but with no ill intent	a) Identify individual – if unthreatening, destroy record b) Check other security databases for red flags
Sub-Low	Case 4: Web activity that indicates no threat	No action

This journal (Mattern et al., 2014) focuses on Cyber Intelligence at the operational level. It claims that cyber intelligence should be at the heart of the cybersecurity mission, which necessitates:

- a proactive security posture;
- a full, accurate, and timely understanding of the threat environment; and
- a commitment to data-driven decisions.

After the malicious actor has penetrated the guarded network long enough to log a discernible pattern of behavior, the Cyber Kill Chain begin shown in Figure 3. This notion is valuable in the event of a known attack, but the primary goal of network defense is to prevent intrusions from occurring in the first place. Defenders must broaden their awareness of the assault pathway or kill chain beyond what occurs on the network in order to do so.

Figure 3.

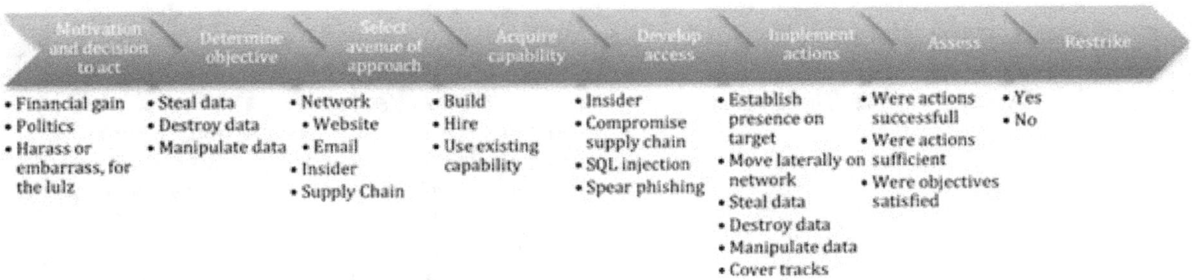

Intelligence efforts can be directed at identifying a succession of processes and behaviors in order to distinguish the following (shown in Table 3):

- Who might be attempting to hack a network?
- What are the malicious actors' aims and capabilities?
- When will they carry out their activity?
- Where will the activity begin?
- How do they intend to infiltrate or influence the network?

Table 3.

Phase	Detect	Deny	Disrupt	Degrade	Deceive	Destroy
Reconnaissance	Web analytics	Firewall ACL				
Weaponization	NIDS	NIPS				
Delivery	Vigilant user	Proxy filter	In-Line AV	Queuing		
Exploitation	HIDS	Patch	DEP			
Installation	HIDS	"chroot" jail	AV			
C2	NIDS	Firewall ACL	NIPS	Tarpit	DNS redirect	
Actions of Objectives	Audit log			Quality of Service	Honeypot	

At this level, cyber intelligence entails watching how malevolent forces plan and prepare assaults depending on what they've learned through gathering their own intelligence and what they think required to achieve their strategic objectives. Attackers must create the infrastructure (such as botnets, malware, and delivery methods like phishing) that will support their tactical operations. They use cyberspace (hop points) to position their capabilities where they'll be needed to carry out their tactics. A hacktivist group, for example, may plan both cyber and physical world operations to promote its aims at the operational level. The following are some instances of operational intelligence:

- An adversary's capabilities are evolving in a technological direction, according to trend analysis.
- Indications that an adversary has decided on a strategy for pursuing an organisation.
- Indications that an attacker is developing the ability to exploit a specific strategy.
- The exposure of enemy tactics, techniques, and operations.
- Understanding the operational cycle of the opponent (i.e., decision-making, acquisitions, and command and control (C2) methods for both technology and manpower).
- Technical, social, legal, financial, and other vulnerabilities of an enemy.
- Information that allows the defender to affect opponents as they go along a path to their attack (via the kill chain).

The MITRE Corporation developed a matrix that aligns levels of cyber preparedness (dubbed "Cyber Prep") with levels of threat. They suggest five Cyber Threat Levels, each of which corresponds to a general cyber defensive strategy/position aimed to enable mission assurance:

- Threat Level 1: Cyber Vandalism, which corresponds to Perimeter Defense.
- Threat Level 2: Cyber Theft/Crime, which correlates to a Critical Information Protection security strategy.
- Threat Level 3: Cyber Intrusion/Surveillance, which correlates to a Responsive Awareness protection strategy.
- Threat Level 4: Cyber Sabotage/Espionage, which corresponds to an Architectural Resilience protection strategy.
- Threat Level 5: Cyber Conflict/Warfare, which corresponds to a Pervasive Agility defense strategy.

According to the author, the Cyber Prep framework provides a hierarchical set of threat levels that may come from different threat vectors; levels in the Early Warning model represent escalation or forward motion toward a given attack in the context of a political conflict or competition, particularly at the nation-state level. The stages of a politically motivated cyber-attack, according to Moran, are as follows:

- *Latent Tensions*: This level describes unaddressed grievances and hostility between political entities that exist (and frequently remain).
- *Cyber Reconnaissance*: This level corresponds to a political entity's attempts to examine and maybe test another entity's cyber infrastructure for vulnerabilities.
- *Initiating Event*: This level represents a "flare up" in inter-party tensions, escalating hostilities and threat perceptions.
- *Cyber Mobilization*: This is the level at which a political entity manipulates the narrative of the "initiating event" in order to instigate action.

- *Cyber Attack:* This level denotes an attack against the cyber infrastructure of another entity, frequently based on intelligence gathered during reconnaissance.

HYPOTHESIS/THEORY

In today's era, where everything moved to the internet from business data to the individual personal information. It is mandate to safe your data. Cyber security is the main evolutionary research to the industry to make their infrastructure, data or network safe i.e., why it attracts various industries worldwide. Threat Modelling is one of the critical aspects of Cyber Security, which is the first phase of the Risk Assessment Cycle. It is widely used in system evaluation and application development in various industries who focuses on security.

According to Wikipedia, Threat modelling is a process by which potential threats, such as structural vulnerabilities can be identified, enumerated, and prioritized – all from a hypothetical attacker's point of view. The purpose of threat modelling is to provide defenders with a systematic analysis of the probable attacker's profile, the most likely attack vectors, and the assets most desired by an attacker as shown in Figure 4.

Figure 4.

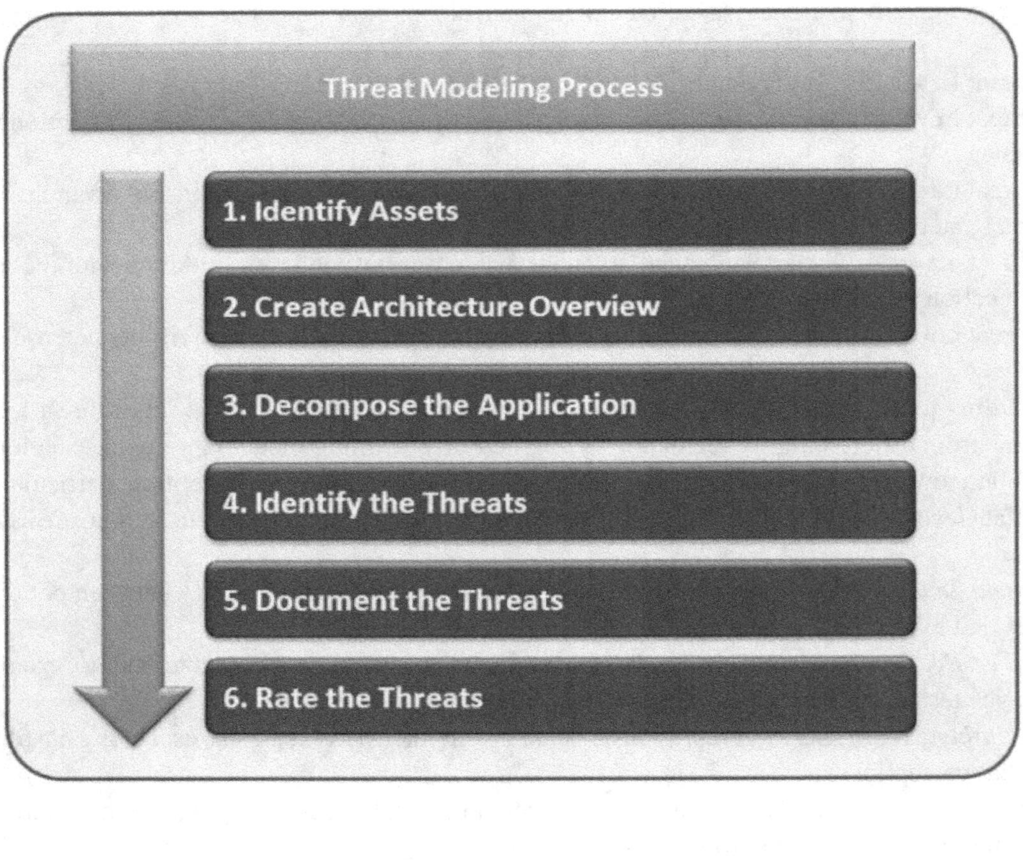

Two types of vulnerability testing are Vulnerability Assessment and Penetration Testing (VAPT). The tests have varying strengths and are frequently combined to provide a more comprehensive vulnerability assessment. In a nutshell, Penetration Testing and Vulnerability Assessments conduct two different jobs within the same area of concentration, usually with different outcomes.

Vulnerability assessment techniques identify which faults exist, but they do not distinguish between those that can be exploited to cause harm and those that cannot. Penetration tests try to exploit system defects in order to discover whether unauthorized access or other malicious conduct is possible, as well as to assess which flaws pose a threat to the application as shown in Figure 5. Rather than finding every flaw in a system, a penetration test is designed to demonstrate how harmful a flaw could be in an actual attack. Penetration testing and vulnerability assessment tools work together to provide a complete view of an application's defects and the risks associated with those issues.

Figure 5.

A computer network IT administrator must perform VAPT, or penetration testing, on a regular basis. This is due to an increase in hacking attempts, regardless of industry. Internal or external attacks might occur with no or limited understanding of the network as shown in Figure 6.

Data security has always been a priority. It is for this reason that people lock their filing cabinets and rent safety deposit boxes at their banks. Data privacy is becoming increasingly important as more of our data is digital and we share more information online. A single firm may hold the personal information of millions of clients—information that must be kept hidden in order for consumers' identities to remain safe and secure, and the company's reputation to remain unblemished. (Did someone say "data breach"?) Data privacy, on the other hand, isn't merely a business problem. When it comes to data privacy, you, as an individual, have a lot on the line. The more you understand it, the better equipped you will be to protect yourself from a wide range of threats.

Figure 6.

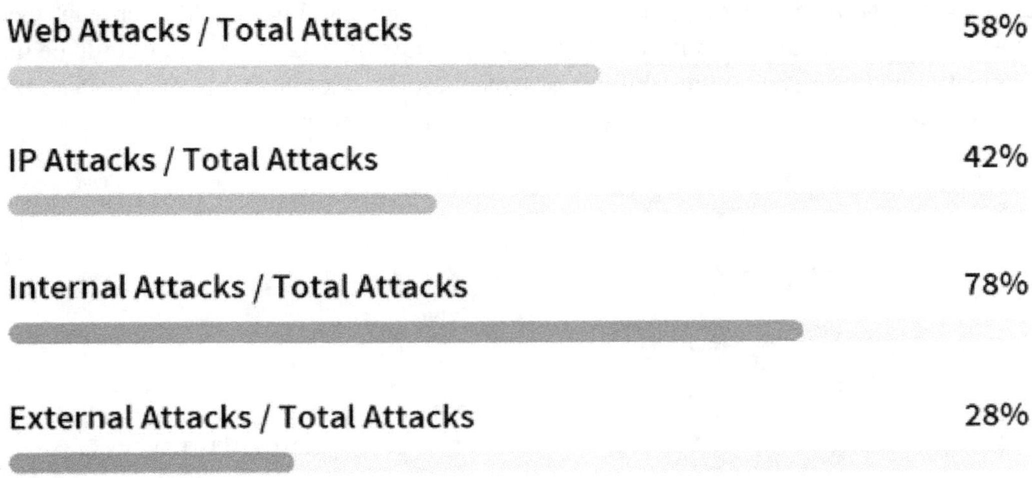

WHAT IS DATA PRIVACY?

Data privacy relates to how a piece of information—or data—should be handled based on its relative importance. For instance, you likely wouldn't mind sharing your name with a stranger in the process of introducing yourself, but there's other information you wouldn't share, at least not until you become more acquainted with that person. Open a new bank account, though, and you will probably be asked to share a tremendous amount of personal information, well beyond your name.

In the digital age, we typically apply the concept of data privacy to critical personal information, also known as personally identifiable information (PII) and personal health information (PHI). This can include Social Security numbers, health and medical records, financial data, including bank account and credit card numbers, and even basic, but still sensitive, information, such as full names, addresses and birthdates.

Why is Data Privacy Important?

When data that should be kept private gets in the wrong hands, bad things can happen. A data breach at a government agency can, for example, put top-secret information in the hands of an enemy state. A breach at a corporation can put proprietary data in the hands of a competitor. A breach at a school could put students' PII in the hands of criminals who could commit identity theft. A breach at a hospital or doctor's office can put PHI in the hands of those who might misuse it.

A cyberattack is a malicious and deliberate attempt by an individual or organization to breach the information system of another individual or organization. Usually, the attacker seeks some type of benefit from disrupting the victim's network. Ten most common cyber-attack types:

1. Denial-of-service (DoS) and distributed denial-of-service (DDoS) attacks: A denial-of-service attack overwhelms a system's resources so that it cannot respond to service requests. A DDoS attack is also an attack on system's resources, but it is launched from a large number of other host machines that are infected by malicious software controlled by the attacker.
2. Man-in-the-middle (MitM) attack: A MitM attack occurs when a hacker inserts itself between the communications of a client and a server.
3. Phishing and spear phishing attacks: Phishing attack is the practice of sending emails that appear to be from trusted sources with the goal of gaining personal information or influencing users to do something. It combines social engineering and technical trickery. It could involve an attachment to an email that loads malware onto your computer. It could also be a link to an illegitimate website that can trick you into downloading malware or handing over your personal information.
4. Drive-by attack: Drive-by download attacks are a common method of spreading malware. Hackers look for insecure websites and plant a malicious script into HTTP or PHP code on one of the pages. This script might install malware directly onto the computer of someone who visits the site, or it might re-direct the victim to a site controlled by the hackers. Drive-by downloads can happen when visiting a website or viewing an email message or a pop-up window.
5. Password attack: Because passwords are the most commonly used mechanism to authenticate users to an information system, obtaining passwords is a common and effective attack approach. Access to a person's password can be obtained by looking around the person's desk, ''sniffing'' the connection to the network to acquire unencrypted passwords, using social engineering, gaining access to a password database or outright guessing.
6. SQL injection attack: SQL injection has become a common issue with database-driven websites. It occurs when a malefactor executes a SQL query to the database via the input data from the client to server. SQL commands are inserted into data-plane input (for example, instead of the login or password) in order to run predefined SQL commands. A successful SQL injection exploit can read sensitive data from the database, modify (insert, update or delete) database data, execute administration operations (such as shutdown) on the database, recover the content of a given file, and, in some cases, issue commands to the operating system.
7. Cross-site scripting (XSS) attack: XSS attacks use third-party web resources to run scripts in the victim's web browser or scriptable application. Specifically, the attacker injects a payload with malicious JavaScript into a website's database. When the victim requests a page from the website, the website transmits the page, with the attacker's payload as part of the HTML body, to the victim's browser, which executes the malicious script as shown in Figure 7.

Figure 7.

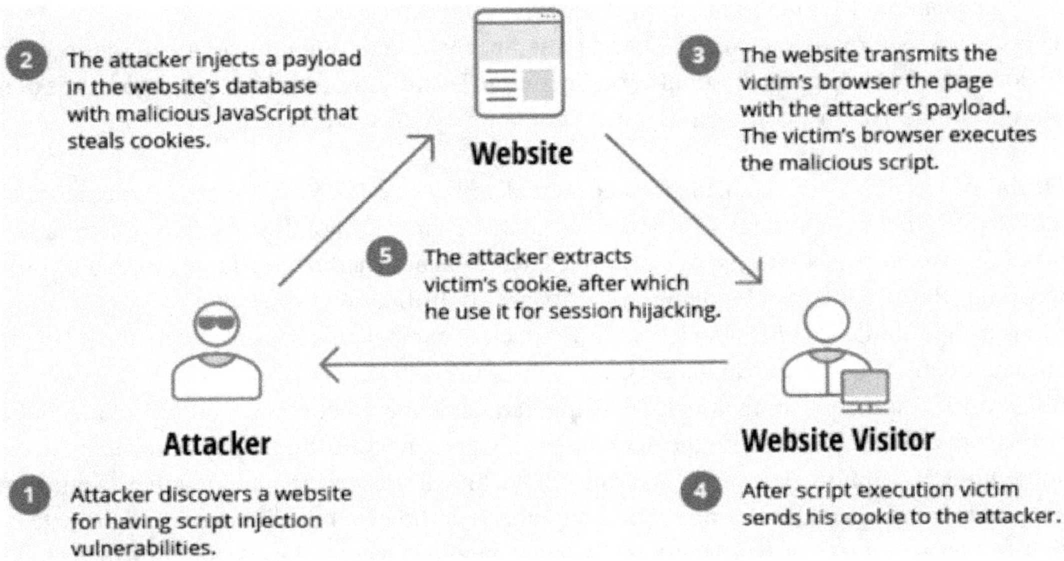

8. Eavesdropping attack: Eavesdropping attacks occur through the interception of network traffic. By eavesdropping, an attacker can obtain passwords, credit card numbers and other confidential information that a user might be sending over the network.
9. Birthday attack: Birthday attacks are made against hash algorithms that are used to verify the integrity of a message, software or digital signature. A message processed by a hash function produces a message digest (MD) of fixed length, independent of the length of the input message; this MD uniquely characterizes the message. The birthday attack refers to the probability of finding two random messages that generate the same MD when processed by a hash function. If an attacker calculates same MD for his message as the user has, he can safely replace the user's message with his, and the receiver will not be able to detect the replacement even if he compares MDs.
10. Malware attack: Malicious software can be described as unwanted software that is installed in your system without your consent. It can attach itself to legitimate code and propagate; it can lurk in useful applications or replicate itself across the Internet.

Cyberspace provides opportunities for innovation, commerce, and societal advancement but also raises significant issues for policymakers in securing cyber vulnerabilities, ensuring privacy and protection of personal data, and considering the use of cyber weapons as a national security asset.

Cybercriminals are going to create 3.5 million new, unfilled cybersecurity jobs by 2021. Compare that with one million openings in 2016. That is an increase of 350 percent in just five years and with that increase comes some serious cybersecurity revenue dedication. Everywhere, businesses are investing a remarkable amount of money into hiring security professionals, maintaining customer privacy and avoiding ransomware attacks. In 2017 alone, all of those protection efforts cost businesses $86.4 billion.

- What can you expect in the future of cybersecurity?
- What do cybercriminals have in store for you in 2018?

- What new threats do you need to be aware of, and how is that going to change — or how is it already changing — the cybersecurity environment?

To answer those questions, here are five cybersecurity challenges and trends that are coming in 2018.

1) Ransomware Evolution

Ransomware is the bane of cybersecurity, IT, data professionals, and executives. Perhaps nothing is worse than a spreading virus that latches onto customer and business information that can only be removed if you meet the cybercriminal's egregious demands. And usually, those demands land in the hundreds of thousands (if not millions) of dollars. Ransomware attacks are one of the areas of cybercrime growing the fastest, too. The number of attacks has risen 36 percent this year (and doubled in cost).

In today's world of evolving ransomware, yesterday's DR strategies no longer work. Data disappears and the business cannot revive it. That is, unless they pay the cybercriminals. In 2018, DRaaS solutions are the best defense that modern technology has to offer against ransomware attacks. With it, you can automatically back up your files, immediately identify which backup is clean, and launch a fail-over with the press of a button when malicious attacks corrupt your data.

2) AI Expansion

Robots might be able to help defend against incoming cyber-attacks. Between 2016 and 2025, businesses will spend almost $2.5 billion on artificial intelligence to prevent cyberattacks as shown in Figure 8.

Figure 8.

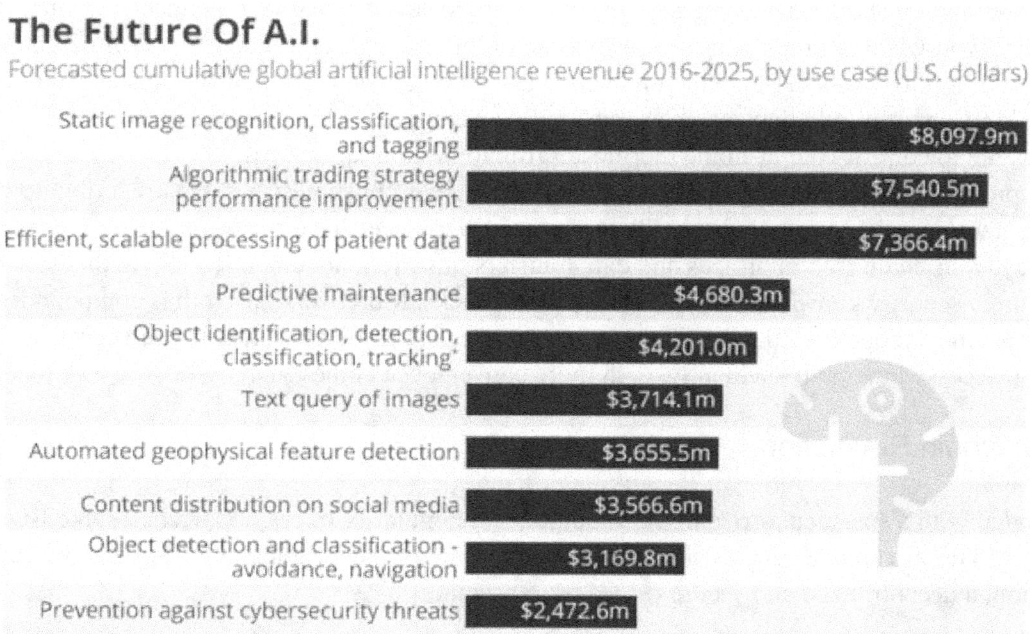

With that, development comes several benefits. First, you do not have to pay robots by the hour. They work free once you have them. (Also, no healthcare or catered lunch.) Which brings us to the next benefit. Namely, that they can work around the clock. (Sans overtime.) Robots do not take breaks. Humans do.

Timing is everything with malware and other vicious data manipulations. If, for instance, you could fight a virus as it was downloading, you would have a much better chance of mitigating its forceful impact. However, when you fight against corrupt data after the fact — after the damage is already done — recovery becomes an uphill battle. Traditionally, IT professionals and cybersecurity experts face these attacks once they have already taken place. Nicole Eagan says,

"These breaches are getting reported as historical events, long after something could have been done about it. That has got to change."

In other words, the biggest problem with recovering from cyber-attacks is that security professionals rarely get the chance to deal with them immediately. Since artificial intelligence does not need to sleep, though, they can set defense systems against malware the moment it begins to download. Developers understand artificial intelligence better than ever and how to manipulate its workings. For that reason, there's no doubt that the future — and 2018 in particular — holds a bit of robot-human cooperation in defense of data everywhere.

3) IoT Threats

The vast majority of humans in first-world countries have an iPhone in their pockets, a computer at work, a television at home, and a tablet in their cars. 84 percent of American households, for instance, have at least one smartphone. 80 percent have at least one desktop or laptop computer. 68 percent have at least one tablet. Moreover, 39 percent have at least one streaming device as shown in Figure 9.

The Internet of Things is making sure that every single device you own is connected. Your refrigerator can tell you when the milk runs out. Alexa can order you a pizza. Of course, all of that connection carries with it massive benefits, which is what makes it so appealing in the first place. You no longer have to log in on multiple devices. You can easily control your TV with your phone. And you might even be able to control your at-home thermostat from other digital devices.

The problem is that all of that interconnectedness makes consumers highly susceptible to cyberattacks. In fact, one study revealed that 70 percent of IoT devices have serious security vulnerabilities. Specifically, insecure web interfaces and data transfers, insufficient authentication methods, and a lack of consumer security knowledge leave users open to attacks. Moreover, that truth is compounded by the fact that so many consumer devices are now interconnected. In other words, if you access one device, you have accessed them all. Evidently, with more convenience comes more risk.

4) Blockchain Revolution

2017 ended with a spectacular rise in the valuation and popularity of cryptocurrencies like Bitcoin and Ethereum. These cryptocurrencies are built upon blockchains, the technical innovation at the core of the revolution, a decentralized and secure record of transactions.

Figure 9.

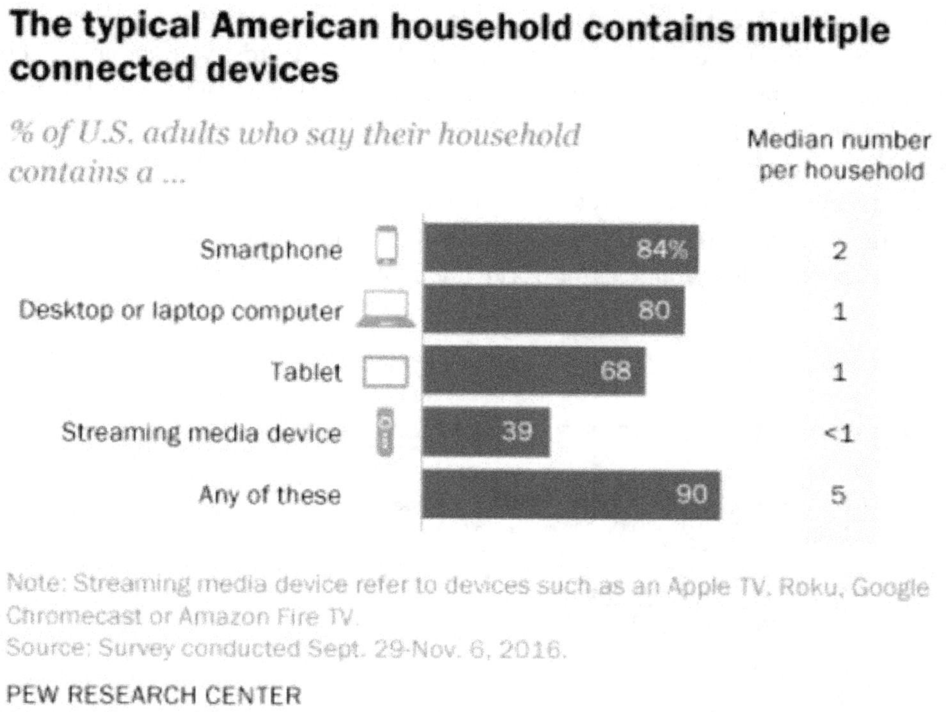

What Does Blockchain Technology Have to Do With Cybersecurity?

It is a question that security professionals have only just started asking. As 2018 progresses, you will likely see more people with answers. While it is difficult to predict what other developments blockchain systems will offer in regards to cybersecurity, professionals can make some educated guesses. Companies are targeting a range of use cases, which the blockchain helps, enable from medical records management, to decentralized access control, to identity management. As the application and utility of blockchain in a cybersecurity context emerges, there will be a healthy tension but also complementary integrations with traditional, proven, cybersecurity approaches. You will undoubtedly see variations in approaches between public & private blockchains. One thing is for sure, though. With blockchain technology, cybersecurity will likely look much different than it has in the past.

5) Serverless Apps Vulnerability

Serverless apps can invite cyber-attacks. Customer information is particularly at risk when users access your application off-server — or locally — on their device.

Why?

Well, on-server — when the data is stored in the cloud rather than the user's device — you have control over that information and the security that surrounds it. In other words, you are able to control what security precautions you take to ensure the user's data remains private from identity thieves and other cybercriminals. With server less applications, however, security precautions are, largely, the responsibility of the user. Of course, you can integrate software into the application that gives the user the best chance of defeating cybercriminals. However, when all was said and done, you, the professional, cannot directly defend the customer. Server less apps are most common as web service and data processing tools, leading uses of serverless architecture worldwide is shown in Figure 10.

Figure 10.

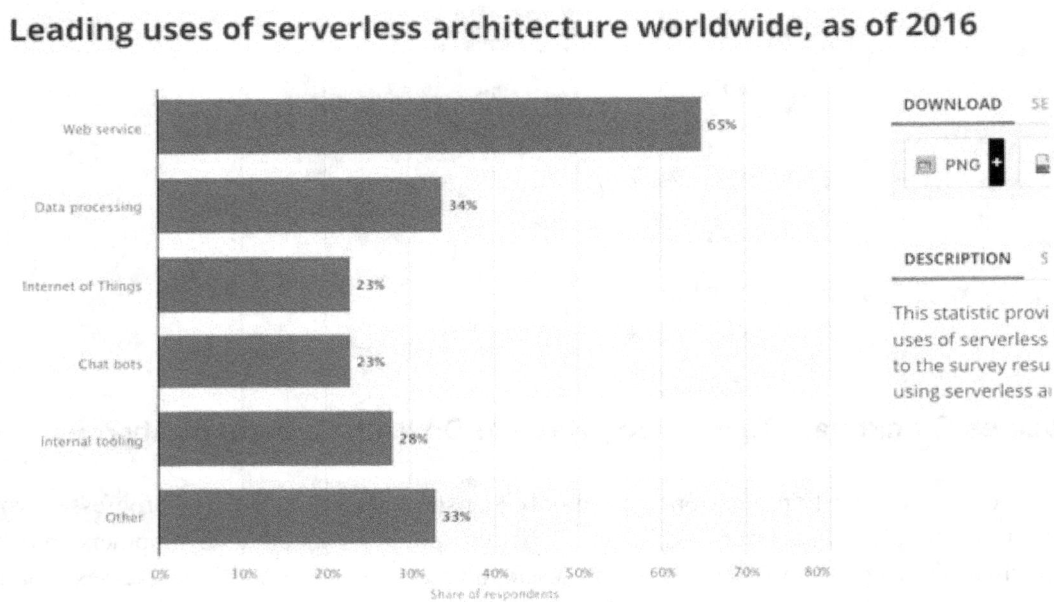

Because AI and ML work in tandem – AI leverages ML capabilities to increase its intelligence and evolve – these systems can be considerably advantageous when it comes to identifying and working to guard against the latest security threats. Emerging threats are very challenging for today's traditional security tools to pinpoint. This is particularly true when it comes to zero-day threats, which pose a significant hazard to enterprise and consumer systems alike. This is an ideal arena wherein AI and ML can shine.

"[Malware and cyber attacks] evolve over time, so more dynamic approaches are necessary". "Cybersecurity solutions that rely on ML use data from prior cyber attacks to respond to newer but somewhat similar risks."

In this way, an AI system powered by ML can leverage what it knows and understands about past attacks and threats to identify other attacks in the same vein or style.

Because hackers are consistently building upon older threats – including new abilities or tweaking previously used samples to build out a malware family – utilizing AI and ML systems to look out for and provide notification of emerging attacks could be incredibly beneficial to stemming the tide of zero-day threats This translates to new dangers being pinpointed more quickly and the necessary updates and patches released in a more streamlined manner, minimizing the number of victim systems impacted by the threats.

In addition, this can also free up considerable time for cybersecurity experts who, without the assistance of AI and ML, would have to identify these threats on their own – or, worse still, wait until a key system is attacked to notice the threat.

A cyber threat is "any circumstance or event with the potential to adversely impact organizational operations (including mission, functions, image, or reputation), organizational assets, individuals, other organizations, or the Nation through an information system via unauthorized access, destruction, disclosure, or modification of information, and/or denial of service." For brevity, this publication uses the term threat instead of "cyber threat". The individuals and groups posing threats are known as "threat actors" or simply actors.

Threat information is any information related to a threat that might help an organization protect itself against a threat or detect the activities of an actor. Major types of threat information include the following:

- *Indicators* are technical artifacts or observables1 that suggest an attack is imminent or is currently underway or that a compromise may have already occurred. Indicators can be used to detect and defend against potential threats. Examples of indicators include the Internet Protocol (IP) address of a suspected command and control server, a suspicious Domain Name System (DNS) domain name, a Uniform Resource Locator (URL) that references malicious content, a file hash for a malicious executable, or the subject line text of a malicious email message.
- *Tactics, techniques, and procedures* (TTPs) describe the behavior of an actor. Tactics are high-level descriptions of behavior, techniques are detailed descriptions of behavior in the context of a tactic, and procedures are even lower-level, highly detailed descriptions in the context of a technique. TTPs could describe an actor's tendency to use a specific malware variant, order of operations, attack tool, delivery mechanism (e.g., phishing or watering hole attack), or exploit.
- *Security alerts*, also known as advisories, bulletins, and vulnerability notes, are brief, usually human readable, technical notifications regarding current vulnerabilities, exploits, and other security issues. Security alerts originate from sources such as the United States Computer Emergency Readiness Team (US-CERT), Information Sharing and Analysis Centers (ISACs), the National Vulnerability Database (NVD), Product Security Incident Response Teams (PSIRTs), commercial security service providers, and security researchers.
- *Threat intelligence reports* are generally prose documents that describe TTPs, actors, types of systems and information being targeted, and other threat-related information that provides greater situational awareness to an organization. Threat intelligence is threat information that has been aggregated, transformed, analyzed, interpreted, or enriched to provide the necessary context for decision-making processes.
- *Tool configurations* are recommendations for setting up and using tools (mechanisms) that support the automated collection, exchange, processing, analysis, and use of threat information. For example, tool configuration information could consist of instructions on how to install and use a

rootkit detection and removal utility, or how to create and customize intrusion detection signatures; router access control lists (ACLs), firewall rules, or web filter configuration files.

While sharing threat information clearly has benefits, certain challenges still remain. Some challenges that apply both to consuming and to producing threat information are:

- *Establishing Trust*. Trust relationships form the basis for information sharing, but require effort to establish and maintain. Ongoing communication through regular in-person meetings, phone calls, or social media can help accelerate the process of building trust.
- *Achieving Interoperability and Automation*. Standardized data formats and transport protocols are important building blocks for interoperability. The use of common formats and protocols enables automation and allows organizations, repositories, and tools to exchange threat information at machine speed. Adopting specific formats and protocols, however, can require significant time and resources, and the value of these investments can be substantially reduced if sharing partners require different formats or protocols. During the standards development process, early adopters need to accept the risk that it may be necessary to purchase new tools if significant changes to formats and protocols take place.
- *Safeguarding Sensitive Information*. Disclosure of sensitive information, such as controlled unclassified information (CUI) and personally identifiable information (PII) can result in financial loss, violation of sharing agreements, legal action, and loss of reputation. Sharing security and event information, such as security logs or scan results, could expose the protective or detective capabilities of the organization and result in threat shifting by the actor. 3 The unauthorized disclosure of information may impede or disrupt an ongoing investigation, jeopardize information needed for future legal proceedings, or disrupt response actions such as botnet takedown operations. Organizations should apply handling designations to shared information and implement policies, procedures, and technical controls to actively manage the risks of disclosure of sensitive information.
- *Protecting Classified Information*. Information received from government sources may be marked as classified, making it difficult for an organization to use. Acquiring and maintaining the clearances needed for ongoing access to classified information sources is expensive and time-consuming for organizations. In addition, many organizations employ non-U.S. citizens who are not eligible to hold security clearances and are not permitted access to classified information.
- *Enabling Information Consumption and Publication*. Organizations that want to consume and publish threat information need to have the necessary infrastructure, tools, personnel, and training to do so. Information sharing initiatives should be carefully scoped, because high-frequency, high volume information exchanges have the potential to overwhelm an organization's processing capabilities. Organizations that are currently unable to support automated indicator exchange can explore other options such as the manual exchange of best practices or summary indicator information. As additional resources become available, an organization may decide to use automated tools and workflows to process and use threat information
- Organizations vital to the security and day-to-day running of the UK attract a certain type of cyber-attack. These organizations are subject, like any other, to the general background noise of constant criminal activity in cyberspace; i.e., largely automated and indiscriminate attacks.

However, because of their critical nature they also attract political activists (hacktivists) and state-backed hackers, intent on destabilizing countries and trading blocs.

Cyber threats vary in their manner and impact, and a concerted campaign will blend different attack types at different stages. Attacks on critical national infrastructure (CNI) organizations primarily seek one or more of the following:

- User accounts for email and other systems
- Access to confidential data – personal information, operating procedures and plans
- Control over operational systems to disrupt services, campaigns or defenses

For example, a campaign will often start with spear-phishing to gain access to a specific person's PC. This is then used to send clone emails to others in the organisation, eventually achieving high-level access to systems. Then data is stolen via exfiltration, or systems are deliberately controlled or crashed, depending on the hackers' goals.

What are all the different attack types and how do they damage CNI organisations?

Types of Cyber Threat

- Phishing and its various off-shoots
- Phishing is the number one way for cyber criminals to get a foothold in your network.

Many of these emails use a scatter-gun approach, sending the same generic email to a large number of email addresses. These will superficially resemble emails from delivery companies, online shopping platforms, banks or other widely-known organisations. They are usually deployed by criminals rather than hacktivists.

- *Spear phishing* is targeted and more sophisticated, using company logos and graphics in emails that target a particular organisation, team or even an individual, perhaps a technical supervisor with privileged system access.
- *Whaling* is a subset of spear phishing where the targets are high level. Targets can include directors and C-suite members, particularly the non-technical ones who may not be as cautious as a CISO or CIO.
- *Clone phishing* is a variant in which a copy of a previously seen message is created and sent from an email address that resembles a company address as closely as possible. Often this is a second stage – the email will be sent from an already compromised but low-level system, to try and gain access to higher level systems.
- The traditional Intelligence cycle is the fundamental cycle of intelligence processing in a civilian or military intelligence agency or in law enforcement as a closed path consisting of repeating nodes. The stages of the intelligence cycle include the issuance of requirements by decision makers, collection, processing, analysis, and publication of intelligence. The circuit is completed when decision makers provide feedback and revised requirements. The intelligence cycle is also called the Intelligence Process by the U.S. Department of Defense (DoD) and the uniformed services.

The intelligence cycle is an effective way of processing information and turning it into relevant and actionable intelligence which is shown Figure 11 and Figure 12

- Direction: Intelligence requirements are determined by a decision maker to meet his/her objectives. In the federal government of the United States, requirements can be issued from the White House or the Congress. In NATO, a commander uses requirements (sometimes called 'essential elements of intelligence' (EEIs)) to initiate the intelligence cycle.
- Collection: In response to requirements, an intelligence staff develops an intelligence collection plan applying available sources and methods and seeking intelligence from other agencies. Collection includes inputs from several intelligence gathering disciplines, such as HUMINT (human intelligence), IMINT (imagery intelligence), ELINT (electronic intelligence), SIGINT (Signals Intelligence), OSINT (open source, or publicly available intelligence), etc.
- Processing: Once the collection plan is executed and information arrives, it is processed for exploitation. This involves the translation of raw intelligence materials from a foreign language, evaluation of relevance and reliability, and collation of the raw intelligence in preparation for exploitation.

Figure 11.

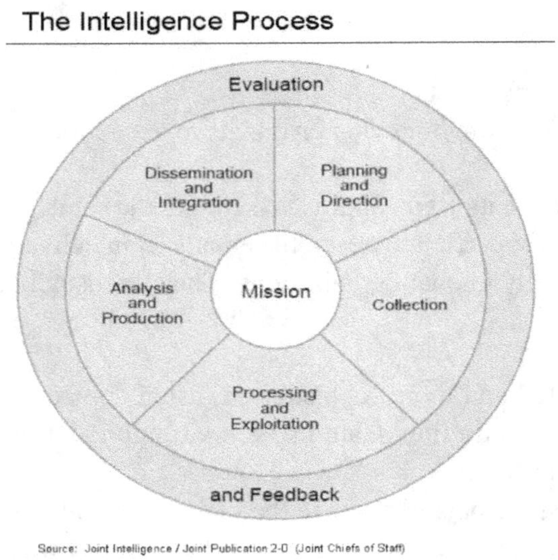

Analysis: Analysis establishes the significance and implications of processed intelligence, integrates it by combining disparate pieces of information to identify collateral information and patterns, then interprets the significance of any newly developed knowledge.

Dissemination: Finished intelligence products take many forms depending on the needs of the decision maker and reporting requirements. An intelligence organization or community typically establishes the level of urgency of various types of intelligence. For example, an indications and warning (I&W) bulletin would require higher precedence than an annual report.

Feedback: The intelligence cycle is a closed loop; feedback is received from the decision maker and revised requirements issued.

Figure 12.

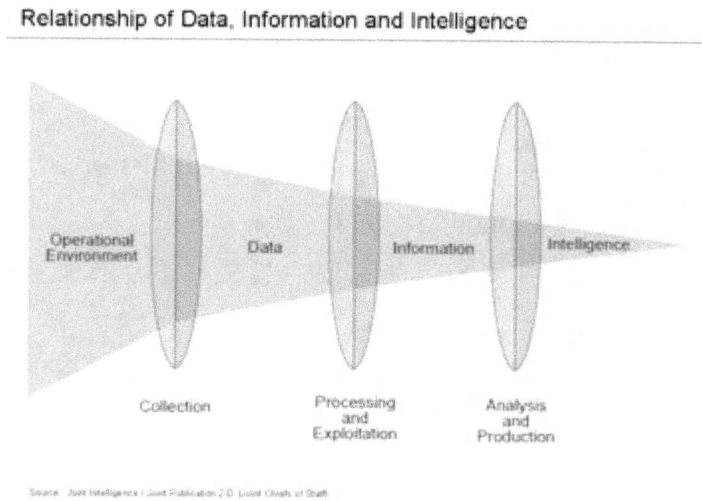

Threat intelligence is what threat data or threat information become when they have been gathered and evaluated from trusted, reliable sources, processed and enriched, then disseminated in a way where it can be considered actionable to its end-user. Source code or JSON files from an underground forum need to be reformatted, for example, or relevant articles need archiving and indexing, in order to make this data and information usable as intelligence. Intelligence means that the end-user can identify threats and opportunities in the cybersecurity landscape, using accurate, relevant, contextualized information. By eliminating the need to sort through thousands of alerts from data, security teams can maximize their own limited resources and accelerate their decision-making processes.

Due to the extraordinarily time-poor nature of their roles, this is where external threat intelligence providers really come into their own. Using automated or manual correlation, internal teams are able to reach out to other organizations to help them prioritize alerts and indicators which is shown in Table 4.

CATEGORIES OF THREAT INTELLIGENCE

Cyberwarfare is computer- or network-based conflict involving politically motivated attacks by a nation-state on another nation-state. In these types of attacks, nation-state actors attempt to disrupt the activities of organizations or nation-states, especially for strategic or military purposes and cyberespionage which is shown in Figure 13.

Although cyberwarfare generally refers to cyberattacks perpetrated by one nation-state on another, it can also describe attacks by terrorist groups or hacker groups aimed at furthering the goals of particular nations. It can be difficult to definitively attribute cyberattacks to a nation-state when those attacks are carried out by advanced persistent threat (APT) actors, but such attacks can often be linked to specific nations. While there are a number of examples of suspect cyberwarfare attacks in recent history, there has been no formal, agreed-upon definition for a cyber "act of war," which experts generally agree would be a cyberattack that directly leads to loss of life.

Table 4.

CATEGORY	ANALYZING	OUTPUT	TIME
Tactical	Security events, IOCs like file hashes, malicious domains, emails, links and attachments, registry keys, filenames, DLLs	MRTI, data feeds	Short-term
Operational	Malware family behavior and profiles, threat actors, human behavior, tactical intel, TTPs, communications and persistence techniques	Reports, lists and trend patterns	Medium-term
Strategic	Operational intelligence, cyberthreats in the context of business objectives, mapping online threats onto geopolitical events	Reports, trends, methodologies	Long-term

Cyberwarfare can take many forms, including:

- *viruses, computer worms and malware* that can take down water supplies, transportation systems, power grids, critical infrastructure and military systems;
- *denial-of-service (DoS) attacks*, cybersecurity events that occur when attackers take action that prevents legitimate users from accessing targeted computer systems, devices or other network resources;
- *hacking and theft of critical data* from institutions, governments and businesses;
- *Ransomware* that holds computer systems hostage until the victims pay ransom.

Objectives of Cyberwarfare

According to Cybersecurity and Infrastructure Security Agency (CISA), the goal of those engaged in cyberwarfare is to "weaken, disrupt or destroy the US." To achieve their goals, "national cyber warfare programs are unique in posing a threat along the entire spectrum of objectives that might harm US interests," says CISA. These threats range from propaganda to espionage and serious disruption with loss of life and extensive infrastructure disruption. A few examples of threats include:

- *Espionage for technology advancement.* For example, the National Counterintelligence and Security Center (NCSC) in its 2018 Foreign Economic Espionage in Cyberspace report notes that China's cybersecurity law mandates that foreign companies submit their technology to the

Chinese government for review and that Russia has increased its demand of source code reviews to approve of foreign technology sold in their country. In 2018, the US Department of Justice charged two Chinese hackers associated with the Ministry of State Security with targeting intellectual property and confidential business information.

- *Disruption of infrastructure to attack* the US economy or, when attacked by the US, to damage the ability of the US to continue its attacks. For example, by controlling a router between supervisory control and data acquisition (SCADA) sensors and controllers in a critical infrastructure, such as the energy sector, an enemy can attempt to destroy or badly damage energy plants or the grid itself.

Figure 13.

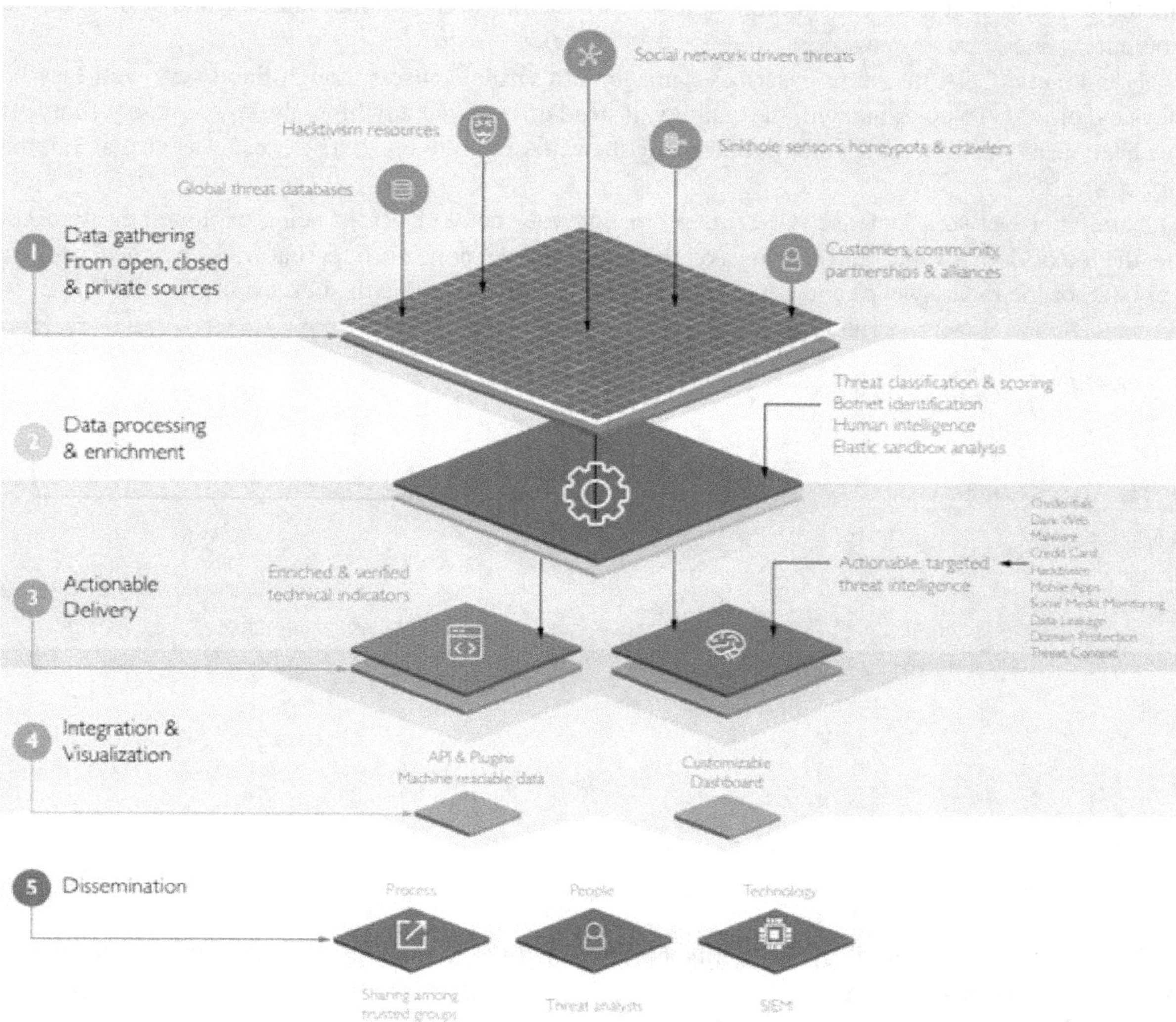

Computer Network Exploitation (CNE) is computer espionage, the stealing of information. It encompasses gaining access to computer systems and retrieving data. An old analogy is that of a cold war spy who picks the lock on a house, sneaks in, takes pictures of documents with his secret camera, and gets out without leaving a trace. A more modern analogy would be a drone that invades a hostile country's airspace to gather intelligence on troop strength which is shown in Figure 14.

Computer Network Attack (CNA) is akin to a traditional military attack or sabotage. It applies the four D's of "disrupt, deny, degrade, or destroy" to computer networks. Now, the cold war spy smashes a few artifacts as he leaves or maybe Fight Club-style, he introduces a gas leak so that the whole place explodes sometime later. Meanwhile, the drone rains hellfire missiles. CNA is the computer equivalent. It describes actions and effects that range from the subtle to the catastrophic.

Non-Kinetic Computer Network Attack is a term this book uses to describe the subset of CNA conducted virtually, that is, any disruption, denial, degradation, or destruction initiated and performed via computers or computer networks.

Non-kinetic CNA therefore describes damage with virtual causes; though there very well may be physical effects. To continue with the analogy, instead of breaking anything, the spy remotely shuts off the heat during an extremely cold night causing the water pipes to burst. The cause was virtual, but the effect was not.

Computer Network Defense (CND) is protecting your networks from being exploited or attacked. It's the locks, doors, walls, and windows on the house and the police officer that walks by once a day on her beat, or the radar sweeps and antiaircraft missile systems that line the border.

Figure 14.

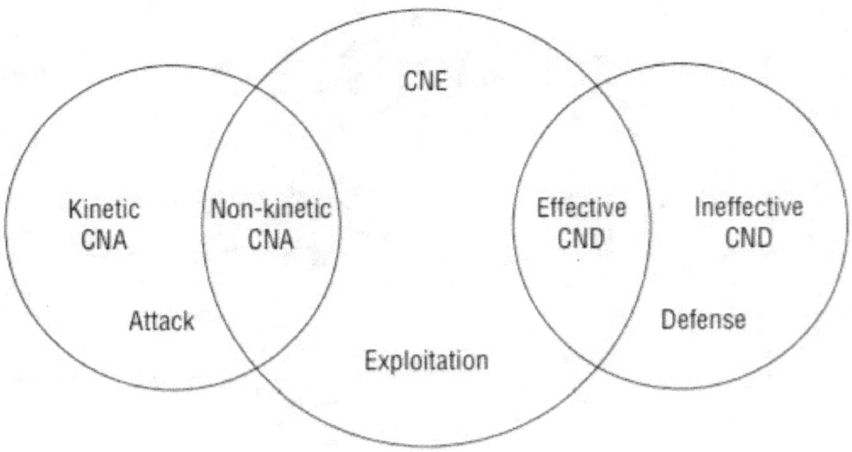

The Operational Levels of Cyber Intelligence, the Intelligence and National Security Alliance (INSA) proposed definitions for the strategic, operational, and tactical levels of cyber activity. While there has been much emphasis on tactical cyber intelligence to help understand the "on the network" cyber attacks so frequently in the news, there has been little discussion about the strategic and operational levels in order to better understand the overall goals, objectives, and interrelationships associated with these tactical attacks. As a result, key consumers such as C-suite executives, executive managers, and other senior leaders in the public, private, and academic sectors are not getting the right type of cyber intelligence to efficiently and effectively inform their organizations' risk management programs. This traditionally tactical focus also hampers the ability of the cyber intelligence function to communicate cyber risks in a way that leaders can fully interpret and understand.

In military operations, a comprehensive view of tactical operations is integral to mission success. Complex intelligence, surveillance, and reconnaissance operations and systems are needed to accomplish this goal.

Cyber reconnaissance is designed to equalize the playing field by providing agencies with a high-resolution picture of its cyber landscape from the adversary's perspective.

In Reconnaissance stage, attackers act like detectives, gathering information to truly understand their target. Detail is everything! From examining email lists to open source information, their goal is to know the network better than the people who run and maintain it. They hone in on the security aspect of the technology, study the weaknesses, and use any vulnerability to their advantage.

The reconnaissance stage can be viewed as the most important because it takes patience and time, from weeks to several months. Any information the infiltrator can gather on the company, such as employee names, phone numbers, and email addresses, will be vital. Attackers will also start to poke the network to analyze what systems and hosts are there. They will note any changes in the system that can be used as an entrance point as shown in Figure 15. For example, leaving your network open for a vendor to fix an issue can also allow the cybercriminal to plant himself inside.

Figure 15.

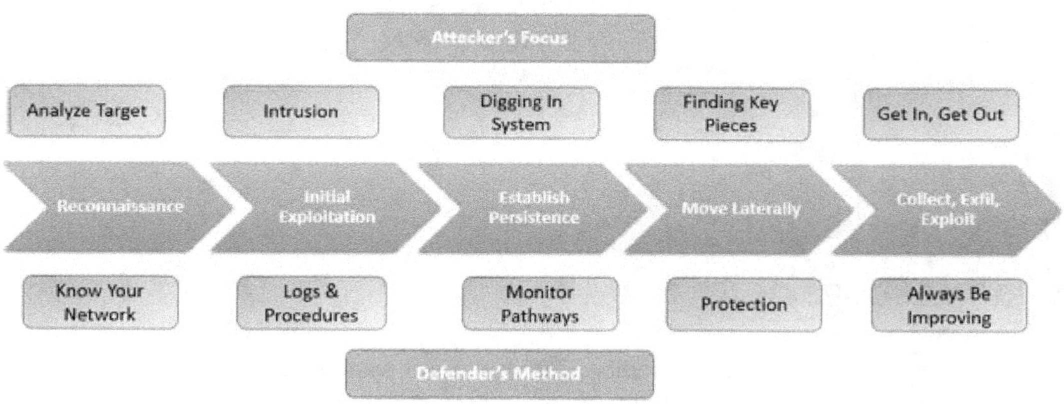

By the end of this pre-attack phase, attackers will have created a detailed map of the network, highlighted the system's weaknesses, and then continue with their mission. Another point of focus during the reconnaissance stage understands the network's trust boundaries. With an increase in employees working from home or using their personal devices for work, there is an increase in areas of data breaches.

The Internet of things (IoT) is the extension of Internet connectivity into physical devices and everyday objects. Embedded with electronics, Internet connectivity, and other forms of hardware (such as sensors), these devices can communicate and interact with others over the Internet, and they can be remotely monitored and controlled.

The definition of the Internet of things has evolved due to the convergence of multiple technologies, real-time analytics, machine learning, commodity sensors, and embedded systems. Traditional fields of embedded systems, wireless sensor networks, control systems, automation (including home and building automation), and others all contribute to enabling the Internet of things as shown in Figure 16. In the consumer market, IoT technology is most synonymous with products pertaining to the concept of the "smart home", covering devices and appliances (such as lighting fixtures, thermostats, home security systems and cameras, and other home appliances) that support one or more common ecosystems, and can be controlled via devices associated with that ecosystem, such as smartphones and smart speakers.

Figure 16.

WHAT IS THE SCOPE OF IOT?

Internet of Things can connect devices embedded in various systems to the internet. When devices/ objects can represent themselves digitally, they can be controlled from anywhere. The connectivity then helps us capture more data from more places, ensuring more ways of increasing efficiency and improving safety and IoT security.

IoT is a transformational force that can help companies improve performance through IoT analytics and IoT Security to deliver better results. Businesses in the utilities, oil & gas, insurance, manufacturing, transportation, infrastructure and retail sectors can reap the benefits of IoT by making more informed decisions, aided by the torrent of interactional and transactional data at their disposal.

How can IoT Help?

IoT platforms can help organizations reduce cost through improved process efficiency, asset utilization and productivity. With improved tracking of devices/objects using sensors and connectivity, they can benefit from real-time insights and analytics, which would help them, make smarter decisions. The growth and convergence of data, processes and things on the internet would make such connections more relevant and important, creating more opportunities for people, businesses and industries

Cyber Kill Chain: A *kill chain* is used to describe the various stages of a cyber attack as it pertains to network security. The actual steps in a kill chain trace the typical stages of a cyber attack from early reconnaissance to completion where the intruder achieves the cyber intrusion. Analysts use the chain to detect and prevent advanced persistent threats (APT) as shown in Figure 17.

Figure 17.

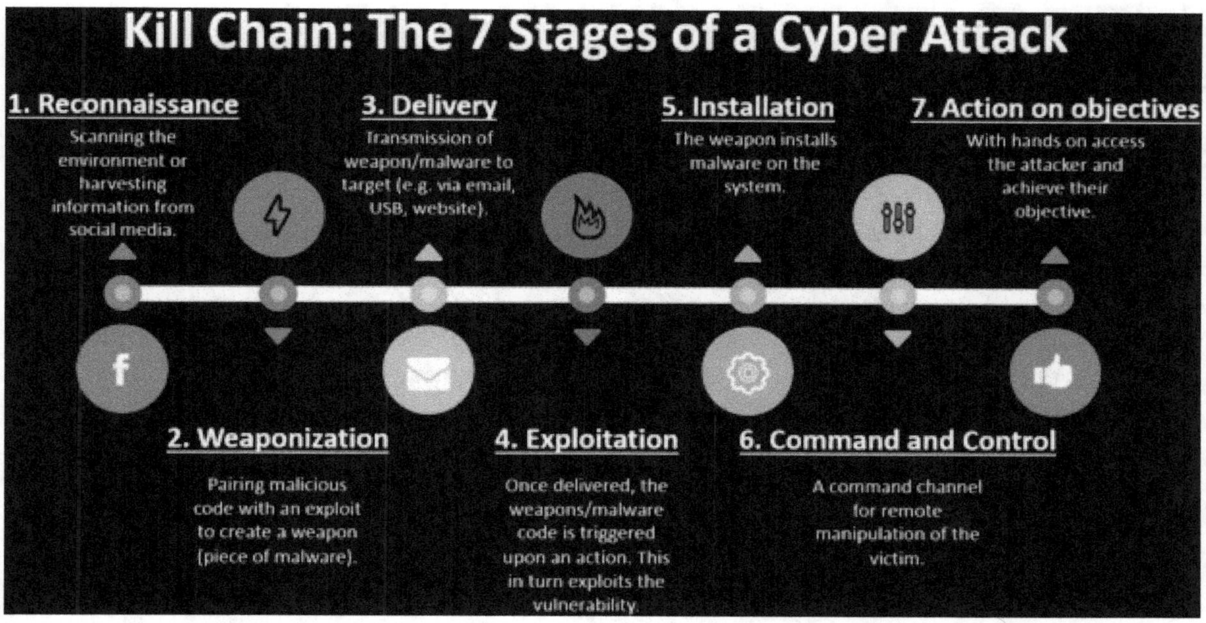

Neural networks help us cluster and classify. You can think of them as a clustering and classification layer on top of the data you store and manage. They help to group unlabelled data according to similarities among the example inputs, and they classify data when they have a labeled dataset to train on. (Neural networks can also extract features that are fed to other algorithms for clustering and classification; so you can think of deep neural networks as components of larger machine-learning applications involving algorithms for reinforcement learning, classification and regression.)

RESULTS AND DISCUSSION

(Xiong & Lagerstrom, 2019), author aims to analyze state-of-the-art threat modeling articles, in terms of general information (including publication year, author affiliations, outlets, and number of citations), type of threat modeling methods it employed, the aimed system, type of threats and attacks it intended to evaluate, the focus and approach, as well as type of methods the article used in terms of validation. The selected 54 articles were classified into three clusters:

Cluster 1 (C1): Applying threat modeling (29 articles)
Cluster 2 (C2): Threat modeling methods (20 articles)
Cluster 3 (C3): Threat modeling process (5 articles)

Studying the focus and approach of the threat modeling articles is important, as it helps us understand what problems were tackled by threat modeling, and how to tackle these issues. The focuses can be classified into "application development", which can be analyzing the security of one or a set of software applications in the design phase, as well as "system evaluation", which takes more than just software into consideration.

Issues, Solutions, and Recommendations

In this chapter, we highlight existing issues in the recent researches. For example, we have Limited AI (Artificial Intelligence) and ML (Machine Learning) based threat and vulnerability analysis techniques exist to provide a better understanding and reduce manual effort, Limited advance techniques are there to reduce the false positives on doing VAPT exercises in application environment with behavioral study and there is only a rudimentary understanding and provision for using CTI to supplement existing cyber security measures.

In this chapter, we also present the previous researches, solutions and recommendations highlighted in Cyber Threat Intelligence area using Artificial Intelligence and Machine Learning. For example, CTI requires the productive, accurate, systematic and robust solutions for new level of threats and vulnerabilities, how to appropriately use CTI in order to augment existing efforts done in cyber security area and AI and ML based techniques which effectively reduce the false positives on doing VAPT exercises in application environment and ML (Machine Learning) and data mining is used to provide efficient malware analysis.

FUTURE TRENDS AND CONCLUSION

While the Intelligent technologies (Artificial Intelligence, Machine Learning and Deep Learning) provides new features and benefits to Cyber Threat Intelligence (CTI), this section presents a research plan with an overarching goal to help ensure that overall Cyber Resilience in CTIshould be maintained. The result of this plan will be a set of tools to improve approaches to CTI, made available to a wider community of academics and practitioners, and, in addition, for the tools to be adaptable to different domains such as cloud security, network and mobile application security and privacy. The chapter concludes with a summary, a balanced assessment of the contribution of AI & ML in CTI, and a roadmap for future directions.

REFERENCES

Bellaby, R. W. (2016). Justifying Cyber intelligence. *Journal of Military Ethics*, *15*(4), 299–319. doi:10.1080/15027570.2017.1284463

Conti, M., Dehghantanha, A., & Dargahi, T. (2018). Cyber threat Intelligence: Challenges and Opportunities. Cyber Threat Intelligence, 1–6. doi:. doi:10.1007/978-3-319-73951-9_1

Mahbub, M. (2020). Progressive researches on IoT security: An exhaustive analysis from the perspective of protocols, vulnerabilities, and preemptive architectonics. *Journal of Network and Computer Applications*, *168*, 1–32. doi:10.1016/j.jnca.2020.102761

Mattern, T., Felker, J., Borum, R., & Bamford, G. (2014). Operational Levels of Cyber Intelligence. *International Journal of Intelligence and CounterIntelligence*, *27*(4), 702–719. doi:10.1080/08850607.2014.924811

Mavroeidis, V., & Bromander, S. (2017) Cyber Threat Intelligence Model: An Evaluation of Taxonomies, Sharing Standards, and Ontologies within Cyber Threat Intelligence. *European Intelligence and Security Informatics Conference (EISIC)*. 10.1109/EISIC.2017.20

Mckinnel, D. R., Dargahi, T., Dehghantanha, A., & Choo, K. R. (2019). A systematic literature review and meta-analysis on artificial Intelligence in Vulnerability Analysis and Penetration Testing. *Computers & Electrical Engineering*, *75*, 175–188. doi:10.1016/j.compeleceng.2019.02.022

Mohanta, B. K., Jena, D., Satapathy, U., & Patnaik, S. (2020). Survey on IoT Security: Challenges and Solution using Machine Learning, Artificial Intelligence and Blockchain Technology. *Journal of Internet of Things*, *11*, 1–32. doi:10.1016/j.iot.2020.100227

Powell, J. L. II. (2016). *Utilizing Cyber Threat Intelligence to Enhance Cybersecurity*. ProQuest LLC.

Preuveneers, D., & Joosen, W. (2021). Sharing Machine Learning Models as Indicators of Compromise for Cyber Threat Intelligence. *J. Cybersecur. Priv.*, *1*(1), 140–163. doi:10.3390/jcp1010008

Ramsdale, A., Shiaeles, S., & Kolokotronis, N. (2020). A Comparative Analysis of Cyber-Threat Intelligence Sources, Formats and Languages. *Electronics (Basel)*, *9*(5), 824. doi:10.3390/electronics9050824

Runder, M. (2015). Cyber-Threats to Critical National Infrastructure - An Intelligence Challenge. *International Journal of Intelligence and Counter Intelligence*.

Shea, D. (2016, May). Preventing Cyber Intelligence Failures by Analyzing Intelligence Failures and Intelligence Reforms. *ProQuest*.

Sherazi, H. H. R., Iqbal, R., Ahmad, F., Khan, Z. A., & Chaudhary, M. H. (2019). *DDoS attack detection A key enabler for sustainable communication in internet of vehicles*. Elsevier.

Shin, M., Cornelius, C., Peebles, D., Kapadia, A., Kotz, D. K., & Triandopoulos, N. (2011). AnonySense: A system for anonymous opportunistic sensing. *Pervasive and Mobile Computing*, *7*(1), 16–30.

Tounsi, W., & Rais, H. (2018). A survey on technical threat intelligence in the age of sophisticated Cyber Attacks. *Computers & Security*, *72*, 212–233. doi:10.1016/j.cose.2017.09.001

Xiong, W., & Lagerström, R. (2019). Threat modeling – A systematic literature review. *Computers & Security*, *84*, 53–69. doi:10.1016/j.cose.2019.03.010

Chapter 6
Cyber Security for Smart Grids

Priyanka Ahlawat
National Institute of Technology, Kurukshetra, India

ABSTRACT

Smart grids use the potential of information technology and advanced analytics to intelligently supply energy via bidirectional communication while also meeting environmental criteria through the facilitation of green technology integration. Communication technology's fundamental vulnerability has exposed the system to a slew of security concerns. Several survey publications have addressed these issues and proposed solutions. The majority of these publications, however, categorized assaults based on confidentiality, integrity, and availability, but they did not include responsibility. Furthermore, present solutions are focused on preventing individual assaults or defending certain components, but there is no comprehensive method to securing the complete system. The authors evaluate the security requirements, describe some serious cyber-attacks, and present a cyber-security approach to identify and counter these assaults in this work.

INTRODUCTION TO SMART GRID

Traditional electric supply systems distribute electrical energy created at a major power plant by gradually rising voltage levels and then distributing it to end consumers by gradually decreasing voltage levels. This power grid, however, has several flaws, including the incapability to accommodate different production sources like renewable energy, time-consuming demand response, costly and heavy assets, outages, and high carbon emissions. It is obvious that the present electrical infrastructure cannot handle these major issues. Smart grid technology promises to provide reliability and flexibility by enabling the addition of different energy supplies (like wind energy, solar energy, and renewable energy), allowing corrective functionalities when failures occur, lowering energy losses within the grid, and minimising the carbon footprint.

An electric power grid is a net of electricity generators, transformers, transmission lines, and distribution systems that supply electricity to its customers (residential, industrial, and commercial). Currently, electricity is produced in centrally controlled electricity production plants and distributed via large transmission networks to distributors before arriving to the end consumers via information exchange

DOI: 10.4018/978-1-6684-3991-3.ch006

and electricity flow is unidirectional, i.e., from electricity production to consumers, which is named to as an power grid. After many years of research, this has been shown that multiple services may link to improve entire power system reliability by correcting for unforeseen disasters as well as interruptions from power equipment such as generators and transmission lines.

Today's electric power infrastructure is ageing and inadequate for the rapidly expanding electricity demand of the twenty-first century due to a lack of monitoring and computerized analysis (Knapp & Samani, 2013). Electricity consumption rate in the United States, for instance, have risen by 2.5 percent each year during the previous 20 years. Furthermore, catastrophic climate change and greenhouse gases from the electrical and transport sectors (Framework, 2012) put further demand on existing power infrastructure. As a result, a novel idea of next generation electricity network is immediately required to address these problems, prompting the smart grid (SG) suggestion.

The Smart Grid may be thought of as a mix of power grid and communication networks. As a result, by combining alternative and renewable energy sources such as solar cells, bioenergy generation, wind energy, wave energy, small hydroelectric generators, and plugin hybrid vehicles, it can increase the efficiency, safety, stability, and protection of power supply to consumers (Rawat & Bajracharya, 2015b).In these four zones of the electric grid, several components are connected together in Smart Grid through bidirectional connectivity and electricity runs to facilitate interoperability between these four sectors of the electric grid. As a result, with the help of smart metres that allow measuring and monitoring of these two-way flows, customers may not only absorb electricity but also transfer excess power to the grid. This new infrastructure has the potential to generate lakhs of alternative small-energy sources and enhance balancing of load via immediate exchanges electric load information, which could assist electricity plants match their production according to demand using data collected by metering, sensing, and monitoring.

Table 1. Smart grid vs traditional grid

Features	Traditional Grid	Smart Grid
Power Generation	Integrated electricity Production	Distributed Electricity Production
Grid Topology	Radial	Network
Sensors	Low grade	High grade
Information Flow	unidirectional	Bidirectional
Monitoring	Not automated	Automated
Testing	Manual	Remote
Environmental Impact	High	Low
Ability Control	Limited	Pervasive

The most significant component to realising the Smart Grid is an advanced metering infrastructure (AMI) built on smart metres. The AMI is an arrangement that receives and analyses information from smart metres through bidirectional connections, and then intelligently manages different power related services and applications depending on that information. The AMI is the placement of a metering system that communicates with the electric metre in both directions. The adoption of AMI is usually regarded as the first step toward the digitization of electric grid control systems. Because of the accuracy increase in live metre analysis and control, AMI has recently attracted a lot of attention in both industry and commerce. The AMI architecture enables smart utility metres and utility suppliers to communicate in real time. The AMI includes smart metres at client locations, access points, a connectivity backbone network connecting consumers and service suppliers, and data supervision systems that track, gather, organize, and evaluate data for further treatment. The smart meter can detect electricity usage in greater accuracy than a typical metre and communicate the gathered data to the service provider on a routine basis for billing purposes and usage checking. Furthermore, information from smart metre readings is required for the control centre to perform Response/Demand programmes.Customers may regulate their power usage and manage how much electricity they use by employing smart metres, particularly by reducing peak load. As a result of customer engagement, utility companies will be able to deliver power at cheaper and more consistent prices to all of their customers, resulting in decreased carbon dioxide emissions. Despite the increased use of AMI, relatively little evaluation or development effort and research has been made to evaluate the safety requirements for this type of systems. Furthermore, security difficulties, main problems, and resolutions in AMI in Smart Grid are given.

Figure 1. Conventional power grid

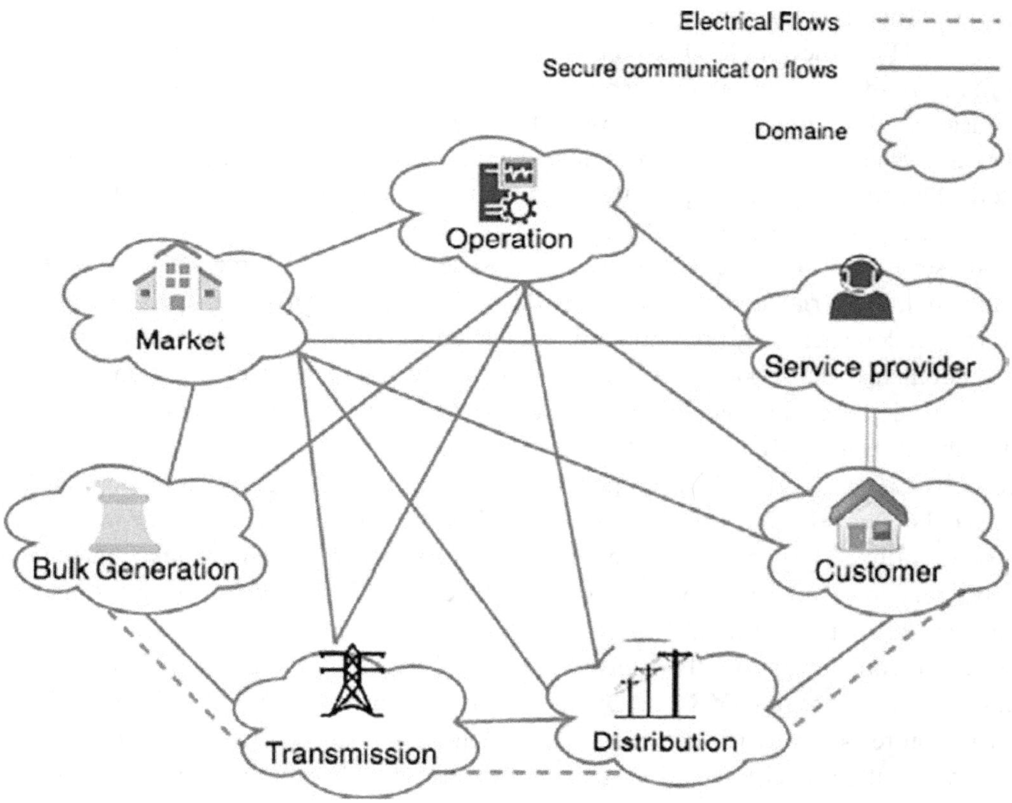

OVERVIEW OF SMART GRID

Features of Smart Grid

The key features of the smart grid are predicted to be improved environmental performance and increased grid resilience. The capacity of a particular entity to survive unforeseen incidents and restore rapidly afterward is referred to as resilience (Knapp & Samani, 2013). Grid resilience is becoming a non-negotiable element, especially since power outages might have a negative economic impact. Smart grid technology promises to increase reliability and flexibility by allowing for more distributed power supply, offering remedial capabilities when errors arise, and simplifying the integration of new resources into the system. Furthermore, smart grid technologies are intended to allow electric cars to replace conventional vehicles, lowering energy losses and energy utilized by consumers within the grid.

Conceptual Model of Smart Grid

In academic and business communities, many frameworks have been established to define the architecture of IoT (), particularly for smart grid applications. Until date, the National Institute of Standards and Technology (NIST) framework has been the most extensively used (). The framework is conceptualized as seven linked domains (Manesh & Kaabouch, 2017) are bulk generation, distribution, transmission, markets, customer, operations, and service provider, each of which contains characters and applications. Characters are systems, devices, and programs, whereas applications are jobs done in each domain by one or more characters. Figure 1 depicts the interaction of characters from many domains over a secure channel and conceptual model of smart grid.

The end user is the primary character in the customer domain. Customers are classified into three types: residential, industrial, and commercial/building. In addition to using power, these characters may create, store, and control energy consumption. This domain communicates with the operation, distribution, market domains, and service provider, and electrically linked to the distribution domain.

In the market domain, this domain is responsible for balancing the demand and supply for electricity. To match demand with supply, the market domain interfaces with generation domains such as distributed energy resources (DER) and bulk generation(Framework, 2012).The service provider domain arrange for services to both utilities and electrical customers. These businesses are in charge of services such as client accounts, billing,and energy use. The service provider and the operation domain interacts witheach other for system control and situational awareness, as well as with the market and customer domains to offer smart services (Framework, 2012).

The operations domain ensures that transmission and distribution activities are efficient and optimum. Energy management systems (EMS) and Distribution management systems (DMS)are used in transmission, distribution respectively (Framework, 2012). Generators are used in the bulk generation domain for large amounts of electricity. The first stage in the process of distributing power to the end customer is energy generating. Coal, oil, nuclear fission, flowing water, and solar output are all used to create energy. The bulk generating domain is electrically coupled to the transmission domain and connects with the transmission domain, market domain, and operations domain via an interface (Framework, 2012).

Figure 2. Conceptual model of smart grid

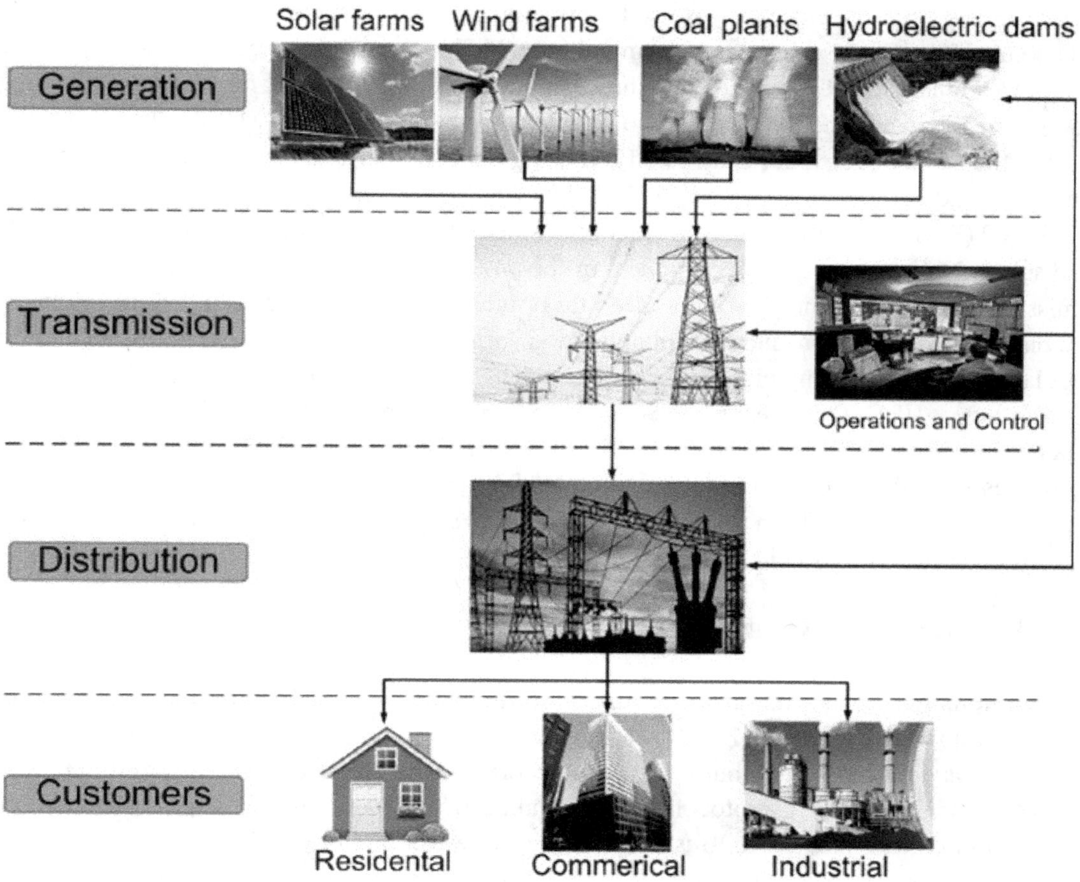

Systems of Smart Grid

Smart grid systems include automated substation, demand response, advanced metering infrastructure (AMI), household energy management (HEM),supervisory control and data acquisition (SCADA), and electrical vehicle (EV). In this part, we will look at three important and sensitive systems: automation substation AMI, and SCADA.

The consumer and distribution domains share advanced metering infrastructure (AMI), which is incharge of measuring, analyzing, and collecting use of energy. It enables bidirectional communication between the utility and the user. It consists of three parts: the communication network, the AMI head-end, and the smart meter(Faisal et al., 2015). An AMI headend is a server that includes a meter data management system (MDMS). Smart meters are digital meters with microprocessors and local memory that are responsible for collecting and monitoring power use from home appliances as well as transferring real time data to the AMI headend on the utility side. Communication between the AMI headend, household appliances,and smart meters take place with the help of several communication protocols, including Zigbee and Z-wave (Faisal et al., 2015).

Supervisory control and data acquisition (SCADA):-It is an operation domain system commonly employed in large-scale scenarios. It is used to monitor,regulate, and measurean electrical power grid. It is made up of three parts: the master terminal unit (MTU), the human-machine interface (HMI),and the remote terminal unit (RTU). There are three components inRTU device: one for data collecting, one for communication, and one for executing instructions from the MTU. The MTU is the device incharge of guiding the Remote Terminal Unit. The HMI is a graphical interface for the operator of the SCADA system. Many industrial protocols, notably IEC 61850 (Knapp & Samani, 2013), and distributed network protocol v3.0 (DNP3) are utilized for interaction inside SCADA systems.

The substation is an important component of the power grid network. It falls within the domains of transmission, generation, and distribution. Its tasks include accepting electricity from the producing plant, controlling distribution, and minimizing power surges (Gungor et al., 2013). It includes equipment for regulating and distributing electrical energy, such as a global positioning system (GPS), intelligent electronic devices (IEDs),remote terminal unit (RTU), and human-machine interface (HMI). The substation transmits operational information to the SCADA with the purpose of controlling the power system. Many processes within the substation are automated in order to improve power system dependability (Knapp & Samani, 2013). The standard IEC 61850 defines the exchange of data between other devices and the automation substation in distribution and transmission.

Network Protocol of Smart Grid

Different communication protocols are required for distributed and heterogeneous smart grid applications. Home appliances utilize the Z-wave and ZigBee protocols in the home area network (HAN) (Faisal et al., 2015). Devices in a neighbourhood area network (NAN) are often linked using IEEE 802.15.4, IEEE 802.11, or IEEE 802.16 protocols (Faisal et al., 2015). Several industrial protocols, particularly Modicon communication bus (ModBus)and distributed networking protocol 3.0 (DNP3) are utilized in supervisory control and data acquisition (SCADA) applications and wide area network (WAN) (Knapp & Samani, 2013). To address the issue of lack of wireless resources,some writers have advocated use of cognitive radio based on IEEE 802.22 and it also enhance smart grid connectivity in the wide area network (Kaabouch & Hu, 2014).We will cover two frequently used yet insecure smart grid protocols: DNP3 and Modbus

Modiconcommunication bus (Modbus) is a7 layer protocol based on OSI model that was developed in 1979 to allow process controllers to connect with computers in real time. Modbus is classified into three types: Modbus RTU, Modbus/TCP, andModbus ASCII. In the Modbus ASCII, encoded messages are in hexadecimal. Despite its slowness, it is perfect for telephone and radio conversations. The messages in the Modbus RTU are encoded in binary and sent through RS232. In the Modbus/TCP, the masters and slaves communicate through IP addresses (Al-Dalky et al., 2014). ModBus is a master-slave protocol that exchanges instructions between one master terminal unit (MTU),master or remote terminal unit (RTU), and numerous slave devices, like programmable logic controllers (PLCs), drivers, and sensors in a SCADA system (Al-Dalky et al., 2014).At one side, Modbus is extensively used in architecture of industrial due to its relative easiness of use in exchanging raw information without encryption, authentication, or any additional overhead (Knapp & Samani, 2013). These characteristics, on either side, render it insecure and easily exploited (Al-Dalky et al., 2014).

Another commonly deployed protocol of communication for vital organization, notably in the electrical industry, is distributed network protocol v3.0 (DNP3) (Rodofile et al., 2016). It started as a serial protocol for managing communication across "Master stations" and slave stations known as "outstations" in 1990 (Knapp & Samani, 2013). DNP3 had been used in electrical stations to connect master stations, like remote terminal units (RTUs), with outstations, like intelligent electrical devices (IEDs). DNP3 was expanded to function across IP networks by encapsulating user data-gram protocol (UDP) or transmission control protocol (TCP) packetsin 1998. DNP3 supports time-stamped (time-synchronized) data and employs various standardized data formats, making data transfer efficient and reliable. DNP3 did not initially have any security mechanisms like as authentication or encryption, however this was corrected with the release of DNP3 secure (Knapp & Samani, 2013; Rodofile et al., 2016; Wang & Lu, 2013).

SMART GRID'SSECURITY REQUIREMENTS

As per the National Institute of Standards and Technology (NIST), the three requirements for maintaining the protection and security of data in the smart grid, namely integrity,confidentiality, and availability (Elmrabet et al., 2016). Accountability is another key security requirement, according to (Liu et al., 2014). All requirement is described in detail here.

Confidentiality: In common parlance, confidentiality ensures that permitted constraints on data access and revelation are maintained. In other terms, the confidentiality requirement necessitates securing both proprietary information and personal privacy from unauthorized entities, persons, or processes. Confidentiality is compromised when information is unlawfully disclosed (Elmrabet et al., 2016). For example, information exchanged between a consumer and other organizations, like meter control, metering consumption, and billing information, must be secret and secured; otherwise, the consumer's information might be exploited, edited, or used for other harmful reasons.

Integrity: In a smart grid, integrity implies safeguarding against unauthorized change or loss of records. Aintegrity loss is defined as an illegal modification, destruction, or alteration of data that occurs undetectably (Elmrabet et al., 2016). Power injection, for example, is a hostile assault conducted by an attacker who cleverly adjusts measured data and transmits them from power injection meters and power flow to the state estimator. To sustain the integrity, both nonrepudiation and accuracy of data are essential. Nonrepudiation refers to the reality that entities, individuals, or organizations are incapable to conduct a certain activity and then disagree it afterwards; authenticity attributed to the idea that information is derived from a valid source.

Availability: The availability of information is described as guaranteeing dependable and timely use of and access to information. It is regarded as being the most critical safety requirement in the smart grid since aavailability loss causes an interruption in information access in a smart grid (Elmrabet et al., 2016). For e.g., a lack of availability can interrupt the function of the control system by preventing data from flowing over the network and, as a result, prohibiting the availability of network to manage the operators of system.

Accountability: Accountability entails assuring the system's tractability and that each and every action taken by an individual, gadget, or even by public authority is admissible as evidence therefore that nobody can disagree withher or his actions. This recordable data can be presented in a court of law as evidence to identify the hacker. Consumers' monthly power bills are an example of an accountability issue. In general, smart meters might calculate the price of power on a daily basis or in real time. Unfortunately, if these meters are attacked, the data they provide is notconsidered as reliable since they have been tampered with. As a consequence, the consumer will get two separate energy bills, one from the company and one from the smart meter (Liu et al., 2014).

SECURITY CHALLENGES AND PROTECTIVE MEASURES IN SMART GRID

Smart Grid Attacks

Malicious hackers often take four phases to hack and gain access of a system: reconnaissance, scanning, exploitation, and access maintenance (Engebretson, 2013). The attacker obtains and gathers information about its target machine during the first phase, reconnaissance. The hacker attempts to find the weaknesses of system in the second stage, scanning. These actions seek to detect the open ports as well as the services operating on each port, as well as their vulnerabilities. During the exploitation stage, he or she attempts to make a compromise in order to gain complete access of the target device. Once the hacker gains administrative control of the target, she or he moves on to the last phase, which is to keep the access. This is accomplished by setting up a covert and undetected application, allowing him/her to quickly return to the target machinein future.

Attackers in smart grids take the same tactics to circumvent security requirements (Knapp & Samani, 2013). During each phase, they employ various tactics to compromise a certain entity in the grid. As a result, depending on these phases, assaults may be categorised. Several forms of assaults might occur throughout the exploitation stage. The malicious actions and assaults that occur throughout each phase are detailed down.

Reconnaissance

The assaults in the first phase, reconnaissance, involve traffic analysis and social engineering. Social engineering (SE) is based on human interaction and social skills instead of technical abilities. An hacker utilises convincing and communication to gain the confidence of a genuine user and get secret information and credentials, such as personal identification numbers (PINs) or passwords, to log on to a specific system. Password pilfering and phishing attacks are well-known Social Engineering tactics. The traffic analysis attack analyses and listens to network traffic with the intention of detecting the connected devices and hosts to the network, as well as their IP addresses. The confidentiality of information is primarily jeopardised by traffic analysis and social engineering.

Scanning

The following phase is a scanning technique, which is used to discover all of the devices and hosts that are still active on the network. Vulnerabilities, services IPs, and ports are the four categories of scanning (Engebretson, 2013). An attacker will often begin with scanning IPs to discover eachconnected hosts to the network as well as their IP addresses. Then she or he digs further to see which ports are open by checking theirports. This scan is carried out on every detected host on the network. Then the attacker does a service scan to determine which system or service is operating behind all opened ports. For example, if the open port 102 on a certain system is found, the attacker may deduce that this system is a messaging system or substation automation control. The target system is a phasor measuring unit (PMU) if port 4713 is open (Knapp & Samani, 2013).

Two industrial protocols, Modbus and DNP3 that may be hacked. Because TCP/Modbuswas built for communication purpose instead of security, hacker uses Modbus network scanning attack. This technique entails delivering a harmless message to all devices connected tonetworkwith the intention of obtaining information about connected devices. Modscan is a scanner for SCADA Modbus network that detects open TCP/Modbus networks and identifies device slave IP addresses as well as their IDs (Al-Dalky et al., 2014).

Exploitation

The exploitation stage comprises malevolent operations that aim to gain control of the smart grid components by exploiting its weaknesses. Trojan horses, Viruses, worms, denial of service (DoS) attacks, man-in-the-middle (MITM) assaults, jamming channels, replay attacks,integrity violations, privacy breaches, and popping the human-machine interface (HMI) are examples of these activities.

In the smart grid, a virus is a software that is used to infiltrate a particular device or system. A worm is a software that replicates itself. It spreads across the network, copies itself, and infecting other devices and systems. A Trojan horse is a software that masquerades as a genuine programme on the target machine. In the background, however, it executes malevolent code. This sort of malware is used by a hacker to install a worm or virus on the victim machine (Knapp & Samani, 2013). In June 2010,Stuxnet, was the first worm attacking supervisory control and data acquisition (SCADA) systems, and discovered by a senior researcher RoelSchouwenber at Kaspersky Lab(Knapp & Samani, 2013).

Various techniques are utilized in denial of service (DoS) attacks, including buffer overflow,SYN attacks, smurf attacks, and teardrop attacks(Knapp & Samani, 2013), time synchronization attack (TSA) (Zhang et al., 2013), and puppet attack (Yi et al., 2014). A SYN attack takes advantage of the three-way handshake needed to setup a session of Transmission Control Protocol (TCP) (SYN, SYN-ACK, ACK). The adversaryoverflows the machine of victim with connection requests while refusing to reply to replays, causing the machine of victimto crash. The TCP/Modbus protocol is susceptible to various assaultsbecause this uses TCP.

In a buffer overflow attack, the adversary provides a large quantity of information to a particular program, depleting its resources (Knapp & Samani, 2013). The ping-of-deathattack uses the internet control message protocol (ICMP) to deliver data of more than 65 K octets. The system then crashes as a result of this. That's why it is classified as a buffer overflow

An attacker uses a teardrop technique to change the size and fragmentation offset fields in successive IP packets (Knapp & Samani, 2013). When the destination device receives these packets, it fails for the reason that the instructions in the packets on how to offset the fragments are conflicting.

A smurf attack not only targets a single system, but it may also congest and overwhelm the traffic of an existing network.The Smurf attack is a type of computer network attack in which a large number of Internet Control Message Protocol (ICMP) packets with the spoofed source IP of the intended victim are broadcast to a computer network using an IP broadcast address. Most network devices will, by default, respond to this by sending a response to the source IP address. If there are a large number of machines on the network that receive and respond to these packets, the victim's computer will be inundated with traffic. This can cause the victim's computer to slow down to the point where it is impossible to work on.

The puppet attack (Yi et al., 2014) aims the network of AMI by leveraging a weakness in the dynamic source routing (DSR) protocol and subsequently draining the bandwidth of communication network. Packet transmission is reduced by 10% to 20% as a result of this assault.

The time synchronization attack(TSA) (Zhang et al., 2013) primarily bull's eye the timing information of smart grid. Because power grid functions like fault diagnosis and event location estimation rely heavily on accurate time information, and because the majority of smart grid measuring instruments are fitted out with a global positioning system (GPS), an assault like TSA that spoofs information of GPS could have a significant effect on the system. DoS is a severe hazard to the smart grid system since control signals and communication in such a system are very time sensitive, and availability of system might be jeopardise with delay a few seconds.

A man in the middle (MITM) attack occurs when a perpetrator inserts himself into a communication between a user and an application, either to eavesdrop or to impersonate one of the parties, giving the impression that a normal exchange of information is taking place. The perpetrator has access to both machines and passes traffic between them. The genuine gadgets seem to interact directly while, in reality, they interact through a third-party device (Wang & Lu, 2013). For instance, a perpetrator may do an MITM by connecting to an Ethernet network and altering or misrepresenting I/O values to the human-machine interface (HMI)and programmable logic controllers (PLC). The MITM might also reroute TCP/IP traffic between the transmission SCADA system and the substation gateway.

Another sort of MITM attack is the intercept/alter attack. It tries to capture, manipulate, and edit data that is being transferred to the whole network or kept in a specific device (Wang & Lu, 2013). for instance, A perpetrator may utilize electromagnetic/radio-frequency interception to interrupt a secure communications in advanced metering infrastructure (AMI). Another sort of MITM attack is active eavesdropping, in which the adversary intercepts secret exchange of data between two authorized devices. Each of these MITM assaults aim to jeopardize accountability, integrity, and confidentiality.

Jamming channel attack, an attacker uses the common feature of the wireless network to broadcast a continuous or random stream of packets with the purpose of keeping the channel occupied, preventing authorized devices from connecting and transmitting information (Reyes & Kaabouch, 2013).,To satisfy quality of service standards, smart grid needs highly available network due to time-critical nature of smart grid, and this kind of assault can substantially impact its performance. The researchers of (Gai et al., 2017) presented a jamming technique called maximal attacking strategy utilising masking and jamming (MAS-SJ) that primarily aims the wireless smart grid network (WSGN) with cognitive radio network (CRN). Since WSGN is necessary for managing power grid in the smart grid with the PMU, which plays a major role in supplying time-synchronized information of power system operational states, cyber-assaults such as MAS-SJ might disrupt the functioning of system or even render it inaccessible.

Replay attack, this type attack occurs as the industrial control message is transferred in plain text,a perpetrator might intentionally collect packets, insert a particular packet, and replay them to authorized destinations, jeopardizing the integrity of communications. The replay attack might have potentially to modify the behaviour of programmable logic controllers (PLCs) (Knapp & Samani, 2013). For interacting and monitoring the SCADA system a device called intelligent electronic device (IED)is built (Wang & Lu, 2013), which might be attacked by replay attacks in order to insert erroneous measurements into a specific register.

Popping the HMI is an assault that leverages a well-known weakness of component/device, particularly one in operating system or software of the device, and then set up a remote shell, letting the perpetrator to establish connection remotely and gain unauthorized access to the server from his computer with the intention of controlling and monitoring the compromised system (Knapp & Samani, 2013).SCADA systems, Substations, and any system running an operating system with a console interface are thought to be possible targets of this assault. Even with the potential consequence of an assault of this kind, doing it does not need extensive networking abilities or substantial knowledge of industrial control and security systems. Because the information for the vulnerabilities of devices is available publicly, this type of hacker or so-called script-kiddies may effortlessly utilize tools available as open source like meterpreter and Metasploit to conduct assault like it and take complete control of the target machine.

A hostile individual may pose as a genuine individual with the intention of obtaining access to a system or get increased freedoms to do unlawful acts in a masquerade attack. This exploit has the potential to interfere with the programmable communicating thermostat (PCT), which helps in minimizing electricity consumption at a domestic location. It jeopardizes integrity, availability, accountability, and secrecy of the system (Knapp & Samani, 2013).

The goal of a privacy violation attack is to breach privacy of consumers by obtaining confidential information (Shapsough et al., 2015). For example, Smart metres gather power usage several times each hour, information on the electricity consumption of users may be retrieved. As a result, if a metre does not display any use of electricity for a stretched period of time, it is likely that the place is unoccupied. Then, this data might be utilised to carry out a physical attack, such as a theft, robbery.

Integrity violation attacks seek to compromise the accountability or/andintegrity of the smart grid by purposefully or inadvertently modifying the information contained in a specific network component. For example, a client may use this attack to change the data on his smart metre with the intention of reducing his power cost. An integrity violation is a false data injection (FDI) attack.This attack might have potentially to target remote terminal units (RTU), causing incorrect information to be relayed to the control centre and increasing outage time. It intends to induce arbitrary mistakes and contaminate the measurements of some devices, so altering the accurateness of the state estimation (SE).Because the SE is critical for monitoring the system and making operation of the power systemmore reliable, as well as to process real-time data collected by the SCADA systemfor the energy management system (EMS).

Maintaining Access

In the last phase, maintaining access, the perpetrator employs a unique technique to get continuous access to the victim's machine, most notably backdoors, Trojan horses, and viruses. A backdoor is an invisible, sneaky application put on the victim's machine to provide easy and rapid access later (Knapp & Samani, 2013). If the perpetrator is successful in planting a backdoor into the servers of the SCADA control centre, she or he can conduct a series of assaults on the system, potentially causing significant damage to the power supply.

In an IT network, security parameter are in the followingorder: confidentiality, integrity, accountability, and availability.In the smart grid, priority order of security parameters aredifferent: availability, integrity, accountability, and confidentiality (Liu et al., 2014).As a result, we may state that severity of assaults that jeopardise the availability of smart grid systems is very high, but assaults that jeopardise confidentiality are of moderate severity. Each assault has a level of likely to be carried out in addition to the amount of severity. For example, Duqu and Stuxnet (Knapp & Samani, 2013) have a high severity since they may vandalise the industrial control system and breach all security barriers; nonetheless, they are complicated and smart. As a result, these viruses are extremely dangerous, yet their chance of being carried out is limited.

The HMI popping attack is another example. It is of high severity and does not need sophisticated networking abilities or extensive knowledge in security and industrial control systems to carry out. Because the information for the vulnerabilities of devices is available publicly, this type of hacker or so-called script-kiddies may effortlessly utilize tools available as open source like meterpreter and Metasploit to conduct assualt like it and take complete control of the target machine.As a result, the severity of this attack is great, and it is extremely likely to be carried out.

Techniques of Detection and Protective Measures

To counter cyber-attacks, a variety of attack detection and countermeasure strategies have been presented in the literature. For example, the authors of (Rawat & Bajracharya, 2015a) suggested a strategy for detecting False detection injection (FDI) attacks.the authors of (Gai et al., 2017; Reyes & Kaabouch, 2013) suggested a method for detecting jamming channel attacks. While these security measures help to secure the smart grid, they are inadequate to combat mixed and sophisticated assaults (Knapp & Samani, 2013).Furthermore, Stuxnet (Knapp & Samani, 2013) demonstrated that strategies such as "security by obscurity" or "defense-in-depth" (Knapp & Samani, 2013) are no longer regarded viable countermeasure.We consider that security cannot be accomplished through a single countermeasure, but rather through the use of numerous approaches as part of a comprehensive plan. We suggest a three-phase cyber-security approach: pre-attack, under-attack, and post-attack. Appropriate published approaches in terms of security technology, encryption,safety protocols, and other cyber-attack solutions are detailed below, and for each step.

Pre-Attack

Numerous published methods are advised at this first phase, pre-attack, to improve the security of the smart grid and to be ready for any prospective assault. Security countermeasures are typically classified into three types: device security,network security, and cryptography.In this chapter, we explore network security protocols and technologies such as intrusion detection systems (IDS), security information and event management systems (SIEM), secure DNP3and network data loss prevention (DLP) (Faisal et al., 2015; Knapp & Samani, 2013). For security of data, authentication, key management, and encryption (Shapsough et al., 2015; Wang & Lu, 2013) are used.Finally, for security of device, compliance checks, host IDS, and diversity techniques are used.

Security of Network

A smart grid's network serves as its foundation. As a result, network security is critical to the overall system's security. To protect the smart grid network, it is suggested that firewalls be used in conjunction with other monitoring and inspection technologies (Knapp & Samani, 2013). A firewall is designed to accept or refuse connections of network based on rules and regulations. However, an undiscovered or sophisticated attack strategy may readily circumvent many firewall mechanisms.Firewalls should thus be used in conjunction with other security technologies such asintrusion detection systems (IDS), security information and event management systems (SIEM), secure DNP3and network data loss prevention (DLP) (Faisal et al., 2015; Knapp & Samani, 2013).An intrusion detection system (IDS) is a system designed to identify malicious activities on a particular host or on a network (Faisal et al., 2015).

SIEMs are information management systems that obtain and collect data from all network devices, such as application logs, network flow, and operating system logs.The information gathered will then be analysed and evaluated by a centralised server with the intention of detecting any possible danger or malicious behaviour in the network (Knapp & Samani, 2013). Network DLP is a system that is in charge of avoiding data loss or theft over a network (Knapp & Samani, 2013).

Secure network protocols like secure DNP3, transport layer security (TLS), secure sockets layer (SSL), and IPsec can be used in addition to these security solutions to improve network security.DNP3 is a smart grid industrial protocol that is extensively used (Rodofile et al., 2016).Originally, the DNP3 protocol lacked any safety features.In other words, messages are sent over the network in plain text and can be readily intercepted.The rising frequency of cyber-attacks affecting power systems and industry in recent years has piqued the interest of a number of experts in both academia and business. As a result, a safe variant of DNP3 protocols known as secure DNP3 has been published.

This protected version introduced a secure layer between the application and TCP/IP layers for authentication and encryption. Numerous attacks may be prevented by using such a protocol. For example, the encryption reduces eavesdropping and replay attempts, while authentication system can guard from MITM attacks.

Data Security using Cryptography

Encryption mechanisms are designed to protect the integrity, confidentiality, and non-repudiation of data. Key encryptions are classified into two types: asymmetric and symmetric. One key is used to encrypt and decode data in symmetric key encryption, also known as single-key encryption. The most often used symmetric encryption algorithms are data encryption standard (DES) and advanced encryption standard (AES).On the other hand, Asymmetric key encryption encrypts and decrypts data using two keys: a public key and a private key.The popular asymmetric encryption algorithm is RSA algorithm (Rivest, Shamir, and Adleman).Different components with varying computing capacities coexist in a smart grid.As a result, both asymmetric and symmetric key encryption can be utilised, with the choice depending on a variety of criteria such as time limitations, computer resources, and data criticality (Shapsough et al., 2015).

Authentication is described as the act of validating the identification of an item, such as the use of a password (Rawat & Bajracharya, 2015b). A smart device, a user, or any equipment linked to the smart grid network might be considered an item.Multicast authentication is a sort of authentication process that is commonly utilised in smart grid applications. The authors of (Shapsough et al., 2015) presented three approaches for multicast authentication: time asymmetry, hybrid asymmetry, and secret-info asymmetry.

Key management is an essential component of authentication and encryption. Shared secret key management (SSKM) or Public key infrastructure (PKI) can be used to assure network communication authenticity. The uniqueness of two parties in PKI infrastructure are confirmed by a certificate issued by a third party known as the certificate authority (CA). This procedure is implemented prior to the establishment of any link between the two parties. To guarantee communication security, four processes are employed in shared secret key management:, key storage, key generation, key distributionand key updating (Shapsough et al., 2015). Because the nature of the smart grid is distributed, some particular-necessities should be pondered when designing a cryptography key management scheme. The authors of (Wang & Lu, 2013) present numerous basic yet related key management scheme necessities, including evolve-ability, efficiency, secure management, and scalability.Furthermore, numerous key management frameworks for the power system have been proposed, including key establishment scheme for SCADA systems (SKE),single-key,advanced key management architecture for SCADA systems (ASKMA), key management architecture for SCADA systems (SKMA), scalable method of cryptographic key management (SMOCK), and ASKMA +, to name a few. The selection of a framework is based on a variety of characteristics, including scalability, computing resource capabilities, and multicast support. The above-mentioned key management strategies were compared by the writers. The comparison was focused on scalability, multicast support, resistance to power system application and key compromise. SMOCK and ASKMA + produce intriguing results. ASKMA + is a good key management method that enables multicast, however it has scalability issue. On the other hand,SMOCK offers strong scalability but certain flaws such as lower computational efficiencyand lack of multicast capability.

Security of Device

The third critical component in the distribution network of smart grid safety is device protection. Several scientific studies and commendation reports have been circulated, all of which contribute to endpoint security assurance. Numerous security technologies, including anti-virus, host IDS, and data loss prevention(DLP), are advocated in (Knapp & Samani, 2013). Furthermore, the authors of (Shapsough et al., 2015) advocated for the use of an automated security compliance check. One tool like this checks all smart grid entities to ensure that configuration of all device, particularly the firmware ofdevice and the current configuration file, is up to date. Because smart grid entities are heavily interconnected and a flaw in one entity can put the whole system at danger, anobedience check is an essential instrument.

Under-Attack

This stage is separated into two jobs: detection of attacks and mitigation of attacks. Through every task, several methodologies and technologies may be employed to identify harmful behaviour and subsequently apply the required countermeasures.All implemented security solutions, including DLP, SIEMS, and IDS, are suggested for attack detection (Faisal et al., 2015; Knapp & Samani, 2013). However, several of these methods, like IDS, have limits and could be improved. IDS is a popular safety technology in IT networks, and this is also utilised in networks of smart grid; nonetheless, it has numerous drawbacks inperformance, most notably a high frequency of false positives. As a result, several research articles were produced in order to increase performance of IDS in the smart environment. The authors of(Faisal et al., 2015) suggested a data stream mining algorithmsbased IDS.They compared seven existing cutting-edge data stream mining algorithms: active classifier, accuracy updated ensemble, leveraging bagging, limited attribute classifier, bagging with adaptive-size hoeffding tree, single classifier drift, and bagging with ADWIN. This comparison was based on numerous measures, together with detection accuracy, memory usage, and execution time. They have used KDD Cup 1999 database for the evaluation. The findings revealed that certain algorithms do not necessitate complex processing resources, making them appropriate for IDS in devices like smart meters. Other techniques are more accurate but demand more processing resources; these techniques can be utilised in AMI headends or data concentrators.

After the attack has been recognised, the following mitigating strategies can be used. The authors of (Shapsough et al., 2015) examined and described numerous approaches for mitigating DoS attacks, including reconfiguration and pushback methods. In reconfiguration mode, the network topology is modified in the reconfiguration approach to separate the attacker. The router is set to prohibit all traffic from the IP address of attacker in the pushback mode. The authors of (Reyes & Kaabouch, 2013) addressed anti-jamming systems based on fuzzy logic for jamming attacks. Other mitigating approaches for man-in-the-middle, buffer overflow, replay attack, CPU exhaustion, false data injection (FDI) and distributed denial of service (DDoS) were thoroughly explored in (Rawat & Bajracharya, 2015a; Rawat & Bajracharya, 2015b).

Post Attack

When an assault remains undetected, such asexample of Stuxnet, the post-attack phase is critical. First and foremost, the situation is vital to detect the entity responsible for the assault. After that, to prevent the smart grid from future similar attacks, anti-virus database, the IDS signature, and security rules must be kept up to speed by learning from assaults. During the post-attack phase, the key approach is forensic analysis. Forensic investigations on smart grids gather, analyse, and capture digital information to find the entity engaged in the occurrence. They are also valuable for identifying and addressing smart grid physical and cyber vulnerabilities in order to forecast prospective attacks.Furthermore, smart grid forensic analysis plays a major role in the investigative process of cyber-crimes like viruses, hacking, cyber terrorism, digital espionage, violating consumer privacy, manipulating smart grid operation, and theftof valuable information such as state secrets and intellectual property (Knapp & Samani, 2013).

DIFFICULTIES AND POTENTIAL SOLUTIONS

Different devices exist and interact simultaneously via diverse network protocols in heterogeneous systems such as smart grid. This heterogeneity is a significant problem and a possible danger to security of smart grid. Communication between devices necessitates protocol translation and data collection. However, this collection might lead to vulnerabilities and accidental breaches merely because one protocol's feature could not be sufficiently transformed into feature of another (Shapsough et al., 2015).

In addition, most of industrial network protocols used in smart grids, like ICCP, Modbus,DNP3, and Profibus were built for connection rather than security. As a result, not only may these protocols not assure security of communication, but they can also be utilised as a surface of attack. Despite the fact that there exist secure variants of several industrial protocols, like secure DNP3, The new version has issue that it is incompatible with legacy setups (Knapp & Samani, 2013).

Furthermore to network protocol concerns, the smart grid's operating systems and physical equipment may uncover the system to anextensive range of assaults. Operating systems lack security features since they are developed for control in automation control components. Furthermore, the majority of physical devices are old, while others lacks in memory space and low processing capability, preventing them from supporting new security procedures. Because smart metres are designed for consumption of lower power, they have limited memory and processing capabilities and hence cannot support several critical security methods like cryptographic accelerators and appropriate random number generators. Even though these mechanisms have a smaller influence on performance of the smart grid, if they are hacked, they provide a possible vector for compromising the entire system.

Safety measures such as intrusion detection systems (IDS), encryption techniques and firewalls play an important part in safeguarding traditional networks. However, these technologies have several restrictions and are not suitable for a dispersed atmosphere with varying application needs likebandwidth and latency (Gungor et al., 2013). Furthermore, these technologies are incapable of mitigating upcoming cyber-assaults. Because cyber-assaults are getting more integrated, smart, and complicated, they can target numerous layers of a communication system at the same time. As earlier stated, Stuxnet (Knapp & Samani, 2013) was capable to vandalise control system of an industrial by circumventing all barriers of security, proving that the safety measures applied in those settings ware incapable of detecting such a powerful virus. In addition, because smart grid has several domains of logic (transmission, distribution, generation, client, service provider, and markets), security needs vary from one domain to the next. Denial of service (DoS) assaults, for example, need quick detection in the generator domain, but not in the domain like market, consumer or service provider. Furthermore, the domain of transmission need key management efficient delay, whereas the domain of market necessitates key management with large-scalability (Wang & Lu, 2013).

As a result, rather than installing a specific security technology or using a single security method, we think that cyber-attacks on smart grid may be more successfully neutralised by integrating numerous security countermeasures via a cyber-security strategy. An approach like this offers various advantages, such as fixing system weaknesses, identifying a variety of cyber-assaults, applying suitable protective measures, and detecting the implicated entity.

CONCLUSION

A smart grid is an arrangement made up of heterogeneous and dispersed components that intelligently supply power while also meeting environmental standards by incorporating renewable technology. This method, however, has a number of security flaws. We offered a detailed review of cyber-security in smart grid and analysed the key cyber-assaults threatening its architecture, protocols for network, and applications in depth. We categorise four stages of cyber-assaults: reconnaissance, scanning, exploitation, and access maintenance. These stages are the procedures that attackers take in order to breach any system. We described the approaches used to obtain sufficient information on the target in the first stage, such as social engineering and traffic analysis. To scan the victim's machine, the techniques used were outlined in the second phase. We demonstrated the strategies utilised to compromise and exploit the victim in the third stage. Virus, DoS, and replay assaults are examples of these tactics. In the previous stage, we discussed the techniques used by the attacker and hacker to get permanent access of victim's machine, like backdoors. In addition, we offered the chance of every assault being carried out, as well as the impact on information security, including integrity, confidentiality, accountability, and availability. Furthermore, we presented a three-step cyber-security strategy: pre-attack, under-attack, and post-attack. We offered a variety of countermeasure and detection approaches for each phase. For example, in the first stage, we discussed a variety of strategies for data security, network safety, and device safety. In the second stage, we provided approaches for detecting and mitigating attacks. The forensic approach for identifying the entity engaged in an assault was described in the last phase. A method like this can address possible vulnerabilities incomponents, improve network and communication safety, and safeguard customer privacy.

Table 2.

Attack Steps	Attack Groups	Compromised security parameter	Possible countermeasures
Reconnaissance	Traffic analysis, Social engineering	Confidentiality	SecureDNP3, SSL, TLS,PKI, Authentication, Encryption(Knapp & Samani, 2013; Wang & Lu, 2013)
Scanning	Scanning Service, Port, IP, Vulnerabilities	Confidentiality	IDS (Faisal et al., 2015), SIEM, Automated (Knapp & Samani, 2013) security compliance checks (Shapsough et al., 2015)
Exploitation	Worms, Virus, Trojan horse	Accountability Integrity Confidentiality Availability	IDS (Faisal et al., 2015), SIEM, DLP, Anti-virus (Knapp & Samani, 2013)
	Denial of service (DoS)	Availability	SIEM (Knapp & Samani, 2013), IDS (Faisal et al., 2015), signal strength, flow entropy, transmission failure count, sensing time measurement, reconfiguration methods, pushback (Shapsough et al., 2015)
	Man-in-the-middle (MITM)	Confidentiality Integrity	SSL, TLS, authentication, encryption (Knapp & Samani, 2013) Secure DNP3, PKI (Wang & Lu, 2013)
	Replay attack	Integrity	SSL, TLS, authentication, encryption (Knapp & Samani, 2013) Secure DNP3, PKI (Wang & Lu, 2013)
	Popping the HMI	Accountability Integrity Confidentiality Availability	SIEM, DLP, Anti-virus (Knapp & Samani, 2013), IDS(Faisal et al., 2015), automated security compliance checks (Shapsough et al., 2015)
	Jamming channel	Availability	Anti-jamming (Reyes & Kaabouch, 2013)
	Masquerade attack	Accountability Integrity Confidentiality Availability	TLS, SIEM, SSL, authentication, encryption, (Knapp & Samani, 2013), IDS (Faisal et al., 2015), SecureDNP3, DLP, PKI (Wang & Lu, 2013)
	Violation of Privacy	Confidentiality	SSL, TLS, authentication, encryption (Knapp & Samani, 2013), PKI, Secure DNP3 (Wang & Lu, 2013)
	Violation of integrity	Availability Integrity	SIEM, SSL, TLS, authentication, encryption (Knapp & Samani, 2013), DLP, Secure DNP3, PKI (Rawat & Bajracharya, 2015a; Wang & Lu, 2013), IDS (Faisal et al., 2015)
Maintaining Access	Backdoor	Accountability Integrity Confidentiality Availability	Anti-virus, SIEM (Knapp & Samani, 2013), IDS (Faisal et al., 2015)

REFERENCES

Al-Dalky, R., Abduljaleel, O., Salah, K., Otrok, H., & Al-Qutayri, M. (2014). A Modbus traffic generator for evaluating the security of SCADA systems. *Proceedings of the international symposium on communication systems, networks digital sign*, 809–14. 10.1109/CSNDSP.2014.6923938

Bouabdellah, M., Kaabouch, N., El Bouanani, F., & Ben-Azza, H. (2018). Network layer attacks and countermeasures in cognitive radio networks: A survey. *J InfSecurAppl*, *38*, 40–49. doi:10.1016/j.jisa.2017.11.010

Elmrabet, Z., Elghazi, H., Sadiki, T., & Elghazi, H. (2016). A new secure network architecture to increase security among virtual machines in cloud computing. *Proceedings of the advances in ubiquitous networking*, 105–16. 10.1007/978-981-287-990-5_9

Engebretson, P. (2013). *The basics of hacking and penetration testing: ethical hacking and penetration testing made easy*. Elsevier.

Essaaidi, M., & Dari, Y. (2015). An overview of smart grid cyber-security state of the art study. *Proceedings of the 3rd international renewable and sustainable energy conference*, 1–7.

Faisal, M. A., Aung, Z., Williams, J. R., & Sanchez, A. (2015). Data-stream-based intrusion detection system for advanced metering infrastructure in smart grid: A feasibility study. *IEEE Systems Journal*, *9*(1), 31–44. doi:10.1109/JSYST.2013.2294120

Fihri, W. F., El Ghazi, H., & Kaabouch, N. (2018). A particle swarm optimization based algorithm for primary user emulation attack detection. IEEE consumer communications and networking conference, 1–6. doi:10.1109/CCWC.2018.8301616

Fihri, W. F., El Ghazi, H., Kaabouch, N., & Abou El Majd, B. (2017). Bayesian decision model with trilateration for primary user emulation attack localization in cognitive radio networks. *Proceedings of the IEEE international symposium on networks, computers, and communications*, 1–6. 10.1109/ISNCC.2017.8071979

Framework, N. (2012). *Roadmap for smart grid interoperability standards, release 2.0* (Vol. 1108). NIST Special Publication.

Gai, K., Qiu, M., Ming, Z., Zhao, H., & Qiu, L. (2017). Spoofing-jamming attack strategy using optimal power distributions in wireless smart grid networks. *IEEE Transactions on Smart Grid*, *8*(5), 1–1. doi:10.1109/TSG.2017.2664043

Gungor, V. C., Sahin, D., Kocak, T., Ergut, S., Buccella, C., Cecati, C., & Hancke, G. P. (2013). A survey on smart grid potential applications and communication requirements. *IEEE Transactions on Industrial Informatics*, *9*(1), 28–42. doi:10.1109/TII.2012.2218253

Kaabouch, N., & Hu, W. C. (2014). *Software-defined and cognitive radio technologies for dynamic spectrum management* (Vol. 1 and 2). IGI Global.

Knapp, E. D., & Samani, R. (2013). *Applied cyber security and the smart grid: implementing security controls into the modern power infrastructure*. Elsevier, Syngress.

Liang, X., Gao, K., Zheng, X., & Zhao, T. (2013). A study on cyber security of smart grid on public networks. *Proceedings of the IEEE green technologies conference*, 301–8. 10.1109/GreenTech.2013.53

Liu, J., Xiao, Y., & Gao, J. (2014). Achieving accountability in smart grid. *IEEE Systems Journal*, *8*(2), 493–508. doi:10.1109/JSYST.2013.2260697

Manesh, M. R., & Kaabouch, N. (2017). *Security threats and countermeasures of MAC layer in cognitive radio networks. J Ad Hoc Netw.*

Manesh, M. R., Mullins, M., Forerster, K., & Kaabouch, N. (2018). A preliminary work toward investigating the impacts of injection attacks on air traffic. IEEE Aerospace Conference, 1–6.

Rawat & Bajracharya. (2015a). Detection of false data injection attacks in smart grid communication systems. *IEEE Signal Process Lett, 22*(10), 1652–6.

Rawat, D. B., & Bajracharya, C. (2015b). Cyber security for smart grid systems: status, challenges and perspectives. *Proceedings of the SoutheastCon*, 1–6. 10.1109/SECON.2015.7132891

Reyes, H., & Kaabouch, N. (2013). Jamming and lost link detection in wireless networks with fuzzy logic. *International Journal of Scientific and Engineering Research, 4*(2), 1–7.

Rodofile, N. R., Radke, K., & Foo, E. (2016). DNP3 network scanning and reconnaissance for critical infrastructure. *Proceedings of the Australasian Computer science week multiconference*, 39:1–39:10. 10.1145/2843043.2843350

Shapsough, S., Qatan, F., Aburukba, R., Aloul, F., & Al Ali, A. (2015). Smart grid cyber security: challenges and solutions. *Proceedings of the international conference on smart grid and clean energy technologies*, 170–5. 10.1109/ICSGCE.2015.7454291

Wang, W., & Lu, Z. (2013). Cyber security in the smart grid: Survey and challenges. *Computer Networks, 57*(5), 1344–1371. doi:10.1016/j.comnet.2012.12.017

Yi, P., Zhu, T., Zhang, Q., Wu, Y., & Li, J. (2014). A denial of service attack in advanced metering infrastructure network. *Proceedings of the IEEE international conference on communications,* 1029–34. 10.1109/ICC.2014.6883456

Zhang, Z., Gong, S., Dimitrovski, A. D., & Li, H. (2013). Time synchronization attack in smart grid: Impact and analysis. *IEEE Transactions on Smart Grid, 4*(1), 87–98. doi:10.1109/TSG.2012.2227342

Chapter 7
The Fifth and the Sixth Order Gopala Hemachandra Representations and the Use of These Representations in Symmetric Cryptography

Çağla Çelemoğlu
Ondokuz Mayis University, Turkey

Ayşe Nalli
https://orcid.org/0000-0002-0489-3649
Karabük University, Turkey

ABSTRACT

We all know that every positive integer has a unique Fibonacci representation, but some positive integers have multiple Gopala Hemachandra (GH) representations, or some positive integers haven't any GH representation. Here, the authors found the first k-positive integer k=(3 2^((m-1))-1) for which there is no Zeckendorf's representation for Gopala Hemachandra sequence whose order m. Thus, the authors formulated the first positive integer whose Zeckendorf's representation can't be found in terms of its order. The authors also described the fourth, the fifth, and the sixth order GH representation of positive integers and obtained the fifth and the sixth order GH representations of the first 26 positive integers uniformly according to a certain rule with a table. Finally, the authors used these GH representations in symmetric cryptography, and the authors made some applications with a method which they construct similar to Nalli and Ozyilmaz.

DOI: 10.4018/978-1-6684-3991-3.ch007

INTRODUCTION

The Fibonacci sequence, $\{F_k\}_0^\infty$, is a sequence of numbers, beginning with the integer couple 0 and 1, in which the value of any element is computed by taking the summation of the two antecedent numbers. If so, for $k \geq 2$, $F_k = F_{k-1} + F_{k-2}$ (Koshy, 2001). There have been many studies in the literature dealing with the quadratic number sequences. One of the studied areas of the Fibonacci sequence is the representations and codes of this sequence.

A universal code transforms positive integers representing source messages into code words of different lengths. There are various universal codes such as the Elias codes, the Fibonacci universal code, Narayana code and non-universal codes such as Rice coding, Huffman coding and Golomb coding (Thomas, 2007), (Platos et., 2007), (Buschmann and Bystrykh, 2013),(Kirthi and Kak, 2016). The best known of them is the Fibonacci code and the Fibonacci code is more useful in comparison with other universal codes. Because Fibonacci universal code is a prefix code of variable size, it is uniquely decodable binary code. Also, this code easily fixs data from damaged parts of code words (Kirthi and Kak, 2016). Fibonacci and Gopala Hemachandra universal codes encode positive integers with binary representations and these code words are obtained based on Zeckendorf representation. Each positive integer has one and only one representation as the summation of non-sequential Fibonacci numbers by Zeckendorf's theorem (Zeckendorf, 1972). Fibonacci sequences and codes can also be defined from higher orders.

Definition 1. The mth order Fibonacci numbers, that are represented by $F_k^{(m)}$, are described with iteration relation as follows:

$$F_k^{(m)} = F_{k-1}^{(m)} + F_{k-2}^{(m)} + \ldots + F_{k-m}^{(m)}$$

for $k>0$ and the boundary conditions $F_0^{(m)} = 1$ and $F_k^{(m)} = 0$ ($-m<k<0$) (Klein and Ben-Nissan, 2010).

One representation can be obtained for each positive integer A with a binary string of length t, $l_1 l_2 \ldots l_{t-1} l_t$, such that $A = \sum_{i=1}^{t} l_i F_i^{(m)}$. The representation is one and only if one uses algorithm to find it as follows: When it is given the integer A, it is detected the largest Fibonacci number $F_t^{(2)}$ equivalent or smaller to A; after that it is continued repeating with $A - F_t^{(2)}$ (Klein and Ben-Nissan, 2010). For instance 17=1+3+13, hence its Fibonacci representation is 101001.

According to above algorithm, Fibonacci numbers aren't used consecutively in any of these summations, that is, in the binary representation, there are no contiguous 1 bits. When generalizing this procedure to higher orders, the same operations are realized as above. Additionally, it is appended (m–1) 1 bits to the mth order variant of Fibonacci representation of k to build the mth order variant of Fibonacci code of any positive integer k. But, unlike the Fibonacci representation, there are no adjacent of m 1 bits in the statement (Klein and Ben-Nissan, 2010).

In this study, the authors found that the first k positive integer $k = (3.2^{(m-1)} - 1)$ for which there is no Zeckendorf's representation for Gopala Hemachandra sequence whose order m. Thus, the authors formulated the first k positive integer whose Zeckendorf's representation can't be found in terms of its order. The authors also described the fourth, the fifth and the sixth order GH representation of positive integers and found the fifth and the sixth order GH representations of the first 26 positive integers uniformly according to a certain rule. The rule the authors determined is that the first 26 positive integers k must exist with the same GH representation for every $u \in \mathbb{Z}^-$. Hence, according to this rule, the fifth and the sixth order GH representations of k for ($1 \leq k \leq 26$) must be independent of u. Finally, the authors used these GH representations in symmetric cryptography and the authors made some applications with a method they constructed in a similar way to the method in (Nalli and Ozyilmaz, 2015).

Table 1.

k	**Zeckhendorf representation for m=2**
1	1
2	01
3	001
4	101
5	0001
6	1001
7	0101
8	00001
9	10001
10	01001
11	00101
12	10101
13	000001
14	100001
15	010001

Background

In this study, the authors made some cryptographic applications by using the Fibonacci codes, the second order Gopala Hemachandra codes and the third order Gopala Hemachandra codes. The definitions of the second and the third order Gopala Hemachandra sequences used in this study are given in this section.

In 2010, it was presented the second order Gopala Hemachandra code in (Basu and Prasad, 2010) and the sequence which used to generate this code is defined as follows:

Definition 2. The second order variant Fibonacci sequence, $GH_u^{(2)}(k)$ is described with the sequence $\{u, v, u+v, u+2v, 2u+3v, 3u+5v...\}$ where $v=1-u$, that is,

$$GH_u^{(2)}(1) = u \ \ (u \in \mathbb{Z}^-); \ GH_u^{(2)}(2) = 1-u;$$

and for $k \geq 3$,

$$GH_u^{(2)}(k) = GH_u^{(2)}(k-1) + GH_u^{(2)}(k-2)$$

(Basu and Prasad, 2010).

Afterwards, in 2015, it was presented the third order Gopala Hemachandra code in (Nalli and Ozyilmaz,2015) and the sequence which used to generate this code is defined as follows:

Definition 3. The third order variant Fibonacci sequence, $GH_u^{(3)}(k)$ is described with the sequence {*u*, *v*, *u*+*v*, 2*u*+2*v*, 3*u*+4*v*, 6*u*+7*v*...} where *v*=1–*u*, that is,

$$GH_u^{(3)}(1) = u \ \ (u \in \mathbb{Z}^-); \ GH_u^{(3)}(2) = 1-u; \ GH_u^{(3)}(3) = 1;$$

and for $k \geq 4$,

$$GH_u^{(3)}(k) = GH_u^{(3)}(k-1) + GH_u^{(3)}(k-2) + GH_u^{(3)}(k-3)$$

(Nalli and Ozyilmaz,2015).

MAIN FOCUS OF THE CHAPTER[1]

The Fourth, The Fifth and The Sixth Order Gopala Hemachandra Sequence and Representations

In this section, the authors described *m*=4,5,6 variant Fibonacci sequences, respectively. Then, the authors found that the first *k* positive integer $k = (3.2^{(m-1)} - 1)$ for which there is no Zeckendorf's representation for Gopala Hemachandra sequence whose order *m*. Thus, the authors formulated the first *k* positive integer whose Zeckendorf's representation can't be found in terms of its order.

Definition 4. The fourth order variant Fibonacci sequence, $GH_u^{(4)}(k)$ is described with the GH sequence {*u*, *v*, *u*+*v*, 2*u*+2*v*, 4*u*+4*v*, 7*u*+8*v*, 14*u*+15*v*...} where *v*=1–*u*, that is,

$$GH_u^{(4)}(1) = u(u \in \mathbb{Z}^-); \ GH_u^{(4)}(2) = 1-u; \ GH_u^{(4)}(3) = 1; \ GH_u^{(4)}(4) = 2;$$

and for $k \geq 5$,

$$GH_u^{(4)}(k) = GH_u^{(4)}(k-1) + GH_u^{(4)}(k-2) + GH_u^{(4)}(k-3) + GH_u^{(4)}(k-4).$$

Definition 5. The fifth order variant Fibonacci sequence, $GH_u^{(5)}(k)$ is described with the GH sequence {*u*, *v*, *u*+*v*, 2*u*+2*v*, 4*u*+4*v*, 8*u*+8*v*, 15*u*+16*v*, 30*u*+31*v*...} where *v*=1–*u*. That is,

$$GH_u^{(5)}(1)=u(u\in\mathbb{Z}^-);\ GH_u^{(5)}(2)=1-u;\ GH_u^{(5)}(3)=1;\ GH_u^{(5)}(4)=2;\ GH_u^{(5)}(5)=4;$$

and for $k\geq 6$,

$$GH_u^{(5)}(k)=GH_u^{(5)}(k-1)+GH_u^{(5)}(k-2)+GH_u^{(5)}(k-3)+GH_u^{(5)}(k-4)+GH_u^{(5)}(k-5).$$

Definition 6. The sixth order variant Fibonacci sequence, $GH_u^{(6)}(k)$ is described as the GH sequence $\{u, v, u+v, 2u+2v, 4u+4v, 8u+8v, 16u+16v, 31u+32v, 62u+63v...\}$ where $v=1-u$, That is,

$$GH_u^{(6)}(1)=u(u\in\mathbb{Z}^-);\ GH_u^{(6)}(2)=1-u;\ GH_u^{(6)}(3)=1;\ GH_u^{(6)}(4)=2;\ GH_u^{(6)}(5)=4;\ GH_u^{(6)}(6)=8;$$

and for $k\geq 7$,

$$GH_u^{(6)}(k)=GH_u^{(6)}(k-1)+GH_u^{(6)}(k-2)+GH_u^{(6)}(k-3)+GH_u^{(6)}(k-4)+GH_u^{(6)}(k-5)+GH_u^{(6)}(k-6).$$

For different values of u, it is obtained different sequences. The authors know that it was demonstrated that solely Fibonacci sequence forms a unique Fibonacci code for all positive integers by Daykin (Daykin, 1960). However, in variants Fibonacci sequences, some integers have more than one Gopala Hemachandra codes while some integers have no Gopala Hemachandra code. For example, for the second order variant Fibonacci sequence while

$$VF_{-5}^{(2)}(k)=\{-5,6,1,7,8,15,23,38,...\}$$

there is no Gopala Hemachandra code for integer k=5,12 (Thomas, 2007),

$$VF_{-2}^{(2)}(k)=\{-2,3,1,4,5,9,14,23,...\}$$

there are two Gopala Hemachandra codes for integer k=8. These codes are 010011 and 1010011. Similarly, for the third order variant of Fibonacci sequence while

$$VF_{-11}^{(3)}(k)=\{-11,12,1,2,15,18,35,68,...\}$$

there is no Gopala Hemachandra code for integer k=11 (Nalli and Ozyilmaz,2015),

$$VF_{-4}^{(3)}(k)=\{-4,5,1,2,8,11,21,40,...\}$$

there are two Gopala Hemachandra codes for integer k=7. These codes are 10000111 and 010111.

In this work, the authors aimed to find the first integer for which the fourth, the fifth and the sixth order GH representation of variant Fibonacci sequences doesn't exist and to formulate the first k positive integer whose Zeckendorf's representation can't be found in terms of its order. In accordance with this purpose, the authors obtained the first integer without Zeckendorf's representation is k=23 for the fourth order GH sequence

$$GH_{-23}^{(4)}(k)=\{-23,24,1,2,4,31,38,75,\ldots\},$$

the first integer without Zeckendorf's representation is k=47 for the fifth order GH sequence

$$GH_{-47}^{(5)}(k)=\{-47,48,1,2,4,8,63,78,\ldots\},$$

the first integer without Zeckendorf's representation is k=95 for the sixth order GH sequence

$$GH_{-95}^{(6)}(k)=\{-95,96,1,2,4,8,16,127,\ldots\}.$$

Hence, the authors obtained by induction that for variant Fibonacci sequence

$$GH_{-(3.2^{(m-1)}-1)}^{(m)}(k)=\left\{-(3.2^{(m-1)}-1),3.2^{(m-1)},1,2,4,8,16,32,\ldots\right\}$$

whose order is m the first integer without Zeckendorf's representation is $k=(3.2^{(m-1)}-1)$. Thus, the authors formulated the first k positive integer whose Zeckendorf's representation can't be found in terms of its order.

In addition, the authors also obtained that the same positive integer exists multiply in variant Fibonacci sequence whose order is m, $GH_{u}^{(m)}(k)$, for some values of u. For example, in the fourth order GH sequence for u=-3, the authors obtained

$$GH_{-3}^{(4)}(k)=\{-3,4,1,2,4,11,18,\ldots\}.$$

In the fifth order GH sequence for u=-3, the authors obtained

$$GH_{-3}^{(5)}(k)=\{-3,4,1,2,4,8,19,\ldots\}$$

and for u=-7

$$GH_{-7}^{(5)}(k)=\{-7,8,1,2,4,8,23,\ldots\}.$$

In the sixth order GH sequence for u=-3, the authors obtained

$$GH_{-3}^{(6)}(k)=\{-3,4,1,2,4,8,16,35,\ ...\},$$

for u=-7

$$GH_{-7}^{(6)}(k)=\{-7,8,1,2,4,8,16,39,...\}$$

and for u=-15

$$GH_{-15}^{(6)}(k)=\{-15,16,1,2,4,8,16,47,...\}.$$

Hence, they obtained by induction that the same positive integer exist multiply in GH sequence whose order is m, $GH_u^{(m)}(k)$, for u=-3, u=-7, u=-15, u=-31,..., and the last integer $u=-(2^{(m-2)}-1)$. So, these columns can't be used while an application of GH representations of positive integers to cryptology is made. Thus, it is important to know these columns in terms of cryptologic.

An Application of The Fifth and The Sixth Order Gopala Hemachandra Representations in Symmetric Cryptography

Cryptography is all of the methods used to make an understandable message incomprehensible to undesirable people (Çimen et al, 2007). The focus of cryptography is privacy. The main purpose of cryptography is to ensure the two in order to get into touch over an insecure canal such that a rival, Oscar, can't understand. The information which is sent is called as plaintext, and the information is selected arbitrary (Stinson, 2002).

Every algorithm uses a key to audit encryption and decryption, and a message may only be decrypted when the key meets the encryption key used. So the most important issue in encryption is the key. According to key structure, cryptography is divided into two as symmetric cryptography and asymmetric cryptography. Examples to symmetric cryptography are Vernam, DES, AES, while examples to asymmetric cryptography are RSA, ElGamal. In addition, there are different cryptographic system built, too. One of them is a system created using source coding. Some examples of these applications are included in (Nalli and Ozyilmaz, 2015), (Das and Sinha, 2019). Here, the authors will make similar cryptographic applications by using new representations the authors described for order m=5 and order m=6 according to the rule they added.

On the other hand, each positive integer has an one and only Fibonacci representation. But GH sequences serve for plural Zeckendorf's representations of the same integer. Thus, some positive integers have plural GH representation or some positive integers haven't a GH representation. In this situation, GH representation of positive integers don't account for a function. Hence, the authors obtained the fifth and the sixth order GH representations of the first 26 positive integers uniformly according to a certain rule so as to do an exercise of these *GH* representations to cryptology. In this part, the rule the authors determined is that the the first 26 positive integers k must exist with the same GH representation for every $u\in\mathbb{Z}^-$. Hence, according to this rule, the fifth and the sixth order GH representations of k for (1≤k≤26) must be independent of u. So, it is necessary for us to find GH representation of 26 positive integers which provide that condition. Since k=23 has no GH representation for the fourth order GH sequence, the authors can't use these sequence by encrypting according to this rule. Then, they will obtain that GH representations of the first 26 positive integers for order m=5 and m=6 according to this rule with a table as follows:

Table 2

k	*The fifth order GH representation*	*The sixth order GH representation*
1	001	001
2	0001	0001
3	0011	0011
4	00001	00001
5	00101	00101
6	00011	00011
7	00111	00111
8	000001	000001
9	001001	001001
10	000101	000101
11	001101	001101
12	000011	000011
13	001011	001011
14	000111	000111
15	001111	001111
16	1000001	0000001
17	1010001	0010001
18	1001001	0001001
19	1011001	0011001
20	1000101	0000101
21	1010101	010101
22	1001101	0001101
23	1011101	0011101
24	1000011	0000011
25	1010011	0010011
26	1001011	0001011

It is clear from the Table 2, according to the rule the authors determined GH representations of both orders of the first 15 positive integers are same.

In addition, in this article, the authors will build a process similar to the method in (Nalli & Ozyilmaz, 2015) so as to use the fifth and the sixth GH representation of positive integers in symmetric cryptography. In this method, they used stream cipher.

In this method, firstly it is represented a number to each letter in message. For convenience, it is expressed these values with a table.

Table 3. The English Alphabet

A	B	C	D	E	F	G	H	I	J	K	L	M
1	**2**	**3**	**4**	**5**	**6**	**7**	**8**	**9**	**10**	**11**	**12**	**13**
N	O	P	Q	R	S	T	U	V	W	X	Y	Z
14	**15**	**16**	**17**	**18**	**19**	**20**	**21**	**22**	**23**	**24**	**25**	**26**

Then, it is obtained GH representation of the number corresponding to each letter in text message from Table 2. And, it is attached '0' (s) to the end of these representations so that the GH representation length of each number in text message is equal to each other. So, it is obtained plaintext (P) .

Later, it will be obtained secret key (K). To do this, firstly, it is obtained the standard Fibonacci representation of the order m (the order of $GH_u^{(m)}$ sequence) from Table 1. And, it is attached '0' (s) to the end of these representation so that the length of standard Fibonacci is equal to the lenght of the GH representation length of each number in the message. Hence, it is obtained K by sender. Then, keystream is generated by repeating the key K in message length. Lastly, it is encrypted by using stream cipher and obtained ciphertext (C).

Now, let illustrates some examples.

Example 1. Let's assume that the message the sender will send using the $GH_u^{(5)}$ sequence is 'SECURITY'. Firstly, it is obtained the following numbers from Table 3.

S	E	C	U	R	I	T	Y
19	5	3	21	18	9	20	25

From Table 2, it is obtained by the sender.

1011001 ...... 00101 .... 0011 1010101
1001001 ...... 001001 ...... 1000101 1010011

Then, it is attached '0'(s) to the end of each representation and the lengths of GH representations are equalized.

1011001 ...... 0010100 ...... 0011000 1010101
1001001 ...... 0010010 ...... 1000101 1010011

In that case, the plaintext is:

10110010010100001100010101011001001001001010001011010011.

Then, it must be obtained the key by sender.

It is known m=5. So, from Table 1, it is obtained 0001. According to the method, the key is 0001000. Hence, the keystream is

00010000001000000100000010000001000000100000010000001000.

So, it is obtained encrypted message (ciphertext) is:

10100010011100001000010111011000001001101010011011011011.

Then, it is investigated how the person who receives the encrypted message will decrypt.

So, the ciphertext is:

10100010011100001000010111011000001001101010011011011011.

and K is 0001000. Hence the keystream is obtained by the receiver.

And, the ciphertext is decrypted. So, it is obtained the message or the plaintext is:

10110010010100001100010101011001001001001010001011010011.

and K is 0001000.

And, He makes sense of plaintext. To do this, firstly, The receiver reserves parts so that the lenght of each part is equal to the lenght of K. Hence, he gets

1011001 ...... 0010100 ...... 0011000 ...... 1010101 1001001
0010010 ...... 1000101 ...... 1010011.

Then he cancels ‘0’ (s) at the end of each part with the inclusion of K. Thus, he gets

1011001 ...... 00101 ...... 0011 ...... 1010101 1001001
001001 ...... 1000101 1010011

and K is 0001.

The standard Fibonacci representation of 5 is 0001 from Table 1. So, he understands that the message was sent by using $GH_u^{(5)}$. He gets what the number value of each part is from Table 2.

So, he gets

19 ...... 5 ...... 3 ...... 21 ...... 18 ...... 9 ...... 20 25

Finally, from Table 3, the message is obtained

S ...... E ...... C ...... U ...... R ...... I ...... T ..... Y.

Example 2. Let's assume that the message the sender will send using the $GH_u^{(6)}$ sequence is 'HELP'.

Firstly, it is obtained the following numbers from Table 3.

H	E	L	P
8	5	12	16

From Table 2, it is obtained by the sender.

000001 ...... 00101 ...... 000011 ...... 0000001.

Then, it is attached '0'(s) to the end of each representation and the lengths of GH representations are equalized.

0000010 ...... 0010100 ...... 0000110 0000001

In that case, the plaintext is:

0000010001010000001100000001.

Then, it must be obtained the key by sender.

It is known m=6. So, from Table 1, it is obtained 1001. According to the method, the key is 1001000. Hence, the keystream is

1001000100100010010001001000.

So, it is obtained encrypted message (ciphertext) is:

1001010101110010011101001001.

Then, it is investigated how the person who receives the encrypted message will decrypt.

So, the ciphertext is:

1001010101110010011101001001 and K is 1001000.

Hence the keystream is obtained by the receiver.

And, the ciphertext is decrypted. So, it is obtained the message or the plaintext is:

0000010001010000001100000001 and K is 1001000.

And, He makes sense of plaintext. To do this, firstly, The receiver reserves parts so that the lenght of each part is equal to the lenght of K. Hence, he gets

0000010 .... 0010100 ...... 0000110 ...... 0000001.

Then he cancels '0' (s) at the end of each part with the inclusion of K. Thus, he gets

000001 ...... 00101 ...... 000011 0000001

and K is 1001.

The standard Fibonacci representation of 6 is 0001 from Table 1. So, he understands that the message was sent by using $GH_u^{(6)}$. He gets what the number value of each part is from Table 2.

So, he gets

8 ...... 5 ...... 12 ...... 16.

Finally, from Table 3, the message is obtained

H ...... E ...... L ...... P

FUTURE RESEARCH DIRECTIONS

New cryptographic applications can be made by using the different sequences. Then these cryptographic applications are comparable with each other.

CONCLUSION

We all know that some positive integers haven't got a GH representation. In (Thomas, 2007), Thomas indicated that for the first integer k=5 there is no Zeckendorf's representation for the second order GH sequence

$$GH_{-5}^{(2)}(k)=\{-5,6,1,7,8,15,23,38,\ldots\}.$$

In (Nalli and Ozyilmaz, 2015), authors obtained that for the first integer k=11 there is no Zeckendorf's representation for the third order GH sequence

$$GH_{-11}^{(3)}(k)=\{-11,12,1,2,15,18,35,68,\ldots\}.$$

In this article, firstly the authors described order m=4,5,6 variant Fibonacci sequences and then the authors obtained the first integer without Zeckendorf's representation is k=23 for the fourth order GH sequence

$$GH_{-23}^{(4)}(k)=\{-23,24,1,2,4,31,38,75,\ldots\},$$

the first integer without Zeckendorf's representation is k=47 for the fifth order GH sequence

$$GH_{-47}^{(5)}(k)=\{-47,48,1,2,4,8,63,78,\ldots\},$$

the first integer without Zeckendorf's representation is k=95 for the sixth order GH sequence

$$GH_{-95}^{(6)}(k)=\{-95,96,1,2,4,8,16,127,\ldots\}.$$

Hence, the authors obtained by induction that for variant Fibonacci sequence

$$GH_{-(3.2^{(m-1)}-1)}^{(m)}(k)=\{-(3.2^{(m-1)}-1),3.2^{(m-1)},1,2,4,8,16,32,\ldots\}$$

whose order is m the first integer without Zeckendorf's representation is $k=(3.2^{(m-1)}-1)$. Thus, the authors formulated the first k positive integer whose Zeckendorf's representation can't be found in terms of its order.

The authors also have obtained that the same positive integer exists multiply in variant Fibonacci sequence whose order is m, $GH_u^{(m)}(k)$, for some values of u. For example, in the fourth order GH sequence for u=-3, the authors obtained

$$GH_{-3}^{(4)}(k)=\{-3,4,1,2,4,11,18,\ldots\}.$$

In the fifth order GH sequence for iu=-3, the authors obtained

$$GH_{-3}^{(5)}(k)=\{-3,4,1,2,4,8,19,\ldots\}$$

and for u=-7

$$GH_{-7}^{(5)}(k)=\{-7,8,1,2,4,8,23,\ldots\}.$$

In the sixth order GH sequence for u=-3, the authors obtained

$$GH_{-3}^{(6)}(k)=\{-3,4,1,2,4,8,16,35,\ \ldots\},$$

for u=-7

$$GH_{-7}^{(6)}(k)=\{-7,8,1,2,4,8,16,39,\ldots\}$$

and for u=-15

$GH^{(6)}_{-15}(k)=\{-15,16,1,2,4,8,16,47,...\}$.

Hence, the authors obtained that the same positive integer exist multiply in GH sequence whose order is *m*, $GH^{(m)}_{u}(k)$, for *u*=-3, *u*=-7, *u*=-15, *u*=-31,..., and the last integer $u=-(2^{(m-2)}-1)$. It is important to know these columns cryptographically, because these columns aren't used in cryptographic applications.

In addition, because GH representation of positive integers don't account for a function. Therefore, the authors obtained the fifth and the sixth order GH representations of the first 26 positive integers uniformly according to a certain rule so as to do an exercise of these GH representations to cryptology. The rule the authors determined is that the first 26 positive integers *k* must exist with the same GH representation for every $u\in\mathbb{Z}^{-}$. Hence, according to this rule, the fifth and the sixth order GH representations of *k* for ($1\leq k\leq 26$) must be independent of *u*. Also, Since *k*=23 has no GH representation, the authors can't use the fourth order GH representation by encrypting according to this rule. Finally, the authors made some applications with a method they constructed in a similar way to the method in (Nalli and Ozyilmaz, 2015).

REFERENCES

Baca, R., Snásel, V., Platos, J., Kratky, M., & El-Qawasmeh, E. (2007). *The fast Fibonacci decompression algorithm.* arXiv preprint arXiv:0712.0811.

Basu, M., & Prasad, B. (2010). Long range variant of Fibonacci universal code. *Journal of Number Theory*, *130*(9), 1925–1931. doi:10.1016/j.jnt.2010.01.013

Buschmann, T., & Bystrykh, L. V. (2013). Levenshtein error-correcting barcodes for multiplexed DNA sequencing. *BMC Bioinformatics*, *14*(1), 1–10. doi:10.1186/1471-2105-14-272 PMID:24021088

Çimen, C., Akleylek, S., & Akyıldız, E. (2007). *Şifrelerin Matematiği Kriptografi*. ODTÜ Yayıncılık.

Das, M., & Sinha, S. (2019). A Variant of the Narayana Coding Scheme. *Control and Cybernetics*, *48*(3), 473–484.

Daykin, D. E. (1960). Representation of Natural Numbers as Sums of Generalized Fibonacci Numbers. *Journal of the London Mathematical Society*, *35*(2), 143–160. doi:10.1112/jlms1-35.2.143

Kirthi, K., & Kak, S. (2016). *The Narayana Universal Code*. arXiv: 1601.07110.

Klein, S. T., & Ben-Nissan, M. K. (2010). On the Usefulness of Fibonacci cCmpression Codes. *The Computer Journal*, *53*(6), 701–716. doi:10.1093/comjnl/bxp046

Koshy, T. (2001). *Fibonacci and Lucas Numbers with Applications*. Wiley. doi:10.1002/9781118033067

Nalli, A., & Ozyilmaz, C. (2015). The third order variations on the Fibonacci universal code. *Journal of Number Theory*, *149*, 15–32. doi:10.1016/j.jnt.2014.07.010

Stinson, D. R. (2002). *Cryptography Theory and Practice*. Chapman & Hall / CRC.

Thomas, J. H. (2007). *Variant of Fibonacci Universal Code*. arXiv: cs/0701085v2.

Zeckendorf, E. (1972). Representations des nombres naturels par une somme de nombres de fibonacci on de nombres de lucas. *Bulletin de la Société Royale des Sciences de Liège*, 179–182.

ADDITIONAL READING

Hoffstein, J., Pipher, J., Silverman, J. H., & Silverman, J. H. (2008). *An Introduction to Mathematical Cryptography* (Vol. 1). Springer.

Hungerfold, T. W. (1987). *Algebra*. Springer-Verlag.

Jones, G. A., & Jones, J. M. (1998). *Elementary Number Theory*. Springer Science & Business Media. doi:10.1007/978-1-4471-0613-5

Niven, I., Zuckerman, H. S., & Montgomery, H. L. (1991). *An Introduction to the Theory of Numbers*. John Wiley & Sons.

KEY TERMS AND DEFINITIONS

Cryptography: It is a process of protecting information and communications such that the only one for whom the information is intended can understand.

Fibonacci Code: It is a universal code which encodes positive integers into binary code words.

Fibonacci Sequence: It is a recurrence sequence beginning with the integer couple 0 and 1, in which the value of any element is computed by taking the summation of the two antecedent numbers.

Gopala Hemachandra Code: It is a code which is a variation of the Fibonacci universal code and has applications in cryptography and data compression.

Gopala Hemachandra Sequence: It is a sequence which generalizes the Fibonacci sequence by allowing the same recursive construction of the Fibonacci sequence to be used with arbitrary starting terms.

Recurrence Sequences: Recurrence sequence is a sequence of numbers indexed by an integer and generated by solving a recurrence equation.

Symmetric Cryptography: Symmetric encryption is a type of encryption where only one key (a secret key) is used to both encrypt and decrypt electronic information.

Universal Code: It is a code transforms positive integers representing source messages into code words of different lengths.

Zeckendorf's Representation: The Zeckendorf representation of an integer is the unique way of representing that integer as a sum of non-consecutive Fibonacci numbers.

Chapter 8
Data Encryption and Hiding for Secured Data Transmission

A. V. Senthil Kumar
Hindusthan College of Arts and Sciences, India

Chennamma H. R.
JSS Science and Technology University, India

Manjunatha Rao
Dr. Ambedkar Institute of Technology, India

Malavika B.
Hindusthan College of Arts and Science, India

ABSTRACT

The secured data hiding for future transfer is an application proposed for a secured data transmission in which the application alters the originality of the file to an encrypted form and embeds that file into a video file which will be transferred. The motive of the application is to provide the ease for the user to pass the information with the encryption standards and algorithms then store the information in an unreadable format. In the next phase, the application has a reverse process where the embedded video can be de-embedded, and the encrypted file can be decrypted to its actual data upon proper user request. In the process of encryption and decryption, the application should confirm the authenticity of the user. The application is generated using the tiny encryption algorithm to encrypt the data. This algorithm is a 64-bit block cipher with a variable length key. The tiny encryption algorithm is used because requires less memory, and as a compliment, it uses only simple operations; thus, it is easy to implement.

INTRODUCTION

The world is revolving around the data, everything is data from the tiny dust to the huge planets every single atom is data. The data have a vital part in the human life. By means of the decade passes the data increases when the data increases the complexity becomes a threat. In earlier days the communication was done face to face as the time moved the technologies grow wider and many more developments were done. In old times, people used to share the information by talking to each other, and by painting pictures, later it changed to letters and books (M. Hussain, 2018). That was the time of learning new languages, people learned to write and read the paper materials (S. Manimurugan, 2020). Then it became a decade of letters and envelopes, people initially made use of pigeons to give the letters to the expected

DOI: 10.4018/978-1-6684-3991-3.ch008

recipient. After that globe was on another pace to used postal addresses to send the letters. It was the method for a long time (A. K. Sahu & G. Swain, 2019).

Subsequently the letter communication changed to the telecommunication, there arrived the telephone then steadily the telephone was developed to the mobile phone. And when the computer arrived the importance of data increased, as computer needs data for processing. In simple words the data is considered as a raw fact. When data is used for communication, there exists a threat or a vulnerability on the security of the data. This security threat needs an appropriate solution (S. Almutairi, 2019). Thus, implementation of a secure transaction is needed. Data security is the exercise of protective virtual data from a not authorised access, exploitation, or robbery during its whole procedure. It's an idea that encompasses each thing of data safety from the bodily safety of hardware and garage gadgets to organizational and admission controls, in addition to the rational safety of software program applications(V. Tyagi, 2012). It additionally consists of administrative guidelines and measures. The data security has a next level as data hiding.

Data hiding is a software program improvement approach specially utilized in item-orientated programming to cover inner item particulars (statistics contributors). Data hiding guarantees extraordinary statistics get right of entry to elegance contributors and defends item truthfulness via way of means of stopping accidental or supposed variations. When a communication is performed, suppose that a confidential document is being passed from person A to person B. If person A is sending that document as a simple word file, anyone in the network can perform social engineering and view, manipulate or change the destination and send that file to anyone the intruder thinks. This is why the data security is important (OUSLIM Mohamed, 2013).

Figure 1 illustrates an unsecured data communication. The intruder can manipulate the data by interrupting inside the network. The straight arrow denotes the actual communication which might have happened if the data communication was secured.

Figure 1.

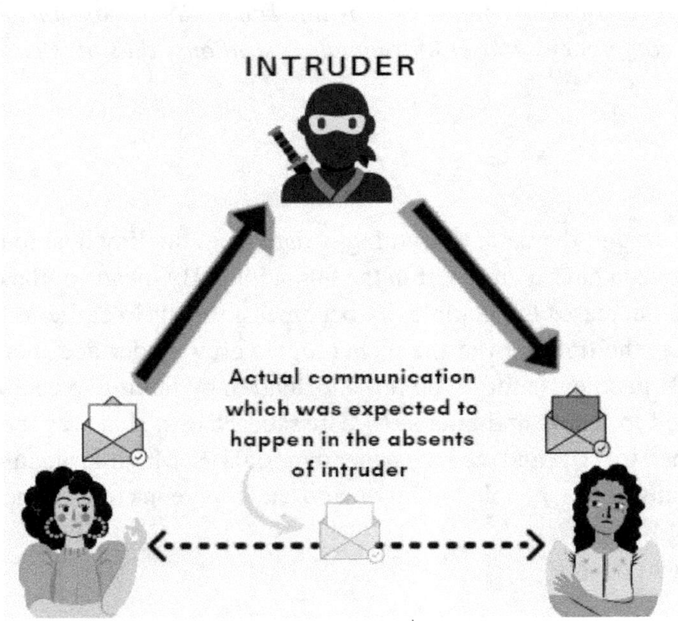

While data hiding, the application must be capable to handle the security in the transaction. Data transfer refers back to the steady exchange of massive documents among structures or companies. In an inner context, data transfer is regularly utilised as an opportunity to a holistic agency application integration structure. There are numerous methods are invented for secured data transfer but when time goes many secured transfers are being cracked by the hacker (S. Manimurugan, 2020).

This application is a combination of two concepts steganography and cryptography. There are many types of steganography such as text steganography, image steganography, audio steganography, video steganography and network steganography, Usually the steganography is done on the images. Since cracking a steganography is a difficult task, thus the steganography is known as the undetectable technique to hide the data (K Anusree& G S Binnu, 2014). And as a statement the cryptography is also undetectable to an extent. So the cryptography alone wont help in secured data transmission (J. A. Kaw, 2019). Thus, this application have combined both these undetectable techniques for secured future data transfer.

In the present decade the concept of data encryption became an essential domain. People started performing encryption to keep their activities confidential from unauthorized persons, so that their data is not altered or tampered. The motive of this project is to provide the secure communication through the network with an authentication of file transferring. The process of encryption plays a significant part in the real time atmosphere to preserve the data out of influence of unauthorized persons, so that it is not transformed and altered and distributing them in video form can be considered as the most protected method to transfer the data through the system (B. Champakamala, 2014).

The whole application must endeavour to accomplish a user friendly Graphical User Interface, that requirement to be in a self-learning mode for the user on the end. The Structure should make available all the purposeful ethics of appropriate course plotting within the atmosphere, that varieties it conceivable for the handlers to have an even movement while working under the atmosphere (J. A. Kaw, 2019).

The complete system ought to make available appropriate menu constructed direction finding for easy operation and navigation. The application must be developed in a certain technique that when it creates a barrier and associates this barrier to some standardized data atmosphere, the application must query the operator for the details of encryption key and should start its operations upon the “logistics” that are on condition that given inside this key. Where this key must prevent the unauthorised people from thieving the material at any opinion of time which is a procedure of “securing the data from third party people”(K. V. Pradeep, 2019).

The data would be premeditated by the means that it avoids the unsanctioned individuals from thieving the material at any opinion of time. This is approximately a fragment of safeguarding the data from third revelrypersons. And the supplementary technique of safeguarding the data is using “Steganography” in which the encrypted file is embedded in to a video file. If anybody trail that file, they will only see the video file not the data file which is embedded. The application of De-embedding and Decryption had better be a converse development at the other end and it would be interpreted only when the recipient of the data delivers the appropriate reversal key (S. A. Parah, 2017).

EXISTING SYSTEM

In the conventional architecture is having only the client and the server. In the utmost approaches the server is the only a database server whom can only provide the data. Thus a large number of the companies had to place the validation phase on the client’s system which leads to an expensive maintenance, these consumers are termed as "fat clients". Which correspondingly means that every single consumer has to be equipped with the procedures of using the application and level the safety in the data transmission is also the feature to be counted (Y. Zhang, 2018).

Since the authentic dispensation of the data occurs in the inaccessible consumer, the data have to be transmitted through the system, which necessitates a secured form in the transmission way. The data communications are measured by the consumer and also while executing, the advance methods of cryptographic standards are being implemented at the time of the data transfer happen. Nowadays transactions are termed as "un-trusted" on basis of security (Saad Al-Mutairi & S. Manimurugan, 2017).

Challenges of Existing System

- There are threats and defenselessness that appear due to the underprivileged strategy of systems.
- Large amount of data transfer might be messy in the network.
- Only a countable number of data is encrypted.
- Low accuracy.
- As a result it has a least efficient performance.

PROPOSED SYSTEM

The application have steganography combined with cryptography. There are many advantages for each concept such as it is difficult to detect, difficult to sense as note and ultimate image data stake identical variety, its tough task to find which is the altered picture and which is the original picture, it is not vulnerable to the hackers, these are the advantages of the steganography. For cryptography there are many other advantages like, confidentiality, data integrity, non-repudiation and Authentication (M. Shanmuganathan, 2020).

Thus sending the video format is considered as a further most protected way to handover the data by the system. The application possess a bundle of processes such as encryption, embedding, de-embedding and finally decryption. When a user logs in with the needed credentials such as user id and password, the authenticated one can encrypt a word file and embed the encrypted file within a video file (K. Rabah, 2004).

This video file is used for the transmission. In the succeeding process that is when the one receives the embedded video, the authorized user can de-embed the video file and then decrypt to the file by providing the appropriate key. When the user fails to provide the appropriate key, it becomes impossible to the user to read the file (S. Almutairi, 2019).

Features of Proposed System

- Better service
- User can encrypt up to 1024 MB file
- Reduce the damages of the machines.
- Have high level of Security
- Comparatively less time needed for the various processing.
- Greater efficiency.

Figure 2. The important phases of methodology

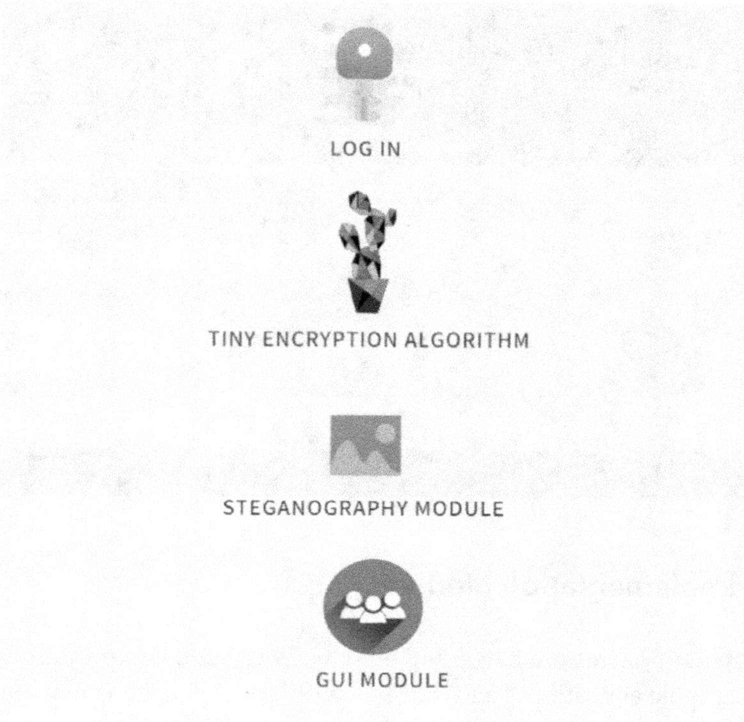

PHASES OF METHODOLOGY

Module Description

Log In

Before entering the system, the user must sign in by providing user credentials to authenticate their identity. In this user login phase, the user can browse a file to encrypt and save that encrypted file in any location of the local system the user need. Next the user can embed the encrypted file to a video file and also can de-embed the encrypted file from the video then decrypt the file which was encrypted initially (R. J. Mstafa, 2017).

The login page contains two credentials such as user id or mail id and a password with sign-in button along with this it has two more button to quit and to clear. This is the initial page which appears first in the application while execution, the user is allowed to proceed to the next step if and only if the user passes this step.

Figure 3.

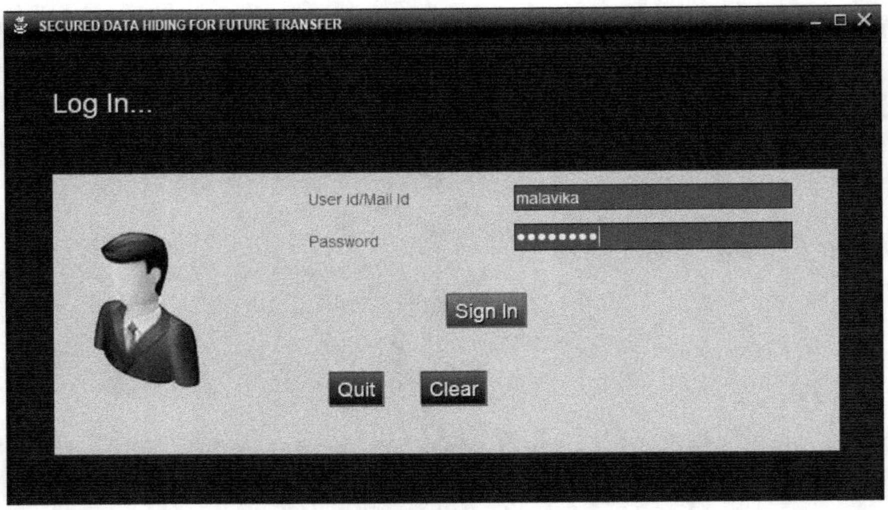

Tiny Algorithm Implementation Module

Day by day the computer systems are becoming more universal and composite, safety is progressively significant. “Cryptographic algorithms” and the “protocols” establish the vital module of structures that defend web transmissions and accumulate data. Where the safety of such structures importantly hinge on the approaches applied to achieve, create, and issue the keys hired by the “cryptographic” procedures. Even if an algorithm for cryptography is found perfectf ormutually in implementation and theory, then the asset of the step by step procedure will be reduced to unusable state if the applicable keys are unwell brought up (Han Yanyan, 2014).

The “Tiny Encryption Algorithm” (TEA) is a “cryptographic algorithm” intended to diminish storage footmark and improvise rapidity. It is a “Feistel type cipher” that utilizes processes from assorted (orthogon) arithmetical collections. As the term proposes, the “Tiny Encryption Algorithm” is small in proportions. All expressed, it can be executed in a limited lines of program writing code. This is significant for the reason that it means that it can be comprised in just about any kind of software package, even those with thoughtful space constrictions (J. F. Dooley, 2013).

TEA appears to be extremely unaffected to differential cryptanalysis, and attains whole dispersal only afterwards 6 rounds (S. A. Parah, F. Ahad, J. A. Sheikh, N. A. Loan, &G. M. Bhat,2017). The time presentation on a contemporary workstation or desktop computer is actually striking. Corresponding to any supplementary encryption algorithm, the Tiny Encryption Algorithm is utilised to hide information. It is substitutable with algorithms like the “Advanced Encryption Standard” (AES)and the “Data Encryption Standard” (DES). These are also call well blocking algorithms, but then again these are larger in size, and classier (Q. Nie, 2018).

The captured screenshot bears that the process called encryption is successful. Before the application ensures the successful encryption, it asks for a strong password or a key to guarantee the security.

Figure 4.

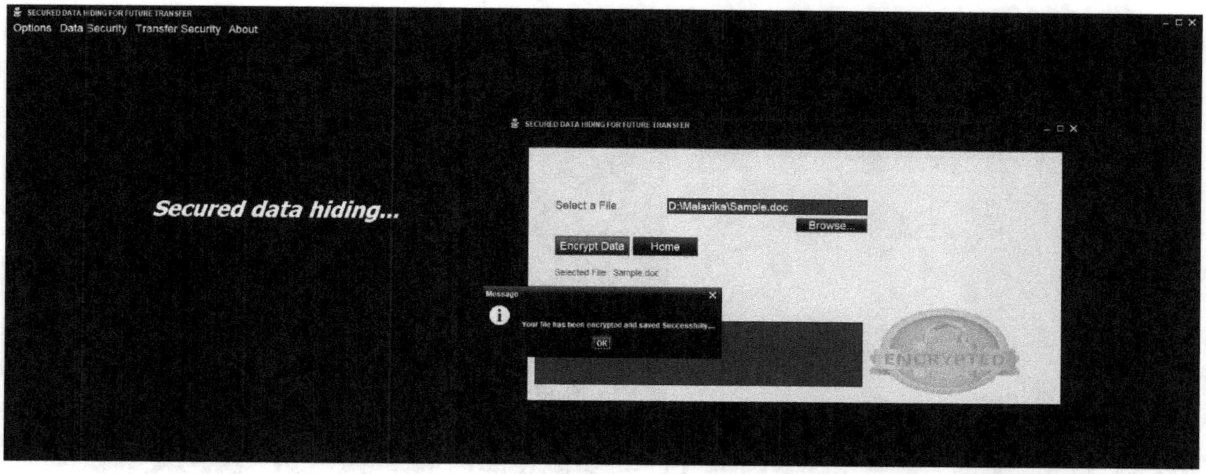

Steganography Module

In Steganography technology, the user at the end classifies a video file, that is occupying to perform as the transporter of information. Then the data file is also designated and before to accomplish better haste of broadcast the data file and the video file are directed. Previous to this the information is embedded within the video file and then directed (Sudip Ghosh, 2015). Then copy if interpreted or hacked by an intruder operator will uncovered up in any video file player but would not be showing the data. Which defends the information from being undistinguishable and hereafter it is protected throughout communication (R. Karakis, 2015). The operator in the getting end usages additional portion of cipher to recover the information from the video.

What is the Purpose of Steganography?

The term "steganography " is inherited from the Greek word "steganos" which means hidden or covered and the graph means to write. The main benefit of utilising steganography concept is to hide data through encryption is that it supports to unclear the detail that there is complex data hidden in the file or other supplementary entity with the hidden text (S. Manimurugan& Saad Al-Mutari, 2017).

In figure 5 the concept called embedding is implemented. Here the user can choose an encrypted file to embed with the video file which can also be chosen by the user itself. The application have the browse button to search the location of the file the user need to embed which is an easy task for the user. Along with that it have an embed button and at the bottom the application displays the file names of the video file selected and the encrypted file selected.

Figure 5.

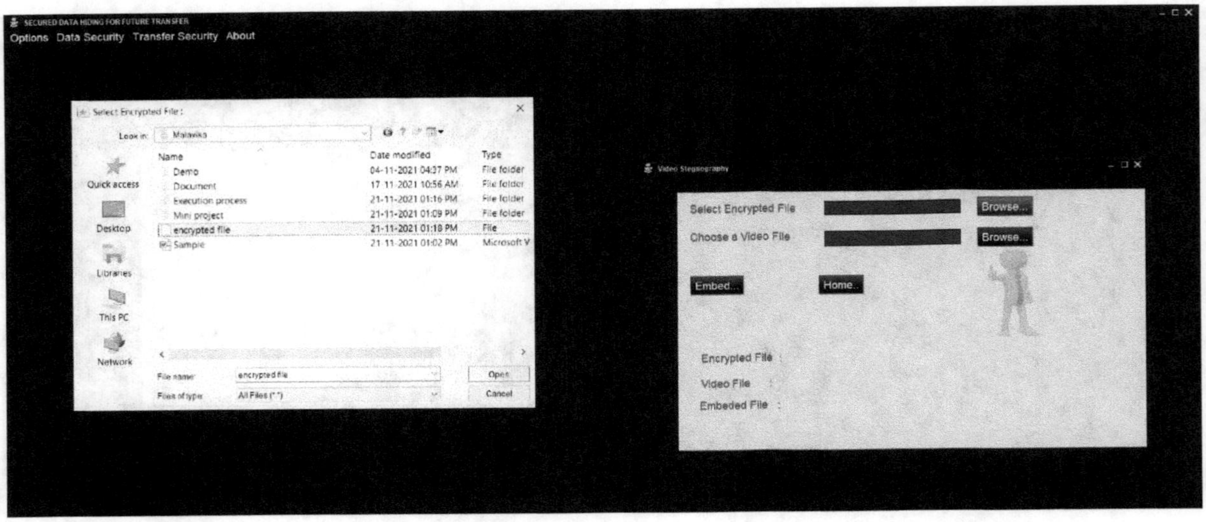

GUI Module

The GUI module or the graphical user interface module is applied in the application for the easy navigation throughout the application. The GUI module is most important for an application because it is not necessary that every application user might have the technical knowledge to operate the application. People might not be familiar with the technical terms(Kumar Sanjay & Dutta Ambar, 2016).This scheme is advanced using visuals in swings in java language(A. K. Sahu & G. Swain, 2019). The choices obtainable are demonstrated in a format of menu, like in an editor available in online. Snapping on somewhat detailed menu item concluded by the mouse or by the keyboard a "dropdown menu" is exhibited, citating all the choices obtainable below that menu element and the operator can choose the desirable actions rendering to their request (X. Weng, 2016).

The Need of GUI

Scheming the pictorial arrangement and time-based behaviour of a GUI is an essential fragment of software application programming (Md. Moniruzzaman, 2016). Its goal line is to improve the effectiveness and comfort to use for the fundamental rational enterprise of a warehoused program, a design correction termed usability(Pooja Rani & Apoorva Arora, 2015).

Figure 6 shows the easy navigation page, it have main menu with four items; options, data security, transfer security and a about item. For each menu item it has a submenu item, in this illustrated figure it has captured with submenu items of data security such as encrypt data and decrypt data.

Figure 6.

WORKING STRUCTURE

The working structure is the way of the performance of the application or the system. In simple words, the work structure is the work flow of the application. The "System Design" contains the preservation of the "secure file transfer" facility with a prearranged encryption form and divided at the attentive side by side of encryption, and embedding process and the receipt provision at the additional end with decryption and de-embedding procedure. The project likewise comprises the delivery of ability to the user to operate the worried info rendering to the user's individual usage and messaging process (Jai Singh, 2015).

The project similarly requests to deliver the network station to the operator to connect with other recorded users concluded the transmitting facilities in a dependable and secured form of the data. Authentication and authorization facilities are favored most for this determination (Meeta Malonia & Agarwal Surendra Kumar, 2016). The "System Design" comprises the preservation of endorsement services, Directory and file facilities with a prearranged encryption arrangement at the absorbed side by side of encryption and the in receipt of facility at the other end with "decryption process". The strategy likewise comprises the delivery of capability to the operator to operate the concerned considerable rendering to his individual use. The enterprise of Effective Privacy Preserving key Generation for Multimedia Application system, essentially encompass the crossing point construction, communication system, and Security services (S. Gunavathy& C. Meena, 2019).

In the interface design this include with the proposal of the user interface with GUI values besides a suitable direction finding structure where the operator prerequisite to arrive into the movement of communications agreement services are checked and supplementary admission is supplied into the structure. Then the operator desires to choose into the processes on condition that, concluded to the GUI where the encryption of the text document, embedding of video file with encrypted document, de-embedding of encrypted file and video file, Decryption of encrypted file, and sending of the text file, these "General Information" and exit are provided. Here the Encryption and decryption and facilities are given for linking the safety services component where the "encryption and decryption" are carried out using the standards of cryptographic algorithms in implementing the Tiny algorithm (N. Varol, F. Aydogan&A. Varol, 2017).

Why to Choose the Tiny Algorithm?

The Tiny Encryption Algorithm (TEA) has been round for simply around more than ten years. It might be the most "minimal" and subsequently comparatively fastest—block cipher always planned and but seems proof against maximum outbreaks. Conventionally, while cryptography changed into the area of the fleet and, more newly, of the lending communal, ciphers had been normally carried out in computer hardware. Every person is acquainted with the superbly fashioned Enigma apparatuses used through the Germans in World War II, besides some of first rate tutelages on them have regarded in those pieces of paper. Anybody who has laboured within side the again workplace of a financial institution has visible the SWIFT stations used for inter-financial institution finances transfer (K. Chachapara&S. Bhadlawala, 2013). These are computer hardware gadgets and are common of the manner wherein cryptography changed into carried out from the most basic civilisations complete to the appearance of the digital laptop. Currently, almost all cipher algorithms are obligatory in software program to be used with packages hitting on PCs (R. Gayathri & V. Nagarajan, 2015).

The cipher is extraordinarily solid and can (nearly) be involuntary from reminiscence in any programming language or CPU coaching set to be used in a wide variety of protection programs inclusive of hashing, suppression and random number generation. Contrasting several different ciphers inclusive of IDEA or the ones from RSA Laboratories, TEA isn't always weighted down with the aid of using copyrights or every other industrial entitlement. It is definitely community area and may be used freely. Its primary gain may be very high speed, that's ideally fitted to trendy programs inclusive of circulation of real-time film through the broadband Internet (J. P. Aumasson, 2018).

What are the Advantages of the Tiny Algorithm?

This category of algorithm can substitute DES in software package, and is sufficiently brief to inscribe into just about in the least program on somewhat all computers. Even though rapidity is not a robust aim with 32 cycles (64 rounds), taking place on one operation it is three times as reckless as a respectable software operation of DES which has a count of 16 rounds. The approaches of usage of DES are all appropriate. The cycle count can willingly be wide-ranging, or even completed part of the key. It is anticipated that security can be boosted by cumulating the count of iterations(A. Gupta &N. K. Walia, 2014).

Figure 7 is a working structure of this application is picturised in this diagram. The process initially initiates with the login page, in the figure the major process flow is structured with which one can study the working of the application.

The process of the application starts from a word file which is predominantly created by the user. This word file is converted to an encrypted file using the Tiny Encryption algorithm. While encrypting the file the application asks for a key which must be strong and must be remembered in the process of decryption. After the successful encryption the user can embed this encrypted file to the video, the application has provided an option to the user to choose the files for the embedding process. Now this file is in that state where this file can be transferred securely (S. B. Sadkhan, 2004).

When the desired recipient receives the embedded file, the receiver has a provision to de-embed the video file and save the encrypted file in an automated file location. And this saved encrypted file is decrypted by the user with the help of the application (Madhuri Rajawat& D S Tomar, 2015). While the decryption process the user is asked to enter the key which was given before for the encryption process. If the user fails to enter the key, then there the user looses their authenticity. For the decryption it is must to know the key otherwise it cannot be decrypted back to the readable format (N. Sharma Prabhjot & H. Kaur, 2017).

Figure 7.

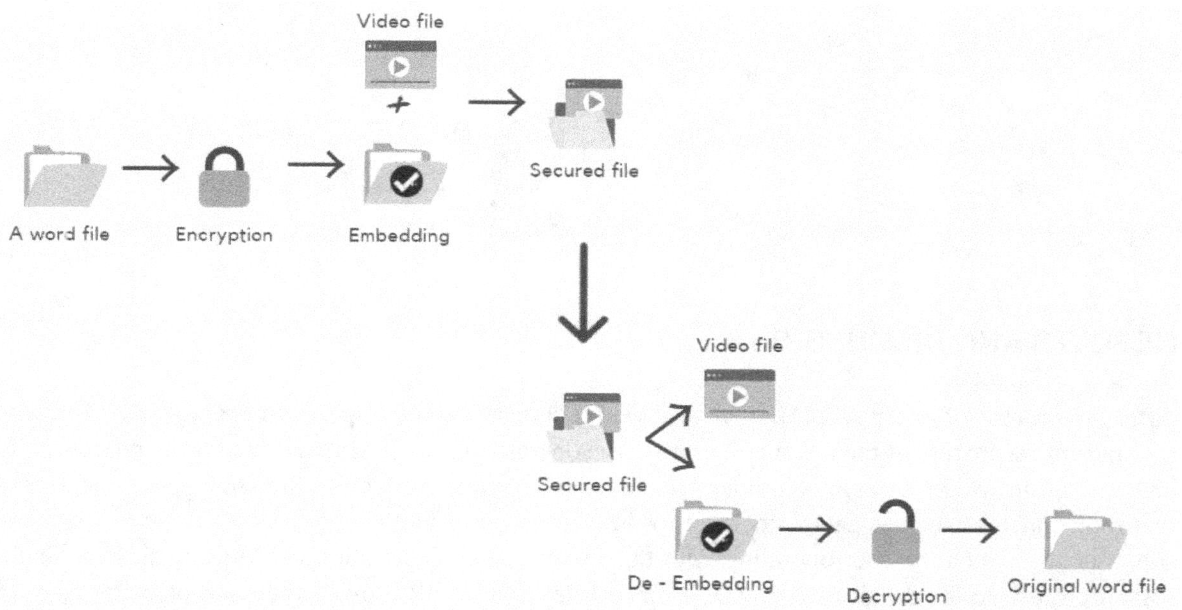

What Happens if the Intruder Breaches the File?

As days pass the security threat increases because if something is made secured there a hacker finds a way to exploit it and take away the data and use it for granted which a worst situation. When the quantity of data increases it craves for security measures. The speciality of the application is that the attacker will never know that the file has an encrypted file within it. Suppose that there exists an intruder in the network, and if the intruder or hacker or some other unauthorised person will open the file and that intruder will only able to find a video which will be a useless entity for that third person (A. K. Sahu& G. Swain, 2019). There exist a least chance of finding the encrypted file which is embedded within the transferred file. This is how the application protects the data from the intruders.

Figure 8 conveys the exact situation when an attacker gets the data. At that time, the hacker's mind will be deviated to the thought that its just a random video file which is not useful for that attacker. And no attacker takes the entity which is not useful for them.

Figure 8.

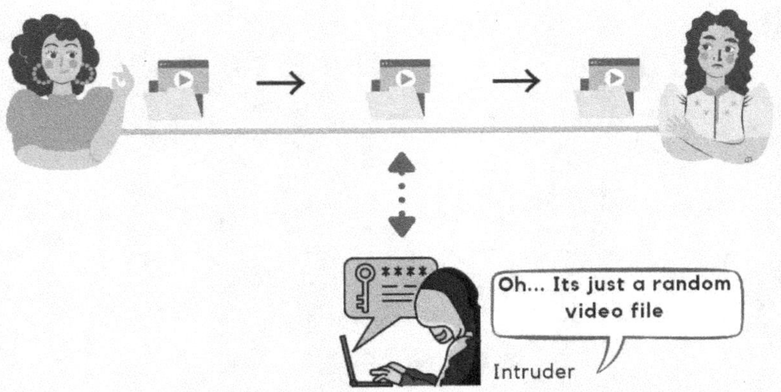

RESULTS AND DISCUSSIONS

After the process of decryption, the operator would choose the file to perform encryption. Afterwards the encryption process of the file finalizes, then the operator has to pick up the file for the process called embedding, in which it requires a video file and sending over the web to the anticipated operator. At that moment the structure gets associated to the battered user and transports the file in video format subsequently which the user operating with the Effective key Generation for Multimedia Application software must go for the decision of De-Embed Files and it will provide the encrypted file from the video to decrypt the file by picking the file pathway by that the file change to the decrypted file and is shown on the structure (J. P. Kaur & R. Kaur, 2014).

Input Design

Input design is the method of adapting a user-focused explanation of the inputs to a computer based commercial system into a programmer-preoccupied with description. The goal line of scheming input data is to create data access simple, reasonable and errors free as conceivable (K. V. Pradeep, 2019).

Input design is a fragment of complete system design, which necessitates cautious attention. If the data is working into the system is improper, then the handing out and the output will enlarge these errors(A. Pradhan, 2016). The key aim throughout the input design is as listed underneath:

- To provide a profitable technique of input.
- To accomplish the maximum conceivable stage of security.
- To guarantee that the input is satisfactory and implicit by the user.

Figure 9 is a screenshot illustrates a word document containing a sample sentence which is saved as sample. This sample document is further used to perform encryption. The figure have the word file which have a short sentence as "Here is a sample word document…"

Figure 9.

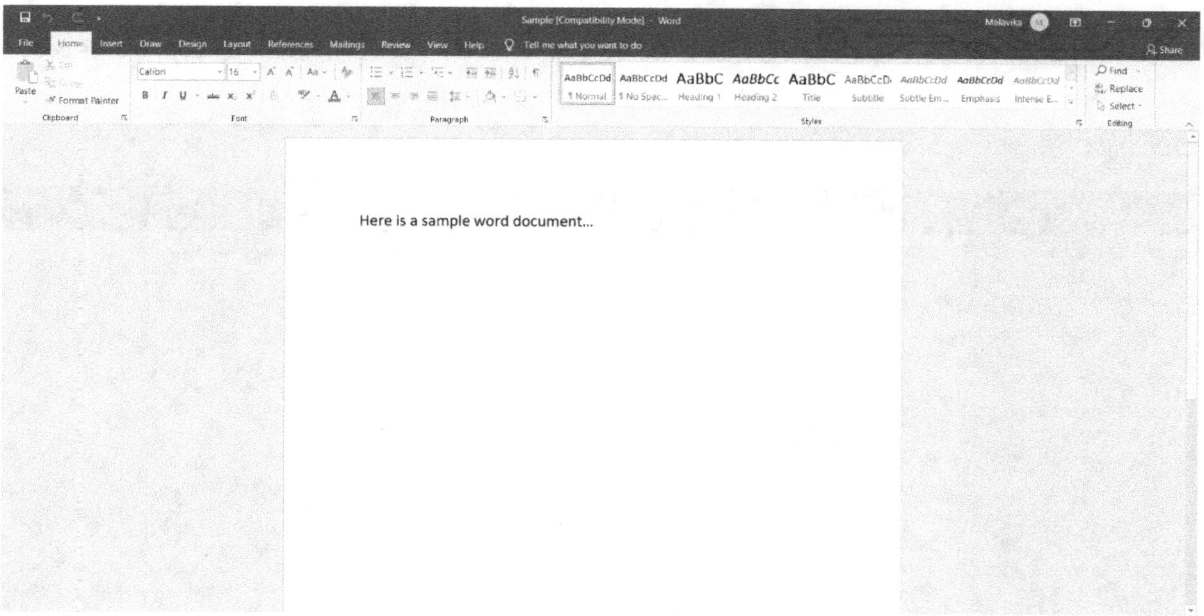

The first and foremost input given to the application is the user id and the user password to ensure the authenticity of the user. Then the user have to give the file which must be encrypted and followed by the other steps or procedures. The next figure have the encrypting process.

Figure 10 is a capture when the user chosen file is encrypted and it is been saved by the user whatever the name the user like. For example, this figure have the filename as encrypted file which was entered by the user.

Figure 10.

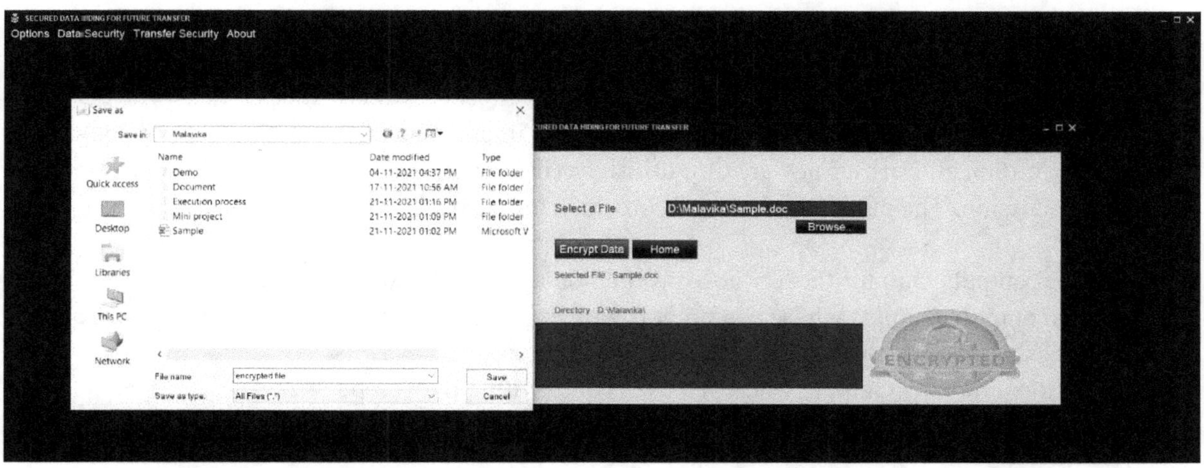

Figure 11 is a screenshot of the encrypted file is shown below. This encryption process is performed using the Tiny encryption algorithm. This encrypted data or the word file is embedded with the user chosen video file for the secured data transmission. The user can save the file in whatever name the user needs.

Figure 11.

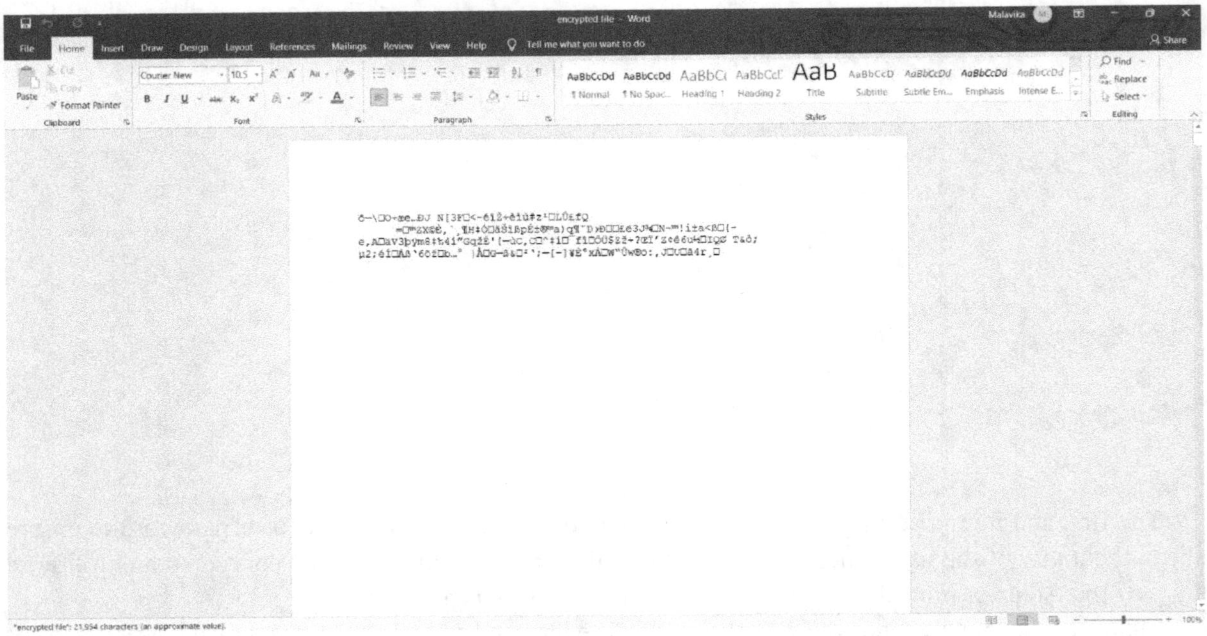

Figure 12 is captured from a video file in which the encrypted data is embedded. This video is in mp4 format. This video is just a random video to hide the encrypted data. In other words, for fool the intruder that it doesn't contain a valuable data, but in reality, it does have an encrypted file embedded to it. This video file will not have any changes before and after the embedding process.

Output Design

Output from computer systems are compulsory mainly to interconnect the consequences of dealing out to operators(M. Pelosi, 2018). They are also utilised to deliver an enduring replica of the outcomes for far ahead discussion. The numerous kinds of outputs in all-purpose are:

- External outputs:Outputs those whose destination are outside the association.
- Internal outputs:Outputs whose end point is with exterior of the association and they are the user's key crossing point with the processer.
- Operational outputs:Outputs where it's use is only within the computer division.
- Interface outputs:Outputs which contain the user in cooperating directly (N. Al-Juaid& A. Gutub, 2019).

Figure 12.

The output section play a vital role in an application because a user always uses an application to get the desired output. If the user doesn't get that expected output, then the user will be disappointed with performance of the application which will lead to the failure of the developed application (S. A. Parah, F. Ahad, J. A. Sheikh, N. A. Loan, &G. M. Bhat, 2017).

Figure 13 shows the de-embedding process is captured an added below. While de-embedding the video file the encrypted file from the video will be automatically saved to the location of the video that too will be automatically named as enc.

Figure 13.

Figure 14 explains the de-embedding process. After de-embedding the encrypted file has to be decrypted back to its initial correct and accurate format in that process it will ask for the key which was given before encryption.

Figure 14.

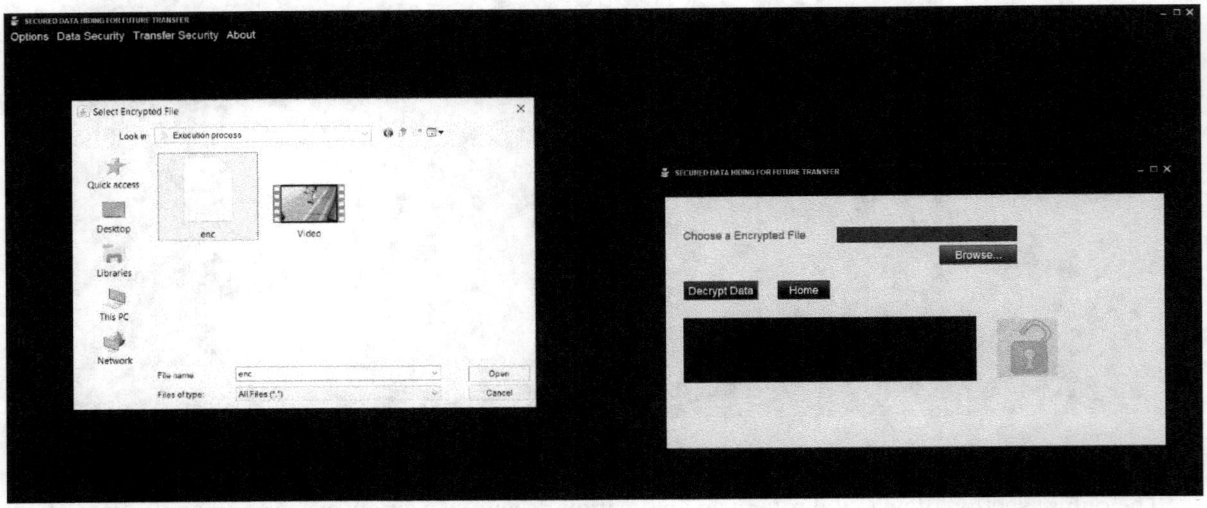

Figure 15 is the document file saved after the decryption process and it contains the same content which was there in the sample document that was mentioned earlier in figure-2.

Figure 15.

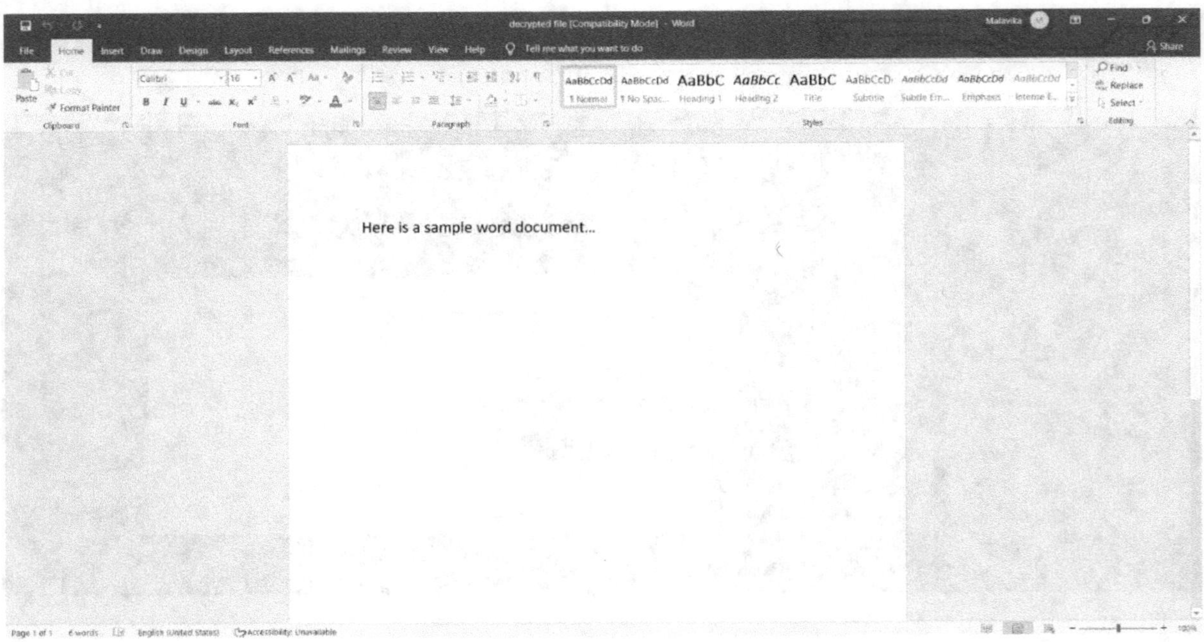

FUTURE ENHANCEMENT

The complete development consumes stayed established and organized as per the necessities of the operator; it is originated to be error free according to the challenging ethics that is applied. In the least requirement-unguided faults will be focused in the upcoming varieties, that are prearranged to be established later on or in forthcoming cases. The structure at contemporary doesn't give a stable support to lesser level check restraints in retrieving the file types in disseminated atmospheres, which is to be measured in the forthcoming progressions (K. Rabah, 2004).

As per the current position the mission established is fine prepared to switch the "Central file system" of an association in a server and deliver admission to the users with numerous rights as agreed by the sophisticated establishments in the password folder (X. Weng, 2016).

It also delivers information about the newest technology used in emerging state in this application and client server technology that will be countless mandate in future. This will deliver improved occasions and supervision in future in developing innovative projects (A. K. Sahu & G. Swain, 2019).

CONCLUSION

The project ensures the security in the state where the user embeds the encrypted file in a video. The user transfers the data by utilising encryption ethics which makes the word document in an unreadable format. In the de-embedding process, the video file and encrypted file is detached (J. F. Dooley, 2013).

By means of decryption process, the illegible data is converted into an initial format that is readable format to the user. The major job is to deliver the operator the suppleness of fleeting the material by employing the encryption ethics as per the algorithms and specification projected and accumulated the evidence in an arrangement that is illegible (Abdullatif Mohammad, 2013). The Application has taken a setback procedure as of that must be in a situation to de embed the statistics file from the video file and decrypt the obtained data file to its initial format upon the appropriate appeal by the operator (M. E. Saleh, 2016). While the "Encryption and Decryption" is executed, the application should authorize the ethics of authorization and verification of the user. The data security is provided by the application when the encryption and decryption is done (Jai Singh, Kamil Hasan & Ravinder Kumar, 2015).

REFERENCES

Al-Juaid, N., & Gutub, A. (2019). Combining rsa and audio steganography on personal computers for enhancing security. *SN Applied Sciences*, *1*(8), 830.

Al-Mutairi, S., & Manimurugan, S. (2017). The clandestine image transmission scheme to prevent from the intruders. *International Journal of Advanced and Applied Sciences*, *4*(2), 52–60.

Ali Al-Haj. (2016). *Combining Cryptography and Digital Watermarking for Secured Transmission of Medical Images*. IEEE.

Almutairi, S., Manimurugan, S., & Aborokbah, M. (2019). A new secure transmission scheme between senders and receivers using HVCHC without any loss. *EURASIP Journal on Wireless Communications and Networking*, *2019*(88), 1–15.

Almutairi, S., Manimurugan, S., & Aborokbah, M. (2019). A new secure transmission scheme between senders and receivers using HVCHC without any loss. *EURASIP Journal on Wireless Communications and Networking*, *2019*(88), 1–15.

Anusree, K., & Binnu, G. S. (2014). *Biometric Privacy using Visual Cryptography Halftoning and Watermarking for Multiple Secrets*. IEEE.

Chachapara, K., & Bhadlawala, S. (2013). Secure sharing with cryptography in cloud. *2013 Nirma University International Conference on Engineering (NUiCONE)*. 10.1109/NUiCONE.2013.6780085

Champakamala, B., Padmini, K., & Radhika, D. (2014). Least significant bit algorithm for image steganography. *International Journal of Advancements in Computing Technology*, *3*(4), 34–38.

Dooley, J. F. (2013). *A Brief History of Cryptology and Cryptographic Algorithms*. Springer.

Gayathri, R., & Nagarajan, V. (2015). Secure data hiding using Steganographic technique with Visual Cryptography &Watermarking Scheme. IEEE ICCSP Conference, 118-123.

Ghosh, Sayandip, Maity, & Rahaman. (2015). A Novel Dual Purpose Spatial Domain Algorithm for Digital Image Watermarking and Cryptography Using Extended Hamming Code. *IEEE Proceedings of International Conference on Electrical Information and Communication Technology (EICT 2015)*, 167-172.

Gunavathy, S., & Meena, C. (2019). A Survey: Data Security In Cloud Using Cryptography And Steganography. *International Research Journal of Engineering and Technology*, *6*(5), 6792–6797.

Gupta, A., & Walia, N. K. (2014). Cryptography Algorithms: A Review. *International Journal of Engineering Development and Research*, *2*(2), 1667–1672.

Han, Y., He, W., Shuai, J., & Qing, L. (2014). A Digital Watermarking Algorithm of Colour Image based on Visual Cryptography and Discrete Cosine Transform. *IEEE Ninth International Conference on P2P*, 527-530.

Hussain, M., Wahab, A. W. A., Idris, Y. I. B., Ho, A. T., & Jung, K.-H. (2018). Image steganography in spatial domain: A survey. *Signal Processing Image Communication*, *65*, 46–66.

Karakış, R., Güler, İ., Çapraz, İ., & Bilir, E. (2015, December). A novel fuzzy logic-based image steganography method to ensure medical data security. *Computers in Biology and Medicine*, *67*, 172–183.

Kaur, J. P., & Kaur, R. (2014). *Security Issues and Use of Cryptography in Cloud Computing*. Academic Press.

Kaw, J. A., Loan, N. A., Parah, S. A., Muhammad, K., Sheikh, J. A., & Bhat, G. M. (2019, April). A reversible and secure patient information hiding system for IoT driven e-health. *International Journal of Information Management*, *45*, 262–275.

Malonia & Kumar. (2016). Digital Image Watermarking using Discrete Wavelet Transform and Arithmetic Progression. *IEEE Students's Conference on Electrical.*

Manimurugan & Al-Mutari. (2017). A Novel Secret Image Hiding Technique for Secure Transmission. *Journal of Theoretical and Applied Information Technology, 95* (1), 166-176.

Manimurugan, S., Al-Mutairi, S., Aborokbah, M. M., Chilamkurti, N., Ganesan, S., & Patan, R. (2020). Effective Attack Detection in Internet of Medical Things Smart Environment Using a Deep Belief Neural Network. *IEEE Access: Practical Innovations, Open Solutions, 8*, 77396–77404.

Manimurugan, S., Al-Mutairi, S., Aborokbah, M. M., Chilamkurti, N., Ganesan, S., & Patan, R. (2020). Effective Attack Detection in Internet of Medical Things Smart Environment Using a Deep Belief Neural Network. *IEEE Access: Practical Innovations, Open Solutions, 8*, 77396–77404.

Mohamed, Ahmed, & Hassan. (2013). Securing biometric data by combining watermarking and cryptography. *IEEE 2nd International Conference on Advances in Biomedical Engineering*, 179-182.

Mohammad, Zeki, Chebil, & Gunawan. (2013). Properties of Digital Image Watermarking. *IEEE 9th International Colloquium on Signal processing and its Applications*, 8-10.

Moniruzzaman, Hawlader, & Hossain. (2014). Wavelet Based Watermarking Approach of Hiding Patient Information in Medical Image for Medical Image Authentication. *IEEE 17th International Conference on Computer and Information Technology (ICCIT)*, 374-378.

Mstafa, R. J., Elleithy, K. M., & Abdelfattah, E. (2017). A robust and secure video steganography method in dwt-dct domains based on multiple object tracking and ecc. *IEEE Access: Practical Innovations, Open Solutions, 5*, 5354–5365.

Nie, Q., Xu, X., Feng, B., & Zhang, L. Y. (2018). Defining embedding distortion for intra prediction mode-based video steganography. *Comput. Mater. Continua, 55*(1), 59–70.

Parah, S. A., Ahad, F., Sheikh, J. A., Loan, N. A., & Bhat, G. M. (2017). A new reversible and high capacity data hiding technique for e-healthcare applications. *Multimedia Tools and Applications, 76*(3), 3943–3975.

Pelosi, Poudel, Lamichhane, Lam, Kessler, & Mac-Monagle. (2018). *Positive identification of lsb image steganography using cover image comparisons.* Academic Press.

Prabhjot & Kaur. (2017). A Review of Information Security using Cryptography Technique. *International Journal of Advanced Research in Computer Science, 8*, 323-326.

Pradeep. (2019). An Efficient Framework for Sharing a File in a Secure Manner Using Asymmetric Key Distribution Management in Cloud Environment. *Journal of Computer Networks and Communications,* 1-8.

Pradhan, A., Sahu, A. K., Swain, G., & Sekhar, K. R. (2016). Performance evaluation parameters of image steganography techniques. *2016 International Conference on Research Advances in Integrated Navigation Systems (RAINS)*, 1-8.

Rabah, K. (2004, March). Steganography—The art of hiding data. *Information Technology Journal, 3*(3), 245–269.

Rajawat, M., & Tomar, D. S. (2015). A Secure Watermarking and Tampering detection technique on RGB Image using 2 Level DWT. *IEEE Fifth International Conference on Communication Systems and Network Technologies*, 638-642.

Rani & Arora. (2015). Image security system using encryption and steganography. *IJIRSET, 4*(6).

Sadkhan, S. B. (2004). Cryptography: current status and future trends. *International Conference on Information and Communication Technologies: From Theory to Applications.*

Sahu, A. K., & Swain, G. (2019, September). An optimal information hiding approach based on pixel value differencing and modulus function. *Wireless Personal Communications, 108*(1), 159–174.

Sahu & Swain. (2019). A novel n-Rightmost bit replacement image steganography technique. *3D Res., 10*(1), 2.

Saleh, Aly, & Omara. (2016). Data security using cryptography and steganography techniques. *International Journal of Advanced Computer Science and Applications, 7*(6), 390-397.

Sanjay, K., & Ambar, D. (2016). A Novel Spatial Domain Technique for Digital Image Watermarking Using Block Entropy. *IEEE Fifth International Conference on Recent Trends in Information Technology.*

Shanmuganathan, M., Almutairi, S., Aborokbah, M. M., Ganesan, S., & Ramachandran, V. (2020, August). Review of advanced computational approaches on multiple sclerosis segmentation and classification. *IET Signal Processing, 14*(6), 333–341.

Singh, Hasan & Kumar. (2015). Enhance security for image encryption and decryption by appling hybrid techniques using MATLAB. *IJIRCCE, 3*(7).

Tyagi, V. (2012). Image steganography using least significant bit with cryptography. Journal of Global Research in Computer Science, 3(3), 53-55.

Varol, N., Aydogan, F., & Varol, A. (2017). Cyber Attacks Targetting Android Cellphones. *The 5th International Symposium on Digital Forensics and Security (ISDFS 2017).*

Weng, Li, Chi, & Mu. (2018). *Convolutional video steganography with temporal residual modeling.* arXiv preprint.

Zhang, Y., Qin, C., Zhang, W., Liu, F., & Luo, X. (2018). On the fault tolerant performance for a class of robust image steganography. *Signal Processing, 146*, 99–111.

Chapter 9
Artificial Intelligence-Based Billing System:
Fingerprint Mechanism

Kathirvel A.
https://orcid.org/0000-0002-5347-9110
Karunya Institute of Technology and Sciences, India

Sabarinathan C.
SRM Institute of Science and Technology, India

Saravanan N.
M. N. M. Jain Engineering College, India

Ramesh S.
Acharya Institute of Technology, India

Meera S.
Agni College of Technology, India

Karnavel K.
Anand Institute of Higher Technology, India

Sudha D.
Mother Theresa Woman's University, India

ABSTRACT

In our daily lives, we conduct billions of payment transactions, yet each payment method requires the transport of a substance. It is typical for users to have a variety of payment materials on hand, such as cash, credit cards, and even mobile phones. Meanwhile, these goods are easily stolen or misplaced. People suffer enormous trauma as a result of these incidents. This chapter details a biometric payment application created to introduce the concept of hardware less payments. It allows users to pay anytime and anywhere by registering their finger without any hardware. This involves registering user information once, and then all subsequent transactions are confirmed and processed by the user's fingerprint, with the application managing the entire process. This solution creates a new payment option and eliminates the risk of shipping valuables abroad. For the company, this application enables an efficient and secure payment system.

DOI: 10.4018/978-1-6684-3991-3.ch009

INTRODUCTION TO PAYMENT TRANSACTIONS

Since the early 16th century, payment transactions have been a part of our lives. It progressed from the barter system, in which products were exchanged in exchange for goods or services, to the modern form of transactions, which includes cash, credit/debit cards, e-wallets, and other methods. The barter system was used for a while until it was discovered that it was inaccurate owing to discrepancies in the value estimation of the items or services exchanged.

Later, coins composed of costly materials such as gold and silver were used in a more standard manner. This approach was convenient since it provided a uniform value for all goods and services, and it was adopted by practically every King or ruler in the world. Currency bills and coins were introduced with the development of governments and the growth of civilisation. Currency bills are still in use today. Many financial institutions, such as banks, have introduced the public to various paper and card-based payment operations. The internet, e-commerce, e-payment, mobile banking, and payment have all grown in popularity as a result of technological advancements and digitisation. All of these transactions have one thing in common: they all require a material to complete the payment.

To identify their payment account, users must carry cash, credit/debit cards, or e-wallets in the form of mobile phones. The transaction must then be authenticated using an OTP, PIN, or password to ensure that it is genuine. The transaction is complete once the server-side verification is completed, and the amount is deducted from the payer. To credit the amount back to the payee, this process must be repeated involving the bank and the payment gateway. People who carry valuables such as wallets are also concerned about keeping them safe and not misplacing them. With technological advancements, this process might be simplified while also being more secure.

The chapter proposes a revolutionary form of material-free payment that simplifies the payment process. Customers are not need to carry cash, debit/credit cards, or mobile phones in order to make a payment. It allows customers to go shopping without worrying about their possessions being in their possession or being protected.

This research established a secure fingerprint-based system that uniquely identifies each user. Yankov et al. (2020) discuss in his paper, it helps to identify the person and functions as a security gateway for the system because fingerprints cannot be the same for more than one person. It uses backend methods for minutiae extraction and minutiae comparison to analyse fingerprints, as outlined in the algorithm portion of this work.

After completing a one-time registration, the app allows the user to enrol their fingerprint to initiate a transaction and pay to any seller. The literature review section of this paper will include a brief discussion of a few alternative payment mechanisms in comparison to the suggested system.

LITERATURE REVIEW

For generations, the desire for possessions has been a part of our existence. People in a society demanded something in return for products. It was known as the Barter system, in which products were swapped for other goods. It was eventually determined that the value and quantity of products differed from one another. It resulted in the creation of currency. Currency notes, cheques, demand draughts, and a few more paper-based transactions were common payment methods a decade ago. Due to their simplicity, currency bills were one of the most simple payment options. However, if the amount was bigger, a large amount of cash had to be transferred, which was inconvenient.

Because cash transactions are difficult to account for, they contribute less to GDP discussed in the paper proposed by Hanzal et al (2019). Cheques and demand draughts alleviate the problem of increased currency volume, but they are a significantly slower process that can take up to three working days.

Recent Trends

In today's world, technological advancements have resulted in a variety of effective payment options being proposed. It was a watershed moment when payment cards were introduced as debit/credit cards. Almost every bank in the world has adopted this system. It allows the user to swipe it through a card reader machine, deducting the money from their account (Sajić et al. 2018). When the card is swiped, a magnetic strip records all of the account information and prompts the user for a secure PIN, which authenticates the transaction. With the use of geo-location, a payment improvement was introduced. It entails using the user's smartphone to track their whereabouts and monitoring their regular geo-locations. As a result, when a user tries to make a payment at a familiar location, the system skips the PIN verification and requests the user to input one only in unusual or unfamiliar locations authors discussed in the work Zolotukhin, O. and Kudryavtseva, M., (2018). This approach may save time and make payment more convenient for the user, but it is vulnerable to unlawful use. Anyone could pose as an imposter and make the payment if the smartphone is lost or stolen.

Drawbacks of Current Payment Method

Payment cards have the disadvantage of requiring the payee to have a card reader machine in order to debit the money. Small-scale merchants may find it difficult to afford these devices because they demand the payment of an annual fee to the bank. Payment by card also has a minimum and maximum limit. Furthermore, while the sleek form of these cards may make them highly useful, they are prone to being misplaced.

Mobile Banking

With the introduction of smartphones, mobile banking has become more convenient for users. Sun, Y. and Havidz, S., (2019) show his work, People find it convenient to pay using their smartphones, but the method employed weighs in here. Banks offer a variety of applications for transferring payments, but in order to complete the transaction, the payer must add the payee's account information as a beneficiary. Even when the transaction is performed online, adding the payee information takes longer in many cases discussed in the paper Tounekti et al (2020). Filling in payee account data while making a payment at the location is not a favoured choice. These transactions also necessitate a persistent internet connection.

e-wallets

Many private businesses have successfully launched e-wallets that allow consumers to load money and pay with QR codes. Because the user must scan the QR code with their smartphone shown in paper Liu et al. (2020), it is a quick way to make a purchase. Users must always have money preloaded in these programmes, which necessitates a constant internet connection because all transactions are conducted online. It also stipulates that the user accepts the terms and trusts the third-party applications with their funds (Islamiati et al. 2019).

Near Field Communication

Near Field Communication (NFC) is an enhanced payment technique widely utilized in industrialized countries (NFC). NFC chips are built into the majority of cellphones produced in industrialized countries. It simplifies financial transactions by requiring users to tap their handset against another device to complete them. It is extremely time-efficient and does not require the user to execute any intermediary steps [9]. To complete the transaction, a management authentication server connects with the device and the application server using an adaptive protocol. It's unlikely to be successful in developing countries because most smartphones don't have NFC chips. Furthermore, no degree of authentication is included in this method. As a result, it might not be the safest way to employ.

Face Recognition

Even the use of facial recognition has been proposed. Users can pay by scanning their face at a camera positioned in the billing counter with this approach deployed as a payment mechanism given in Al-Haj, A. and Al-Tameemi, M. (2018). The issue is registration, which takes a minimum of a thousand photos of input to reliably identify a single person. The database and server required to process and store these data would be prohibitively expensive. Furthermore, the technique fails to distinguish identical twins. Although the findings of iris recognition are accurate, the technology required to detect iris is expensive shows by author Caron (2018), and installing it in every shop would not be a financially viable option for most businesses.

Biometrics Method

Biometrics is efficient and produces precise findings, but the implementation costs are a disadvantage. As a result, a reduced biometric method, such as a fingerprint, has been proposed in payment technology.

QR Method

A novel payment technique combining both QR codes and face recognition is discussed. It employs QR to identify the user and account information, as well as face recognition for authentication discussed in Ximenes et al. (2019). It is extremely safe because it adds up the benefits of both ways while also adding up the disadvantages. There are no ways given in the research to avoid the extensive use of a server to process the facial data or to avoid the need for a constant internet connection in order for the QR code to work. It also lengthens the total process.

IoT Method

IoT components were used to create a distinct device. A fingerprint sensor, a WIFI module, a keypad, and an LCD are all connected to a Raspberry Pi that serves as a processor discussed in Hualin et al (2018).

It is a self-contained, portable device that may be used at any shop's billing counter. The biggest disadvantage is that IoT devices have yet to produce significant results in fingerprint matching. Employees must have proper training in order to use the equipment. Even a single component failure would cause the entire system to malfunction.

Finger Print Method

Many inventors were motivated by the use of fingerprints to come up with new ideas. The use of a fingerprint sensor on an existing contactless debit/credit card was discussed in a paper. It assists the user in registering their fingerprint on their card. You can use it to make payments without having to enter your PIN. The method self-authenticates when the user holds the card with the right finger on the sensor shown in the paper Suwald, and Rottschafer, (2019), ensuring that only the account owner may make the payment. The difficulty with this technology is that the sensor requires power to function effectively. Power is required for the card to function properly. It eventually leads to the purchase of an adaptor at an additional cost.

E-Transaction

A unique and exciting payment method based on offline E-cash transfer has been proposed. With the use of blockchain technology, the emphasis is on using virtual cash instead of real hard cash. The concept entails using E-cash for payment chores such as withdrawal, transfer, and so on discussed in Luo and Yang (2018). It needs the user to register with their favourite bank and obtain a card as well as an e-cash mobile application. The transaction was authenticated using the payer's signature on the smartphone itself, according to the document. However, it was just explored conceptually, with no actual model supplied.

Advantages and Disadvantages of Biometric

Taking into account the benefits and drawbacks of all of the above-mentioned methodologies, a biometric payment application based on fingerprints is created, with a mix of possible benefits and minor drawbacks.

BILL PAYMENT SYSTEM USING BIOMETRIC MECHANISM

This section introduces a fresh method to the payment transaction. The notion of material-free payment with the use of a fingerprint has been implemented as a smartphone application. Carrying wallets and protecting them is no longer necessary because users are encouraged to conduct transactions with just their fingertip. This approach necessitates a one-time registration in which users enter their information, including their fingerprint and a secure PIN. Sensitive data, such as a fingerprint, is transformed into numerical values and stored in a byte array for data privacy concerns. All data is safely kept in a real-time cloud database and may be accessed without the use of a server. All the information gets securely stored in the real-time cloud database, function seamlessly without a server. Following this, the user can use this programme to enter any shop and pay their debts using their fingerprint and PIN.

Architecture Diagram

The payer and the payee are both involved in the architecture. The payer is the one who makes a purchase and pays for it. The shopkeeper or shop owner who must receive the money is the payee. The payer does not need to carry any materials or a wallet to make a payment with this approach. As shown in Figure 1, the shopkeeper has the application connected to a USB fingerprint scanner. After the payee logs into the application, the amount due is entered. To finish the payment, the customer provides his information.

Figure 1. The biometric payment application

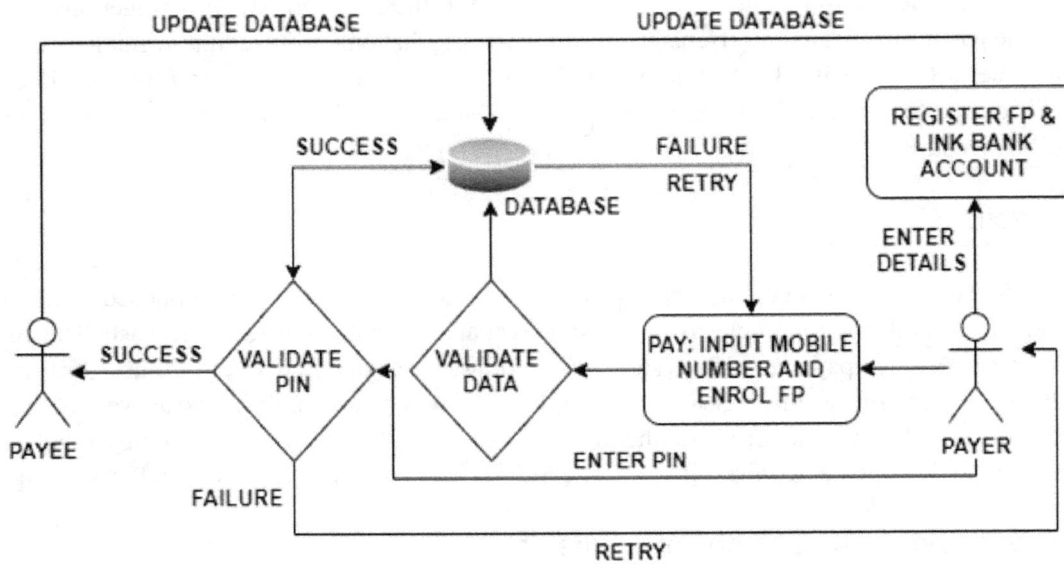

For simplicity, the application opens right into the make payment action, with the option to register. The app also offers a registration feature that asks the user for their information as well as the connection to their preferred payment account. The user's actions on the programme are depicted in Figure 2. This is a one-time registration process in which the user's information is securely saved in a real-time database, following which the user can make a payment using their fingerprint and PIN. By connecting the database, the application verifies the data. The amount is debited from the customer and credited to the shopkeeper's account after proper validation. The fact that the application communicates directly with the database without the use of a middleman such as a payment gateway makes it a useful tool.

Methodology for Minutiae Extraction and Comparison

Fingerprint capture, fingerprint pre-processing, fingerprint enhancement, feature extraction, and minutiae matching are the five processes that make up the algorithm's operation given by Liban and Hille (2018).

Figure 2. Biometric payment application

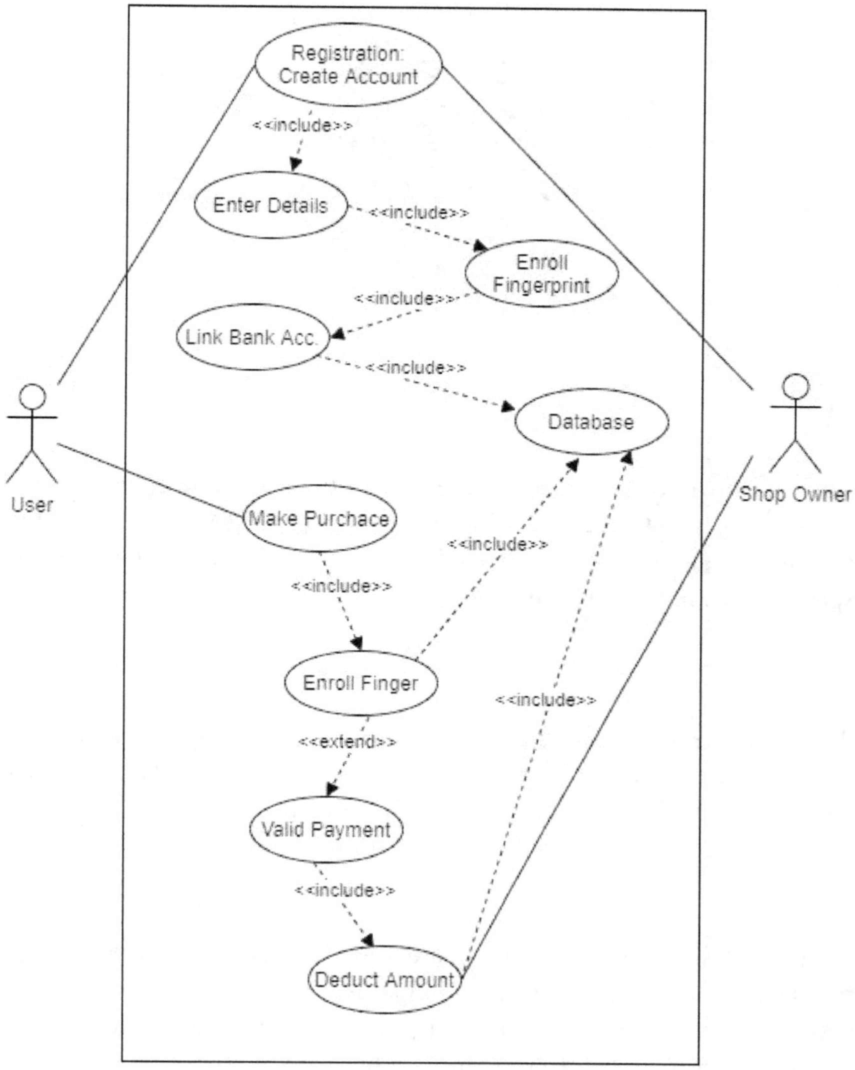

Acquisition

It is the procedure through which the application obtains the user's fingerprint input via the fingerprint scanner. A fingerprint is placed on the scanner's surface, and it takes an image of it. The application uses device drivers to interface with the scanner and retrieves unique fingerprint image for further processing.

Pre-Processing

The acquired image is transformed to pure gray scale, or black and white, during pre-processing. As seen in Figure 3, it ensures that the ridge, ridge endings, bifurcations, valleys, and whorls are different.

Figure 3. Representation of fingerprint minutiae

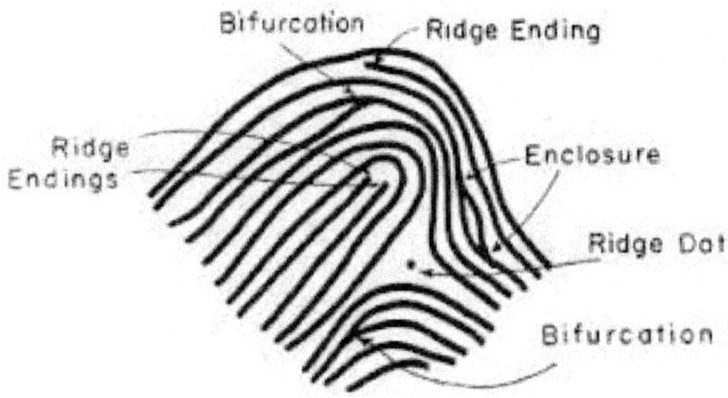

Enhancement

The picture quality is increased to its highest level at this point in order to better distinguish between details and white space. A quality check is performed on the image to confirm that it is suitable for the extraction process. Binarization assigns a value of 0 to every black ridge inside an image and a value of 1 to the white space. It aids in the extracting process by identifying the fingerprint's features.

Extraction

For binary bits having the value 0 as features, the image becomes entirely rooted. The value of each pixel determines the value of each feature. 1 is used to compute these binary and pixel values, which are then saved in a byte array given in Gudkov and Lepikhova., (2018).

Matching

The byte arrays of the desired two fingerprints are compared because the byte array represents the whole fingerprint. Due to the fact that fingerprint input is not continuous and tends to change orientation over time, even with the same finger, a matching score is generated, as illustrated in Figs. 5 and 6. This matching score indicates the degree of similarity between the fingerprints under consideration. It determines whether the fingerprints are identical. If the resulting result is greater than 100, the biometrics are similar according to a pre-determined standard. It is regarded as a contest. Any result that falls below that threshold is called a no-match.

BIOMETRIC PAYMENT MECHANISM

Biometric payment and register are the two key functions of the programme. The signup activity begins when the customers click the Register button. The user is prompted to provide information for the registration procedure in this module. Details such as a person's name, mobile number, or email address are gathered.

Figure 4. Payment registration activity

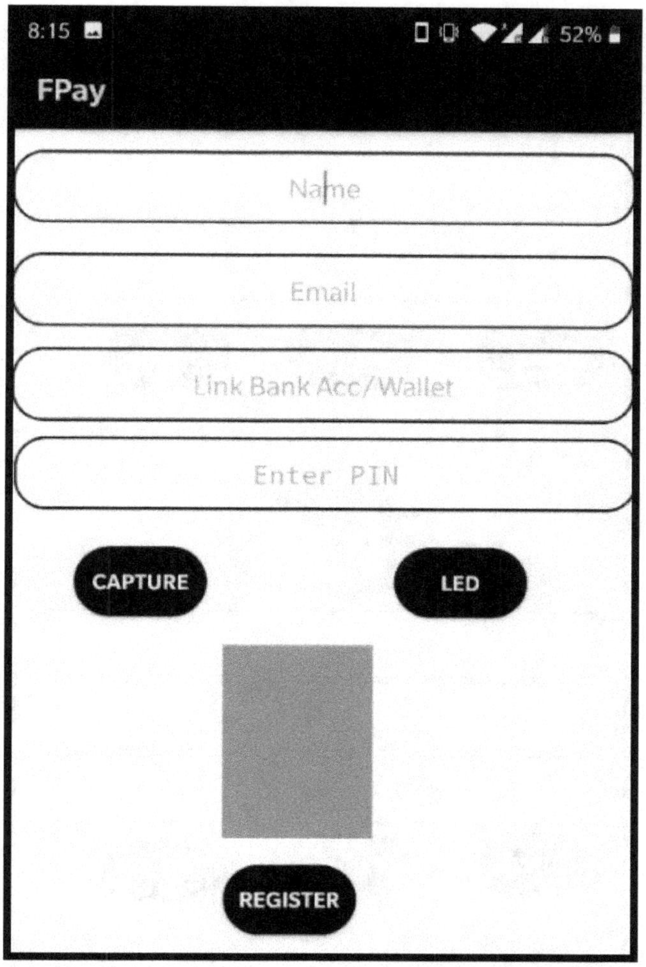

Users must enter their bank account information and choose a preferred payment account to integrate with the app. Finally, the user's fingerprint is taken for identification for security purposes. To use the external fingerprint scanner, the user must push the record fingerprints button and place their choice fingertip on it. It takes a picture of the palm to be processed later. In order to authenticate upcoming payments, the user must also provide a secure PIN.

The minutiae extraction process converts the fingerprint to a byte array. The Password is also hashed for security. As a result, sensitive data is not stored. All of this information is safely kept in a single cloud database. It is an each registration, following which the user could use the application at any time and in any location. After signing up, the user receives an instant success notification. Users of this programme are free using this payment option anywhere it is offered after registering.

The user does not need to bring any payment materials to the store because they can finish their shopping while scanning their fingerprint, authenticating the payment, and completing the transaction. The shopkeeper enters the bill amount into the application. To complete the transaction, the user must input their mobile number, verify their fingerprint, and enter their PIN.

Figure 5. Payment activity

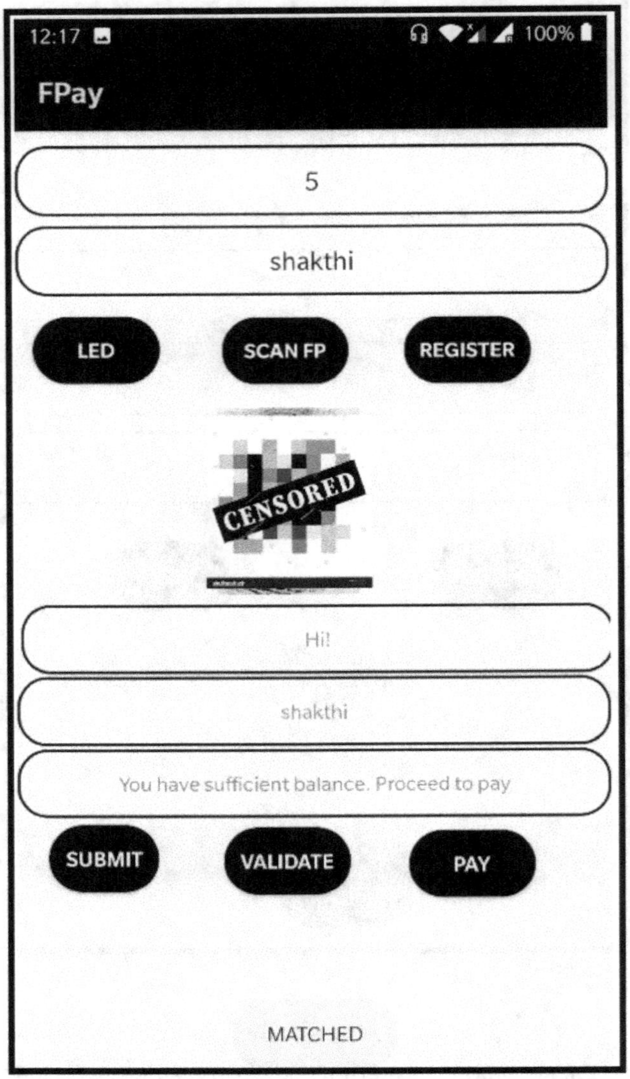

Identification, validation, and authentication are the three sub-processes of the procedure. The user's mobile number is utilised as a key to select details from the database throughout the identification process. The minutia matching algorithm is used to match the byte array holding the fingerprint with current byte array in the repository during the validation phase. Whenever a successful match is found, the programme checks the account balance for sufficient. When enough credit is given, the user is prompted for their PIN for authentication; otherwise, the user is informed that they do not have enough credit to complete the transaction.

Whenever the PIN is entered correctly, the application connects the system (see Figure 7) and debit cards the amount provided. The debited amount is credited to the payee's account at the same time. Finally, the user is shown a transaction success message. Every person's fingerprints are distinct. This approach is more secure because it cannot be modified by anyone.

Figure 6. Transaction message

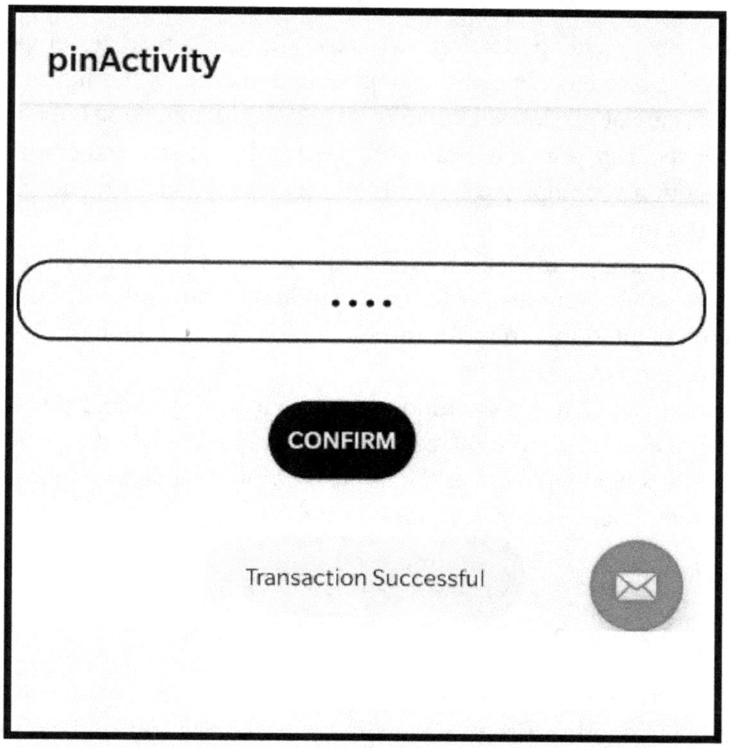

Figure 7. Cloud schema

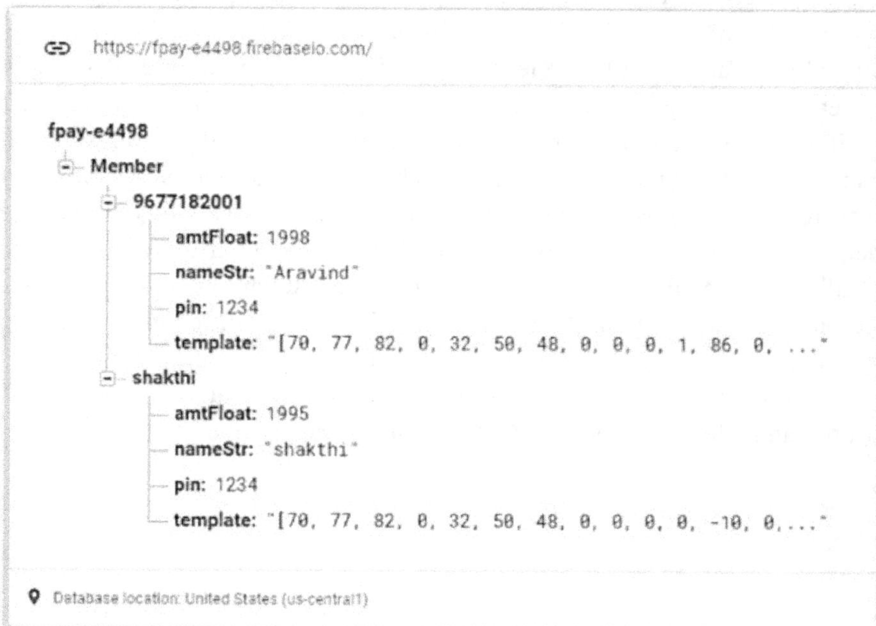

EXPERIMENTAL RESULTS AND DISCUSSION

As a result, the biometric payment application was successfully designed, developed, and implemented. A new method of material-free payment was also presented to society. It relieves the strain on consumers to keep their wallets safe at all times. Unlike OTP and 2 authentications, the programme combines fingerprint and PIN for security, which is straightforward and produces less confusion for clients. With fewer inputs, the transaction is completed faster. It eliminates the need for consumers to type large account numbers and speed up the procedure.

Because fingerprints are unique to each user, they make it easy to prevent unauthorised use. Regression testing yielded 100 percent accurate results for the minutia matching algorithm. In term of the system, real-time database implementation allows for instant data retrieval without a server. It guarantees that there will be no downtime on the server.

Fingerprints and other sensitive information are translated to bytes and saved in a byte array. The programme is secure to use in terms of information security. No critical information is ever lost. Lastly, the payment contract was concluded without the use of any intermediaries, such as online payment. It's also a cost-effective alternative to a card-swiping machine that requires an annual membership.

SUMMARY

Finally, the biometric application sets the door for material-based payment to be phased out. The application incorporates all of the important benefits and eliminates the majority of the disadvantages identified in the literature. To tackle the problem of material reliance, the programme employs automation. In addition, it offers a number of other benefits, which are detailed in the paper's conclusion. Users don't have to worry about losing their purse or carrying exact change when they use the app. Both the client and the payer have a pleasant overall user experience. People easily adapt to this strategy because there are few items to carry or processes to complete on the user's end.

A few additional innovations will be released as updates in future sprints. The application will be changed to allow each user to register multiple fingerprints. It ensures that the user can access the system even if their previously registered fingerprint has a problem. To deal with any unexpected failures in the fingerprint sensor, a backup payment mechanism will be added. The Aadhar Enabled Payment Service (AEPS) allows customers to pay with their Aadhar card, which is already linked to their bank account. To make payment simple, the proposed biometric application might partner with India Stack, company developers of Aadhar. It also allows the user to bypass the enrolment process because their fingerprint is already stored on their Aadhar card. It also gives you the option of linking a bank account or changing an existing bank accounts linked with biometric payment app all online. Based on the needs of the investor and the convenience of the users, a few additional main characteristics such as forgotten PIN, a bug report feature, and a feedback system could be introduced as an update.

REFERENCES

Al-Haj, A., & Al-Tameemi, M. (2018). Providing security for NFC-based payment systems using a management authentication server. *2018 4th International Conference on Information Management (ICIM).*

Caron, F. (2018). The Evolving Payments Landscape: Technological Innovation in Payment Systems. *IT Professional*, *20*(2), 53–61.

Gudkov, V., & Lepikhova, D. (2018). Fingerprint Model Based on Fingerprint Image Topology and Ridge Count Values. *2018 Global Smart Industry Conference (GloSI).*

Hanzal, P., & Homan, J. (2019). Electronic Exchange SAF-T Standard of Data from Organizations to Tax Authorities or Auditors - Situation in the Czech Republic. *2019 9th International Conference on Advanced Computer Information Technologies (ACIT)*, 405-408. 10.1109/ACITT.2019.8780001

Hualin, Z., Qiqi, W., & Yujing, H. (2018). Design Fingerprint Attendance Machine Based on C51 Single-chip Microcomputer. *2018 IEEE International Conference of Safety Produce Informatization (IICSPI).*

Islamiati, D., Agata, D., & Anom Besari, A. (2019). Design and Implementation of Various Payment System for Product Transaction in Mobile Application. *2019 International Electronics Symposium (IES).*

Liban, A., & Hilles, S. (2018). Latent Fingerprint Enhancement Based On Directional Total Variation Model With Lost Minutiae Reconstruction. *2018 International Conference on Smart Computing and Electronic Enterprise (ICSCEE).*

Liu, W., Wang, X., & Peng, W. (2020). State of the Art: Secure Mobile Payment. *IEEE Access: Practical Innovations, Open Solutions*, *8*, 13898–13914. doi:10.1109/ACCESS.2019.2963480

Luo, J., & Yang, M. (2018). Offline Transferable E-Cash Mechanism. *2018 IEEE Conference on Dependable and Secure Computing (DSC).*

Sajić, M., Bundalo, D., Bundalo, Z., Stojanović, R., & Sajić, L. (2018). Design of Digital Modular Bank Safety Deposit Box Using Modern Information and Communication Technologies. In *7th Mediterranean Conference on Embedded Computing MECO 2018. 10-14 June 2018* (pp. 107-112). 10.1109/MECO.2018.8406014

Sun, Y., & Havidz, S. (2019). Factors Impacting the Intention to Use M-Payment. *2019 International Conference on Information Management and Technology (ICIMTech).* 10.1109/ICIMTech.2019.8843758

Suwald, T., & Rottschafer, T. (2019). Capacitive Fingerprint Sensor for Contactless Payment Cards. *2019 26th IEEE International Conference on Electronics, Circuits and Systems (ICECS).*

Tounekti, O., Ruiz-Martinez, A., & Skarmeta Gomez, A. (2020). Users Supporting Multiple (Mobile) Electronic Payment Systems in Online Purchases: An Empirical Study of Their Payment Transaction Preferences. *IEEE Access: Practical Innovations, Open Solutions*, *8*, 735–766. doi:10.1109/ACCESS.2019.2961785

Ximenes, A., Sukaridhoto, S., Sudarsono, A., Ulil Albaab, M., Basri, H., Hidayat Yani, M., Chang Choon, C., & Islam, E. (2019). Implementation QR Code Biometric Authentication for Online Payment. *2019 International Electronics Symposium (IES).*

Yankov, M. P., Olsen, M. A., Stegmann, M. B., Christensen, S. S., & Forchhammer, S. (2020). Fingerprint Entropy and Identification Capacity Estimation Based on Pixel-Level Generative Modelling. *IEEE Transactions on Information Forensics and Security*, *15*, 56–65. doi:10.1109/TIFS.2019.2916406

Zhang, W., & Kang, M. (2019). Factors Affecting the Use of Facial-Recognition Payment: An Example of Chinese Consumers. *IEEE Access: Practical Innovations, Open Solutions*, *7*, 154360–154374.

Zolotukhin, O., & Kudryavtseva, M. (2018). Authentication Method in Contactless Payment Systems. 2018 International Scientific-Practical Conference Problems of Infocommunications. *Science and Technology (PIC S&T).*

Chapter 10
Emerging Computing Platforms for Solving Complex Engineering Problems

Ankita
Birla Institute of Technology, Mesra, India

Sudip Kumar Sahana
https://orcid.org/0000-0002-2493-3695
Birla Institute of Technology, Mesra, India

ABSTRACT

The basic utilities such as electricity, gas, and water are the types of services that can be commercialized and delivered at the doorstep of a user. The user is not concerned about the location of the service provider or the method of the service delivery. Computing is a new utility whose nature has been intensely transformed to a traditional utility where users can utilize the services as per their requirement and irrespective of their location. The significant advancement in information and communication technology needs a type of computing, which is required to meet the everyday demands of the generic computing class of users. There are many computing models such as cluster computing, grid computing, cloud computing, and many more which have been used in solving complicated problems of science and engineering. This chapter discusses the latest computing platforms which can provide fundamental computing services to users for fulfilling their computational needs.

INTRODUCTION

The basic utilities such as electricity, gas, water are the type of services that can be commercialized and delivered at the doorstep of a user. The user is not concerned about the location of the service provider or the method of the service delivery. Computing is a new utility whose nature has been intensely transformed to a traditional utility where users can utilize the services as per their requirement and irrespective of their location. The modern times have seen a substantial increase in computing needs of the users. One

DOI: 10.4018/978-1-6684-3991-3.ch010

of the chief scientists of Advanced Research Projects Agency Network (ARPANET), Leonard Kleinrock, (Kleinrock, 2005) in 1969, stated "'As of now, computer networks are still in their infancy, but as they grow up and become sophisticated, we will probably see the spread of 'computer utilities' which, like present electric and telephone utilities, will service individual homes and offices across the country". The computing vision of Kleinrock came into light in the 21st century, which has seen a revolution in the computing industry. The users can easily access computing services on demand without making a heavy invest in IT organizations. Though a seamless, timely, hassle-free and ubiquitous computing service has proven really helpful in the modern times, it presents new challenges to the software developer for building software that can be provided as a service to the users without installing them in their individual systems. The significant advancement in Information and Communication Technology needs a type of computing, which is required to meet the everyday demands of the generic computing class of users. There are many computing models such as cluster computing, grid computing, cloud computing and many more which have been used in solving complicated problems of science and engineering. These are the key terms of this chapter as given below:

(i) *Distributed Computing*: It is a computing model in which software components are shared among multiple computing devices within a fixed area.
(ii) *Cluster Computing:* Cluster computing is a way of computing where a group of computers is combined together through some wired network in a manner, such that it behaves like a single entity.
(iii) *Grid Computing:* The grid computing model involves a distributed computing environment which involves multiple systems, including heterogeneous resources to find solution for complex engineering problems.
(iv) *Cloud Computing:* It is a service based computing model which involves the distribution of services (private or public) through mode of Internet.
(v) *Fog Computing:* It is an extended version of cloud computing model which places resources in logical locations between clouds and sources of data.
(vi) *Edge Computing:* The edge computing model attempts to reduce the bandwidth and latency by bringing neighboring the computing and data source together. The computing can be brought to any IoT device or user's computer which minimizes the cost of long distance relationship between customer and provider.
(vii) *Utility Computing:* This computing model provides resources and other services to the customer on demand basis. The cost of computation depends on the type of usage, which varies from service to service.

The chapter will discuss the latest trends in the computing platforms which can provide fundamental computing services to the users for fulfilling their computational needs. Also, it will explain various limitations or issues related to these computing platforms along with their possible solutions.

LITERATURE REVIEW

The ability to interconnect computers by making a worldwide network has made communication possible among different computer systems irrespective of their geographical location. With this computing ability, one can harness the seamless computing potential of distributed computing resources which is managed by several organizations. The author in (Buyya, Yeo, Venugopal, Broberg, & Brandic, 2009) has discussed latest trends in computing platforms and has revealed the importance of utility computing in the recent times. The authors in (Yousefpour et al., n.d.) has reviewed one of the latest computing platform, Fog computing and other computing platforms. The author has also compared the similarities and differences of different computing platforms and various challenges as well as future possibilities of research in fog computing. The author in (Q. Zhang, Cheng, & Boutaba, 2010) has provided a detailed review report on cloud computing paradigm – its benefits, challenges, limitations and also the prominent area of research in the cloud computing. The new computing platforms has a wide range of application areas. The author in (Ujjwal, Garg, Hilton, Aryal, & Forbes-Smith, 2019) has proposed a cloud computing model which provides natural hazard modeling systems to cope up with the natural disasters and address the challenges of the traditional Information Communication Technology (ICT). In (Ali, Mazen, & Hassanein, 2018), the author has proposed a hybrid computing model based on cloud computing which can be used in dealing with different challenges of e-government such as setting up new services and managing the increased inflation rate of data and application. The above application areas of the new computing paradigms has paved the way for the future research and defining new possibilities in the discussed subject. This chapter presents a deep insight into the current computing platforms that provide computing service as a basic utility to the users at minimum cost and time. It also compares the similarities between the computing platforms and explains the significance of emerging computing platforms over traditional computing platforms.

Types of Computing Platforms

Computing is an organized way of studying the algorithmic processes that can be used in displaying and converting information. The research trends show that the latest computing platforms have spanned across a wide variety of application areas of worldwide importance. The computing platforms discussed in this chapter are given below:

1. Distributed Computing

The distributed computing (Attiya & Welch, 2004) deals with the study of modeling and constructing of algorithms for the management of distributed systems. It is a way of computing where multiple programs are running across various computing systems on a single network and matches the needs for better performance in the scientific community as well as generic applications. The primary reasons behind the acceptance of this computing technology has been its performance, resource sharing, availability of computing systems and fault tolerance. The distributed computing allows the combination of multiple devices together to avail increased computation power, bandwidth and memory. There are several ways to implement distributed computing. A high-speed network can be used to interconnect a class of desktop computers, which can support computational power equivalent to supercomputers.

In a distributed system (van Steen & Tanenbaum, 2016), the autonomous computers are collected together, which seems to be a single intelligible system to its users. It has been in the computing industry for more than four decades now. The distributed computing (Belfiore et al., 2006) is strongly concerned with the aggregation of distributed resources and consolidated usage. Communication is an essential attribute of distributed computing because the exchange of data and information is required for computing systems to work together which is provided by the network. The communication among the networked computers in a distributed system occurs through message passing.

Working in a distributed computing environment has many advantages. Some of them are listed below:

(i) Availability of computing resources: A company working within a distributed computing framework is not restricted or dependent on a single server. Even if one server fails, there are other servers which will be in a working state and allow smooth functioning of the organization.
(ii) Data availability: Similarly, data or information exchange is common in a distributed computing platform, the failure of one storage device will not lead to loss of data.
(iii) Avoid system overloading: Many user requests can be redirected to several servers in a distributed computing rather than hitting a single server (machine).
(iv) Scalability: The design of a distributed computing network can be easily modified by adding new machines to the distributed network.
(v) Parallelism: During scientific computation, it is often required to break the complicated simulations into sub parts and run them in parallel to achieve faster results rather than performing the entire computation in the series mode.

Limitations of distributed computing network:

(i) The security and privacy become a non trivial issue in a distributed computing network because the information of users is placed in multiple locations (or servers).
(ii) Synchronization is difficult to achieve in the case systems attempt to read or write the same data which can cause possible errors.

2. Cluster Computing

The cluster computing is a form of computing where several working machines combine together to run an application or perform different jobs. An illusion of a single virtual machine (resource) is created for the user, such that the user believes that it is responding to a single system. A High Performance Cluster (Buyya, 1999) consists of a large number of computing devices called nodes (computing nodes and master node).

There are two types of cluster models- High availability cluster models (Gadir et al., 2005) and load balancing cluster models. The high availability cluster model uses implicit redundancy to grant service availability and resources in a continuous manner. In the case of a failure of a single cluster node, the application or services will be available in the other cluster nodes. This part of migrating an application from one node to another to continue the client application is called failover of an application to another node in the clustered network. The important application areas are binding of important missions, files and servers. A high availability cluster aims at providing uninterrupted services by duplicating computing servers and services through excessive restructuring of software and hardware.

Cluster load balancing (Werstein, Situ, & Huang, 2006), (Overeinder, Sloot, Heederik, & Hertzberger, 1996) is the second model where the systems or nodes are integrated in a manner such that all the client requests are channeled across the systems in an even fashion. The requests are redirected to the various nodes independently on the basis of a scheduler and a scheduling algorithm, but the computing devices are not jointly in a single operation. They are used in real time applications such as resolving cargo disparities from several input requests.

Some of the benefits of working in a cluster computing environment are given below:

(i) This model of computing is scalable such that any new node can be added to the clustered network for expanding the network size.
(ii) The cluster computing model also assures availability of a node within the cluster in the case of a node failure. It can also manage the abrupt increase in the number of jobs.

The drawbacks of the cluster computing model are given below:

(i) The initial set up of a cluster computing environment is costly because it requires installation of workstations and other computing devices.
(ii) Since there are a large number of systems in a clustered network, it requires proper monitoring and maintenance for proper functioning of these systems which expands the infrastructure.

3. Grid Computing

The grid computing (Berman, Fox, Hey, & Hey, 2003) has reformed the computing style of the scientists and engineers all over the world and the management of data and information services have also been modified by the user community. The recent technological advancements have made possible to inter-connect a large number of resources such as workstations, servers, storage devices and other services. These connected devices together can be treated as a single united resource to the user community. Further, the users are allowed to access these resources, in a seamless manner without any adherence to the physical location of the resource. This new model of computing is commonly known as "Grid Computing" (Foster & Kesselman, 2003b). The area of distributed computing has been broadened with the evolution of grid computing. During early 1990s, the grid computing (Buyya & Venugopal, 2005) emerged as a new working platform and a lot of research has been done on distributed computing and parallel programming.

The current innovation in the field of science and engineering has been boosted by the unparalleled growth of computing and communication technologies. The main aim is to provide a ubiquitous computing (Lyytinen & Yoo, 2002) platform using standardized protocols which allow universal access to the resources and coordinate the sharing of resources. (Software and hardware). The protocols which enable access and discovering of resources as well as communication among the resources are developed by a grid computing forum called Global Grid Forum (GGF) and Grid community.

In future, this technology can help the scientists and researchers to perform real time and practical analysis of complicated problems of science and engineering.

The advantages of the grid computing platform are given below:

1. The cost of deploying parallel machines has always been a concern for the developers. With grid computing (Foster, Kesselman, & Tuecke, 2001), we can have a collection of aggregated heterogeneous or homogeneous devices with tremendous computation power at low cost.
2. The intergrids are used to provide security in the grids which is similar to the level of the security provided by the LAN network.

3. It is a fault-tolerant computing platform (Foster, 2003), (Foster & Kesselman, 2003a). If a single machine fails or collapses, the grid system identifies it quickly and the job is redirected to a working machine which completes the idea of a flexible and invulnerable computing frameworks.

There are some limitations which cannot be overlooked while working on a Grid Computing platform.

(i) There are several small servers distributed across multiple administrative domains. Maintaining a synchronization between all the servers in a large and dynamic environment like Grid is really challenging and requires efficient tools for handling such situations.
(ii) Sometimes, political indifferences between countries restrict sharing of resources.
(iii) Building a safe and trusted environment for users is one of the primary concerns of service providers in the grid computing platform.

4. Cloud Computing

A cloud computing platform (Buyya et al., 2009), (Weiss, 2007) contains the features of both grid and cluster, but it has its own special characteristics such as virtualization, storage and compute services which makes it more powerful and popular computing platform. It (Q. Zhang et al., 2010), (Varghese & Buyya, 2018) is a next generation computing, data center whose nodes is virtualized using "hypervisor" technologies to fulfill the explicit service-level agreements. The cloud environment is formed using "negotiation" and can be accessed using SOAP (Simple Object Access Protocol) and REST (Representational State Transfer) Both of them are Web Service Technologies. The set up of the service model is based on the user requirements only which can be public, private, community or hybrid.

There are three types of service models (Kavis, 2014) provided by the cloud computing model: The first one is SAAS (Cusumano, 2010) which stands for Software as a Service. It is an on-demand software delivery model where the other (third) party hosts the applications for the user over the internet. A customer is given a network-dependent access to an application copy which is created by the supplier for the SAAS distribution. The second one is PAAS(Pahl, 2015) which stands for Platform as a Service in which the other (third) party provider provides the owned APIs to the developers for running the application in the given environment. It is totally based on the theory of virtualization, such that the resources can easily be increased or decreased according to the changes in business.The applications developed using PAAS are cost effective and sometimes free which also allow to take the advantage of a previously evolved cloud organization for transferring or moving a current application.

Once an application is generated, it is bound to that platform while the developers are allowed to generate any new application on the platform. With PAAS, one can work with the functions and programming language only provided by the chosen platform.

The last service model is IAAS (Laniepce et al., 2013) which stands for Infrastructure as a Service. They are composed of highly expandable computational resources which is totally a self-service model, providing several features liking handling of computers, networking and repository services. The applications, middleware and data are handled by the IAAS clients while IAAS is bound to handle the servers, repository, networking and virtualization. Out of the three computing models, the most flexible is the last one i.e. IAAS. The consumption of the hardware devices decides its purchase where several users can work on a single hardware equipment. IAAS are a boon to small companies and start ups which do not have enough funds to spend on the purchase of the hardware and software.

The advantages of cloud computing platform are given below:

In the recent years, the cloud computing has proved its influence on the computing industry. The survey report shows that companies which are investing in the technologies such as big data and cloud computing have a rapid revenue rate (approximately 53% faster) than the other companies which are still not investing in these technologies. Some of the advantages of using a cloud computing environment are listed below:

(i) The cloud environment provides scalability to the business or the organization implementing it. A business may scale high or low which will require changes in the operational and repository needs to meet the changing environment of the organization. Rather than getting bothered for installing or upgrading the requisite resources (given by the cloud provider), one can use their time on running their businesses effectively.
(ii) The cloud computing platform facilitates collaboration among the employees who are working on a single project across multiple places to access the same file required for the project.
(iii) "Work from home"- this flexibility is granted to the employees who are off-site or wants to work from home for some reasons. The only requirement is to have a good internet connection.
(iv) The cloud computing always makes its users aware with the latest technologies and upgrades to computing devices.
(v) The cloud computing environment also provides quality data being used at multiple locations by several employees working on a single project in an organization. While many people are accessing the same data, the consistency of the data can still be maintained and human errors can be avoided.
(vi) The cloud services are really helpful in providing recovery from the disaster situations (Ujjwal et al., 2019) such as power outages or even natural disasters.
(vii) Data loss can be prevented using a cloud-based server. The information which are saved locally to the computing systems are vulnerable to system failure or malfunctioning. The information which is uploaded to the cloud server are relatively safe and easy to access from anywhere provided you are having a working computing device and an internet connection.
(viii) Last but not the least, the cloud environment is environment-friendly and results in less usage of carbon imprints.

The cloud computing platform has some key issues (Branco Jr, de Sá-Soares, & Rivero, 2017) discussed below:

(i) The service provider has to keep low operational cost, make quick allocation and de-allocate the resource from the cloud as soon as the user finishes its job. Moreover, the resource provisioning decisions should be made online to satisfy the fast demand variations.
(ii) The improvement of energy-efficiency is a matter of concern for all the infrastructure developers in cloud computing.
(iii) Proper management and analysis of traffic is important so that network operators can make quick planning decisions by viewing the flow of traffic in the network.
(iv) Security (Subramanian & Jeyaraj, 2018), (Zissis & Lekkas, 2012) is an inevitable subject in cloud computing platform. The service providers solely depend on the infrastructure providers for achieving data security. Confidentiality and data integrity are the concerning issues of the cloud computing environment.

(v) The threats (Brohi, Bamiah, Brohi, & Kamran, 2012) in the virtual networking cannot be overlooked and it is necessary to establish a safe network among all the objects of the cloud.
(vi) A temporary downtime can happen in a cloud network because of several factors such as low internet connection, power loss and failure of maintenance of data computing center.
(vii) The stored data on cloud poses threat of theft because data is available online and security breach happens to be a potential hazard of cloud computing.

5. Fog Computing

The data have been the most crucial element of today's computing industry. The generation of data has been rising unexpectedly and it is estimated by CISCO that 2020 will see approximately 50 billion connected machines (Evans, 2011). Since there has been a tremendous amount of data, it is very difficult for the network designers to manage the speed, magnitude and movement of data. The bandwidth constraints,sometimes, do not allow to move large amount of data from the IoT machines to the cloud efficiently. Also, there are privacy concerns in some situations where it is not possible or feasible to send data to the cloud. Such issues of bandwidth constraints and privacy concerns can be handled using a new computing platform called fog computing platform (Hu, Dhelim, Ning, & Qiu, 2017), (Upadhyay, 2018). The void between the cloud and IoT devices is reduced if storage, computing, data management and networking is enabled on the nodes of the network within the close proximity of IoT devices. So the computation is not restricted to the cloud only, the computation also happens on the path of data movement from IoT to the cloud.

There are two platforms, horizontal and vertical platform in a fog computing model (Mouradian et al., 2017). The horizontal function allows the distribution of computing functions among several computing platforms and companies. A vertical platform is enabled for a specific type of applications, but there is no interaction between the platform to platform interaction in the different vertically based platforms. It also provides a flexible platform to fulfill the data-driven requirements of the developers as well as users. It is also aimed at providing support for the Internet of Things (IoT).

Fog computing (Yousefpour et al., n.d.) is very much similar to cloud computing, but there are still many dissimilarities that are addressed below:

(i) The mechanism of security in fog computing differs from that of cloud computing. In cloud computing, the structure of the security system is centralized located in the specified buildings for data centers of cloud whereas in fog computing, the security should be given at the edge or in the specified areas of fog nodes.
(ii) As mentioned above, the data centers of cloud are centralized where as nodes of the fog are mostly arranged in less centralized areas. Since the fog computing follows a decentralized approach, they can be used as fog computing nodes as well as the resources of the fog which act as a fog client.
(iii) The power utilization as well as the availability of the resources is high in the case of the cloud computing paradigm while both power utilization and resource availability is comparatively low in the case of fog computing (Mukherjee, Shu, & Wang, 2018).
(iv) There are large data computing centers in the cloud environment while fog computing operates on small gateways, switches and routers

(v) Internet connectivity is essential requirement to work in a cloud computing environment while fog computing does not require stable or continuous internet connection. (Can work with or without internet connection)

The advantages of fog computing are listed below:

(i) The fog nodes at the edge are densely populated because the data produced by the machines and sensors are acquired by the fog nodes at the network edge which helps in reducing the movement of data across the internet and helps in achieving quality services.
(ii) It also saved bandwidth by performing the computation of data and storage between the cloud and edge nodes.
(iii) The fog computing model facilitates mobility of various devices such as smart watch and smart phones while devices like traffic cameras at the network end remains static. A fog node can work as a static resource (the setup of a fog node in a coffee shop) or a mobile resource (placed on a moving train). The mobile devices can communicate with each other without transmitting the data to the cloud or any base station.
(iv) There are multiple nodes distributed across the geographical boundaries which have the capability to record and extract the position of the nodes to provide support for mobility. It helps in making real time decisions, performing faster data analysis and achieving improved location-oriented services.
(v) It also supports heterogeneity of nodes in the fog computing model. A wide variety of fog nodes are available such as servers, switches, gateways, routers and base stations. Virtualization is also supported by the fog computing model such that virtual network nodes and computational nodes are also operational here and can be treated as fog nodes.
(vi) It also secures data by providing means of data encryption, integrity check and measures of isolation to prevent security breaches.

The Fog Computing platform is basically designed to support Internet of Things. The IoT devices help the computing platform to reduce the gap between cloud and end devices. The major challenges of the fog model are written below:

(i) The two major issues in fog computing platform are- trust and authentication. A fog node is the leading element of fog computing model which is responsible for assuring secrecy and obscurity for the users. The end users must be assured of the authorization of the fog node because the fog node holds the responsibility of carrying out the global concealing process and non vicious actions. Hence it is highly required for the fog nodes to maintain a trust on each other.
(ii) It is very important to detect unauthorized access inorder to prevent the real data from hacking (Mandlekar, Mahale, Sancheti, & Rais, 2014).
(iii) A fog node is vulnerable to malicious attacks (Mahmud, Kotagiri, & Buyya, 2018) because fog nodes are loacated in different regions with some regions having weak security arrangements. In that case, a malicious user can misuse and alter readings or may cause IP Spoofing. Security issues (P. Zhang, Zhou, & Fortino, 2018) such as IP address spoofing and eavesdropping are needed to be addressed in fog computing platform.

(iv) It requires potential efforts to achieve data consistency. Every user is concerned about the confidentiality of their data (Koo, Shin, Yun, & Hur, 2016). Hence preservation of privacing is an important issue in fog computing model.

(v) The communication channel used for transmission between (a) IoT equipments and fog nodes and (b)among the fog nodes, must be a secure one. It may happen that fake messages are being comunicated because of the false information being sent by the intruder during the time of communication in the network (Soleymani et al., 2017).

(vi) Scheduling of task is complex here.

(vii) Currently, Service Level Agreements (SLAs) are defined for cloud computing infrastructure and not for the fog network.

(viii) The design of fog computing model considers only a few objectives such as offloading and load balancing and overlook other important objectives.

6. Edge Computing

Edge computing (Tu, Pop, Jia, Wu, & Iacono, 2019) is a recent computing paradigm that has gained immense popularity in the recent years and serves as a computing paradigm for latest technologies such as Internet of Things (IoT), vehicle to vehicle communication and provide services to the user community. The rise of IoT applications has led to the introduction of number of communicative devices. The data produced by these IoT devices are tremendous and hence processing of data is required to execute several IoT services. The edge computing (Ahmed & Rehmani, 2017) model is meant to provide computation of these data at the network edge rather than sending the unprocessed data to data centers. This leads to a reduction of the bandwidth and computational complexity required by the clouds. The edge computing forms a combination of IoT devices and cloud by cleaning, preprocessing and collecting the IoT data together using the services of the cloud which are located near to the IoT devices. The opportunistic edge computing (Olaniyan, Fadahunsi, Maheswaran, & Zhani, 2018) aims at creating scalable foundations using the resources given by the end users.

The advantages of edge computing are given below:

(i) It is able to handle important issues like latency and confidentiality. The cyber attacks are highly reduced and data security is enhanced because of the decentralized structure of the edge computing model.

(ii) Since connected devices need not to wait for the centralized platform to implement a service, the availability of service is relatively higher than cloud computing.

(iii) The operational cost is reduced because the data is preprocessed at the network edge rather than sending it to the cloud data centers. Hence, it will also reduce the infrastructure cost.

(iv) The edge computing model is highly scalable and can form hybrid structures with cloud computing infrastructure.

There are still open issues or challenges (Cao, Zhang, & Shi, 2018), (Varghese, Wang, Barbhuiya, Kilpatrick, & Nikolopoulos, 2016) that need to be addressed in Edge Computing.

(i) The resources and the services located at the edges of the network demands efficient resource discovery mechanisms to locate them because of the abrupt amount of devices present on the edges of the network.
(ii) The data processed at the network edge is only a subset of data and not the complete data. Hence it becomes important for the organizations decide the acceptable level of information loss.
(iii) The hardware requirements of edge computing infrastructure is high.
(iv) Efficient fault recovery mechanisms are needed to deal with improper faults on the edge node.
(v) Partitioning of tasks is not a new task in a distributed computing environment, but it becomes challenging in edge computing platform. A scheduler is required, such that it can schedule partitioned tasks over edge nodes.
(vi) The risk factor associated with the service provider and the organizations owning these devices must be coherent. It is important to build a cost efficient computational model such that edge nodes will be accessible by the intended user community.
(vii) The conventional authentication protocols are no longer suitable in emerging computing platforms because of the heterogeneity of the computing nodes

7. Utility Computing

Nowadays, computing (Fortino & Palau, 2012) has become a necessary part in carrying out our daily activities such as use of computers or mobile phones to communicate with other people, reading national and international news, managing finances and many more. The unexpected need for computing has posed challenges to the scientists and service providers to provide secure, available as well as an economical service to the every user. The computing is required, just like other utilities such as water and electricity because of the ease of the present utility services. A type of computing is needed where the complexity as well as the cost of maintenance is divided equally among the users (customers). The utility computing (Mondal & Sarddar, 2015), (Adhikari, Das, & Mukherjee, 2016) aims at providing a technology which will allow the organizations to deliver as well as employ resources and other functionalities as per the demand. It supports a type of infrastructure which intends to provide the services of information technology as and when demanded. Utility computing (Canali, Rabinovich, & Xiao, 2005) is different from outsourcing because outsourcing deals with the location and the manager of the resources. But the utility computing is concerned with the management, utilization and consumption of the resource. The utility computing infrastructure is applicable to the corporate as well as outsourcing data center. It helps in breaking down the massive IT foundation into multiple independent segments. A classification is made in context of the business process that are supported by these independent segments. It will help those resources which are not being used (idle) because they have been supporting those business operations which are outdated and currently not functional. Now these resources can be provided to other business operations as needed. These business resources can be turned on or off as per requirement.

Utility computing offers several benefits which are listed below:

(i) It is cost effective. The enhanced resource utilization helps the owners of the IT infrastructures by decreasing the cost of management of the resources through mutual sharing of resources among the users in the company.
(ii) The need for buying resources is mitigated because of the improved capability to match the resource specification with the changing space specifications over a time period.

(iii) The complexity of a utility model is reduced because of the improved management and maintenance of the system.

Some of the limitations of the utility computing model are given below:

(i) The utility computing machines are vulnerable to hacking. A hacker could try to access the important files of the clients which is a potential threat towards the privacy of the clients.
(ii) Reliability is also a concern in the utility computing model because an organization facing a financial crisis may curtail its services and clients will be deprived of the services for which they are spending money.

SOLUTIONS AND RECOMMENDATIONS

Providing a hassle free computing services to millions of users is not easy and hence poses several challenges for the providers and the software developers. Some of the solutions and recommendations in dealing with such challenges, issues or problems are presented below:

(i) It is required to define appropriate SLAs along with SLAs management techniques for fog computing network.
(ii) A multi-objective fog computing model that can cover many objectives such as latency, security, availability, bandwidth and energy should be designed to incorporate several objectives together.
(iii) In order to facilitate high-speed users, it is required to develop fast protocols and machine learning algorithms.
(iv) A scheduler is required, such that it can schedule partitioned tasks over edge nodes. Also, a mechanism is needed for the receivers to verify the exactness of the partitioned tasks.
(v) There must be some new design of authentication and trust protocols that can handle the heterogeneity of the nodes.

CONCLUSION

The emerging computing platforms are transforming the view of information technology and turning the vision of utility computing into reality. The rapid growth of the Internet technology and Internet of Things (IoT) has increased the significance of the latest computing paradigms such as fog, cloud and edge computing. This chapter discusses some of the emerging computing platforms and their application areas. The challenges of these computing paradigms are also articulated in the chapter, which covers the challenges and limitations of individual computing platforms. The current computing platforms are serving applications of national and international importance, such as disaster management, e-governance and many more. The technologies are still in their growing stage and researchers have opportunities to make new advancements that will bring promising benefits to mankind in the field of medicine, energy utilization, gaming and many more.

FUTURE WORK

There is yet a lot to be explored in the area of emerging computing paradigms. The IoT is being supported by the latest computing platforms such as fog computing and edge computing. In future, billions of users and devices will be connected through the IoT that will benefit the entire working community. A potential computing platform (such as fog computing) is required to manage the big data, generated by Internet of Things, are often time-sensitive and critical terms of security. The clouds are present everywhere and hence it is expected to connect them and use them for trading services by forming market-based global clouds. In future, the emerging computing platforms such as fog, cloud, edge and other computing paradigms will be required to assemble together to form a unified, secure and interoperable computing platform for providing services to the individual users and organizations.

REFERENCES

Adhikari, M., Das, A., & Mukherjee, A. (2016). Utility Computing and Its Utilization. In *Emerging Research Surrounding Power Consumption and Performance Issues in Utility Computing* (pp. 1–21). IGI Global. doi:10.4018/978-1-4666-8853-7.ch001

Ahmed, E., & Rehmani, M. H. (2017). *Mobile edge computing: Opportunities, solutions, and challenges*. Elsevier.

Ali, K. E., Mazen, S. A., & Hassanein, E. E. (2018). A proposed hybrid model for Adopting Cloud Computing in E-government. *Future Computing and Informatics Journal*, *3*(2), 286–295. doi:10.1016/j.fcij.2018.09.001

Attiya, H., & Welch, J. (2004). *Distributed computing: Fundamentals, simulations, and advanced topics* (Vol. 19). John Wiley & Sons. doi:10.1002/0471478210

Belfiore, J., Campbell, D., Capps, S., Cellini, S., Fitzgerald, C., Gundotra, V., … Rudder, E. (2006). *Distributed computing services platform.*

Berman, F., Fox, G., Hey, T., & Hey, A. J. (2003). *Grid computing: Making the global infrastructure a reality* (Vol. 2). John Wiley and Sons. doi:10.1002/0470867167

Branco, T. Jr, de Sá-Soares, F., & Rivero, A. L. (2017). Key issues for the successful adoption of cloud computing. *Procedia Computer Science*, *121*, 115–122. doi:10.1016/j.procs.2017.11.016

Brohi, S. N., Bamiah, M. A., Brohi, M. N., & Kamran, R. (2012). Identifying and analyzing security threats to virtualized cloud computing infrastructures. *2012 International Conference on Cloud Computing Technologies, Applications and Management (ICCCTAM)*, 151–155. 10.1109/ICCCTAM.2012.6488089

Buyya, R. (1999). High performance cluster computing: Architectures and systems (volume 1). *Prentice Hall.*

Buyya, R., & Venugopal, S. (2005). A gentle introduction to grid computing and technologies. *Database*, *2*, R3.

Buyya, R., Yeo, C. S., Venugopal, S., Broberg, J., & Brandic, I. (2009). Cloud computing and emerging IT platforms: Vision, hype, and reality for delivering computing as the 5th utility. *Future Generation Computer Systems*, *25*(6), 599–616. doi:10.1016/j.future.2008.12.001

Canali, C., Rabinovich, M., & Xiao, Z. (2005). Utility computing for Internet applications. In *Web Content Delivery* (pp. 131–151). Springer. doi:10.1007/0-387-27727-7_6

Cao, J., Zhang, Q., & Shi, W. (2018). Challenges and opportunities in edge computing. In *Edge Computing: A Primer* (pp. 59–70). Springer. doi:10.1007/978-3-030-02083-5_5

Cusumano, M. A. (2010). Cloud computing and SaaS as new computing platforms. *Communications of the ACM*, *53*(4), 27–29. doi:10.1145/1721654.1721667

Evans, D. (2011). The internet of things: How the next evolution of the internet is changing everything. *CISCO White Paper, 1*(2011), 1–11.

Fortino, G., & Palau, C. E. (2012). *Next Generation Content Delivery Infrastructures: Emerging Paradigms and*. Academic Press.

Foster, I. (2003). The grid: Computing without bounds. *Scientific American*, *288*(4), 78–85. doi:10.1038cientificamerican0403-78 PMID:12661319

Foster, I., & Kesselman, C. (2003a). *The Grid 2*. Morgan Kauffman.

Foster, I., & Kesselman, C. (2003b). *The Grid 2: Blueprint for a new computing infrastructure*. Elsevier.

Foster, I., Kesselman, C., & Tuecke, S. (2001). The anatomy of the grid: Enabling scalable virtual organizations. *International Journal of High Performance Computing Applications*, *15*(3), 200–222. doi:10.1177/109434200101500302

Gadir, O. M., Subbanna, K., Vayyala, A. R., Shanmugam, H., Bodas, A. P., Tripathy, T. K., … Rao, K. H. (2005). *High-availability cluster virtual server system*. Academic Press.

Hu, P., Dhelim, S., Ning, H., & Qiu, T. (2017). Survey on fog computing: Architecture, key technologies, applications and open issues. *Journal of Network and Computer Applications*, *98*, 27–42. doi:10.1016/j.jnca.2017.09.002

Kavis, M. J. (2014). *Architecting the cloud: Design decisions for cloud computing service models (SaaS, PaaS, and IaaS)*. John Wiley & Sons. doi:10.1002/9781118691779

Kleinrock, L. (2005). A vision for the Internet. *ST Journal of Research*, *2*(1), 4–5.

Koo, D., Shin, Y., Yun, J., & Hur, J. (2016). A Hybrid Deduplication for Secure and Efficient Data Outsourcing in Fog Computing. *2016 IEEE International Conference on Cloud Computing Technology and Science (CloudCom)*, 285–293. 10.1109/CloudCom.2016.0054

Laniepce, S., Lacoste, M., Kassi-Lahlou, M., Bignon, F., Lazri, K., & Wailly, A. (2013). Engineering intrusion prevention services for iaas clouds: The way of the hypervisor. *2013 IEEE Seventh International Symposium on Service-Oriented System Engineering*, 25–36. 10.1109/SOSE.2013.27

Lyytinen, K., & Yoo, Y. (2002). Ubiquitous computing. *Communications of the ACM*, *45*(12), 63–96.

Mahmud, R., Kotagiri, R., & Buyya, R. (2018). Fog computing: A taxonomy, survey and future directions. In *Internet of everything* (pp. 103–130). Springer. doi:10.1007/978-981-10-5861-5_5

Mandlekar, V. G., Mahale, V., Sancheti, S. S., & Rais, M. S. (2014). Survey on fog computing mitigating data theft attacks in cloud. *Int. J. Innov. Res. Comput. Sci. Technol*, *2*, 13–16.

Mondal, R. K., & Sarddar, D. (2015). Utility Computing. *International Journal of Grid and Distributed Computing*, *8*(4), 115–122. doi:10.14257/ijgdc.2015.8.4.11

Mouradian, C., Naboulsi, D., Yangui, S., Glitho, R. H., Morrow, M. J., & Polakos, P. A. (2017). A comprehensive survey on fog computing: State-of-the-art and research challenges. *IEEE Communications Surveys and Tutorials*, *20*(1), 416–464. doi:10.1109/COMST.2017.2771153

Mukherjee, M., Shu, L., & Wang, D. (2018). Survey of fog computing: Fundamental, network applications, and research challenges. *IEEE Communications Surveys and Tutorials*, *20*(3), 1826–1857. doi:10.1109/COMST.2018.2814571

Olaniyan, R., Fadahunsi, O., Maheswaran, M., & Zhani, M. F. (2018). Opportunistic edge computing: Concepts, opportunities and research challenges. *Future Generation Computer Systems*, *89*, 633–645. doi:10.1016/j.future.2018.07.040

Overeinder, B. J., Sloot, P. M., Heederik, R. N., & Hertzberger, L. O. (1996). A dynamic load balancing system for parallel cluster computing. *Future Generation Computer Systems*, *12*(1), 101–115. doi:10.1016/0167-739X(95)00038-T

Pahl, C. (2015). Containerization and the paas cloud. *IEEE Cloud Computing*, *2*(3), 24–31. doi:10.1109/MCC.2015.51

Soleymani, S. A., Abdullah, A. H., Zareei, M., Anisi, M. H., Vargas-Rosales, C., Khan, M. K., & Goudarzi, S. (2017). A secure trust model based on fuzzy logic in vehicular ad hoc networks with fog computing. *IEEE Access: Practical Innovations, Open Solutions*, *5*, 15619–15629. doi:10.1109/ACCESS.2017.2733225

Subramanian, N., & Jeyaraj, A. (2018). Recent security challenges in cloud computing. *Computers & Electrical Engineering*, *71*, 28–42. doi:10.1016/j.compeleceng.2018.06.006

Tu, W., Pop, F., Jia, W., Wu, J., & Iacono, M. (2019). *High-Performance Computing in Edge Computing Networks*. Elsevier. doi:10.1016/j.jpdc.2018.10.014

Ujjwal, K. C., Garg, S., Hilton, J., Aryal, J., & Forbes-Smith, N. (2019). Cloud Computing in natural hazard modeling systems: Current research trends and future directions. *International Journal of Disaster Risk Reduction*.

Upadhyay, N. (2018). Fogology: What is (not) Fog Computing? *Procedia Computer Science*, *139*, 199–203. doi:10.1016/j.procs.2018.10.243

van Steen, M., & Tanenbaum, A. S. (2016). A brief introduction to distributed systems. *Computing*, *98*(10), 967–1009. doi:10.100700607-016-0508-7

Varghese, B., & Buyya, R. (2018). Next generation cloud computing: New trends and research directions. *Future Generation Computer Systems, 79*, 849–861. doi:10.1016/j.future.2017.09.020

Varghese, B., Wang, N., Barbhuiya, S., Kilpatrick, P., & Nikolopoulos, D. S. (2016). Challenges and opportunities in edge computing. *2016 IEEE International Conference on Smart Cloud (SmartCloud)*, 20–26. 10.1109/SmartCloud.2016.18

Weiss, A. (2007). Computing in the clouds. *Networker (Washington, D.C.), 11*(4), 16–25.

Werstein, P., Situ, H., & Huang, Z. (2006). Load balancing in a cluster computer. *2006 Seventh International Conference on Parallel and Distributed Computing, Applications and Technologies (PDCAT'06)*, 569–577. 10.1109/PDCAT.2006.77

Yousefpour, A., Fung, C., Nguyen, T., Kadiyala, K., Jalali, F., Niakanlahiji, A., … Jue, J. P. (n.d.). *All One Needs to Know about Fog Computing and Related Edge Computing Paradigms*. Academic Press.

Zhang, P., Zhou, M., & Fortino, G. (2018). Security and trust issues in Fog computing: A survey. *Future Generation Computer Systems, 88*, 16–27. doi:10.1016/j.future.2018.05.008

Zhang, Q., Cheng, L., & Boutaba, R. (2010). Cloud computing: State-of-the-art and research challenges. *Journal of Internet Services and Applications, 1*(1), 7–18. doi:10.100713174-010-0007-6

Zissis, D., & Lekkas, D. (2012). Addressing cloud computing security issues. *Future Generation Computer Systems, 28*(3), 583–592. doi:10.1016/j.future.2010.12.006

Chapter 11
Data-Driven Android Malware Analysis Intelligence

Ketaki Anandkumar Pattani
https://orcid.org/0000-0002-7942-8637
Institute of Advanced Research, Gandhinagar, India

Sunil Gautam
Institute of Advanced Research, Gandhinagar, India

ABSTRACT

Android OS powers the majority of the market share. Malware acts as stimuli to the vulnerabilities in Android devices as it affects a huge amount of user data. Users' data is at high risk when it comes to attacks through varied types of malware. Also, mutations in malware have brought up newer variants in them. Malware families have been expanding, thereby making analysis and classification diverse. Mainly classified into static, dynamic, and alternative or hybrid analysis, the field of malware analysis is facing many repercussions. The development of malware is endless and hence calls for intelligent and self-learning approaches in this regard. However, more distinct techniques are in need and can be served by integrating intelligent and analytical capabilities. This chapter involves a fourfold approach with major contributions to review existing Android malware analysis techniques, intelligent techniques for Android malware detection, determination of future challenges and need of security in this direction, and finally, analyzing possible defense mechanisms possible in this regard.

INTRODUCTION

In current scenario, Android Operating System (OS) prevails over the global mobile market. By 2021, Android devices influenced the global mobile market with a share of around more than 83% according to the statistical information provided by International Data Corporation (IDC) corporate. It is also predicted to notice a rise by 2025 (Ceci, 2021). These Android devices may store highly confidential information of users like SMS, contacts banking information, Personal Identification Number (PIN), location specific information, photographs, videos and many others (Faruki et. al, 2014, p. 998). The extensive capabili-

DOI: 10.4018/978-1-6684-3991-3.ch011

ties along with numerous connectivity options have made Android devices popular amongst users and so an ideal target of attackers. A boom in the usage of Android thereby implies a growing concern for security and privacy of users and their data. Security in Android becomes a prime concern as bridging day-to-day communication; it also experiences threats, penetration and insecurities (Lu et.al, 2016, p. 3). Figure 1 shows a high rise in the number of Android applications (Statistica et.al, 2020).

Figure 1. Number of android applications 2009-21
(Statistica, 2020)

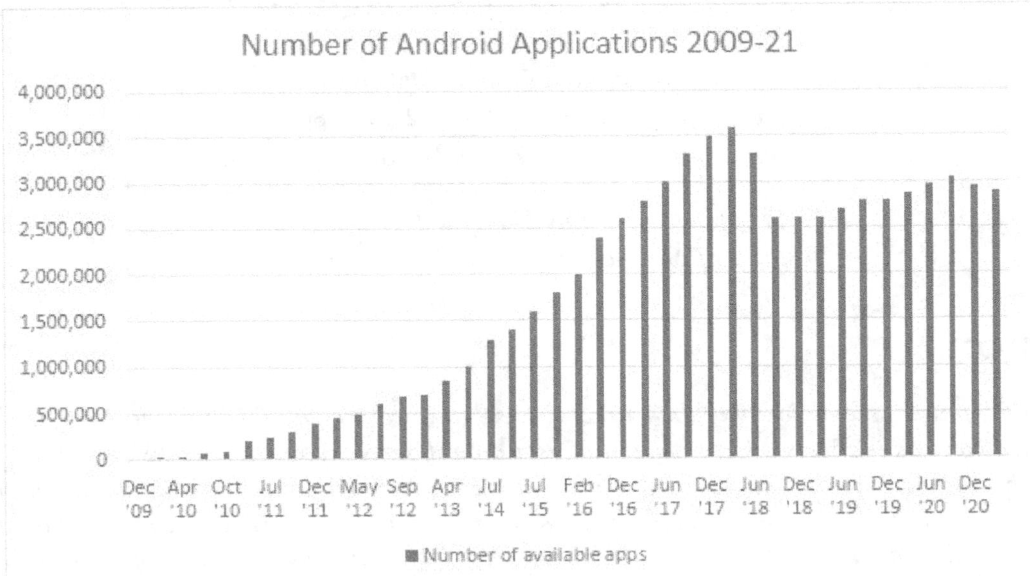

On the other hand if we observe the status of malware development, it has noticed a huge rise of more than 46% detections and 400% rise since 2010 (Li et. al, 2018, p. 3216). Several researchers and the private anti-malware industries have also realized the vulnerabilities in Android devices and attack scopes. Mobile malwares are expected to evolve along with increasing exposure to these devices. As an example more the people will install loosely checked mobile applications from various sources, higher are the chances of them being infected by malicious code. Malwares come in different forms like Trojans, spywares, risk wares, ransom wares, viruses, worms etc. and may be grouped into certain malware families based on their characteristics. It is likely that newer malwares may also be uncategorized or are developed through mutations. Malware mutations are not novel malwares but are created by automatic grouping, collection and management of various types of malwares (Yen et. al, 2019, p. 109). There are several state-of-art tools and techniques that are being used for detection of these malwares such as Taintdroid (Enck et. al, 2014, p. 1), Mockdroid (Beresford et. al, 2011, p. 49), Crowdroid (Gordan et. al, 2015, p. 110), Droidscope (Yan et. al, 2012, p. 569) and IntelliDroid (Wong et. al, 2016, p. 21) to name a few. These detection approaches are mainly classified based on their nature. However, time demands futuristic fine-grained and robust detection mechanisms for the ever changing threats.

Figure 2. Development of new android malware worldwide June 2016-March 2020 (Faruki et. al, 2014)

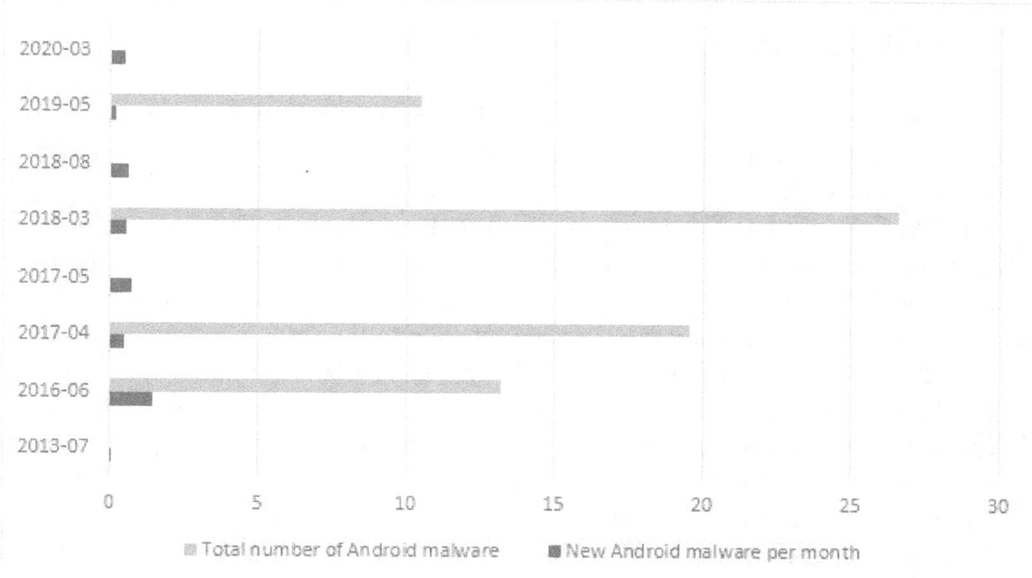

Data is a principal concern when it comes to mobile security or malware analysis. Now, the mobile devices are more presumably holding the privacy and security based information. The user's information may be personal, professional or financial when it comes to privacy. Attackers have an intent to breach the privacy and observe or leak this information. In either of the cases, user may face critical issues. Figure 2 shows the number of Android malwares and new malwares encountered every month. International standards determine the information security management and requirements at organizational level also. However, Android being open source and due to its popularity, becomes a significant base of such information. In these days, where every few seconds a new malware is entering the cyber space, the need for intelligent data-driven malware analysis becomes utmost essential. The characteristic of malwares, detection mechanisms and their impact on existing systems needs to be studied. Also, tools and technologies supporting detection and mitigation has to be focused.

This chapter entails a fourfold approach detailing Android malware analysis, existing work in this direction and intelligent analysis systems, challenges and need for security and detection mechanisms on whole. The current work begins with the role of data security in Android malwares and various existing analysis techniques along with its classification including static, dynamic and alternative or hybrid approaches section 2. Static malware analysis includes permission based analysis, Dalvik byte code based analysis and hybrid static analysis technique. Whereas, dynamic analysis includes anomaly based analysis, taint analysis and emulation based analysis techniques. Hybrid analysis involves both static code analysis as well as dynamic behavioral analysis. Thereafter, in section 3 machine learning based defense mechanisms are explained. In section 4, the chapter entails the need of intelligent analysis systems and its detection capabilities determine the requirement of automated detection systems in this data-driven world. Following this, are the real time challenges and issues faced to maintain data privacy and security in the current scenario. Finally, concluding section entails possible developments in the field through intelligent analysis and its scope. The chapter provides an overview of the current order and evolution of data-driven Android malwares and its future projection.

Traditional Android Malware Analysis Techniques

The term malware in generic way refers to the malicious paradigms involving programs or files that can be infectious or detrimental to the user privacy and security. Malware is represented by its capacity to replicate, propagate, execute itself, and damage a computer system. Confidentiality, Integrity, and Availability (CIA) form the basis of privacy and security in any system. The confidentiality, integrity, and availability of information can all be harmed if a computerized system is corrupted. G. McGraw and G. Malware is defined as "any code added, changed, or removed from a software system in order to intentionally cause harm or subvert the intended function of the system" (Wong et. al, 2016, p. 21) (Kramer et. al, 2010, p. 105). Malwares characteristics are categorized into 'Malware Families' and the malwares that have same characteristics belong to the same malware family. This encompasses varied malware types including worms, Trojans, virus, backdoors, etc. The purpose of each malware may vary based on the intention of its creation as fraud, privacy leak, ransom, (Distributed Denial of Service) DDoS victim or send spam emails. The penetration techniques are used by the attackers to bypass the various detection techniques and state-of-art tools. Following are the different types of penetration techniques (Statistica, 2020):

- Repackaging with malware: This technique develops he view of some popular app by disassembling the application and then adds its malware content to the app and again reassembles it and puts it on less monitored 3rd party market. For example: Amazon Application Store.
- Drive by Download: Here, as user tries to download any resource unintentionally a malware is downloaded in the background without the user being notified. Developers use the 'Non-Compatible Android Trojan' to perform this task.
- Dynamic Download of Payloads: Here, the encrypted payloads are executed at runtime to perform malicious activities. Certain malwares are also used to download these payloads at run time in order to fool static analysis tool.
- Malware with Stealing Techniques: Since direct analysis of android app causes battery and resources issues, certain techniques are used to obfuscate anti-malwares as key permutation obfuscation approach, dynamic loading of data for obfuscation, native code execution and stealth of data etc. to attack victim's device.

Malware Analysis is therefore required to determine the scope and behavior of these malwares. The analysis techniques based on software engineering or reverse engineering approaches are termed as traditional malware analysis and broadly categorized as static analysis, dynamic analysis and hybrid analysis.

STATIC ANDROID MALWARE ANALYSIS

Here, the maliciousness of malware is checked by analyzing source code without executing it. Static Malware Analysis just analyzes the code through all possible paths without considering the dynamically loaded code (Nath et. al, 2014, p. 440). Example: Certain behavior of the system seen on occurrence of particular event. They are further categorized as signature based technique, permission based technique, Dalvik bytecode-based technique, and the hybrid technique.

A. Signature Based Malware Analysis

Signature based analysis is used highly by commercialized services to have malware detection. Each malware has its own unique signature based on its syntactic and semantic features using which it can be recognized and its characteristics can be determined. So, if new application comes having similar type of malware, it is detected using its signature. However, such methods are very easy to obfuscate if there occurs new or unforeseen variants then it becomes difficult to recognize it. Further, due to pre-defined database of signatures there may be results with undetected outcome.

AndroSimilar (Faruki et. al, 2013, 152) uses Signature based analysis wherein malware is introduced to existing apps and then repackaged. The aforementioned tool detects such obfuscating malware-infected gadgets. It is statistical feature signature based method to detect zero-day variants of the already known malware. More than 60% of genuine results with proper detections are provided by the instrument. Droid Analytics (McGraw & Morrisett, 2000, p. 33), a tool that follows the same signature-based methodology as before, but it extracts and analyses each application at the operation-code or instruction-syllable level. It accomplishes this by creating three different levels of signatures. Using code tracing at the Application Programming Interface, class level, and application level, the levels are identified as Method Level, Class Level, and Application Level. The detection score based on resemblance does not prove to be completely accurate, and there may be false positives in the data.

B. Permission Based Malware Analysis

Android's fundamental structure is contained in AndroidManifest.xml, which contains all of the permissions that an application requires. Most of the time, the permissions provided to an application are in excess of what is necessary. The issue with this technique is that various applications ask for innocuous permissions and with those permissions perform malicious activities such as in SoundComber (Schlegel et. al, 2011, p. 17) and Sonic Evasion (Pattani & Gautam, 2021, p. 1). Permission analysis, which depicts a list of susceptible permission combinations and then checks for their presence, is the most extreme way for malignity analysis. Since the border between the permissions sought and those really necessary for benign and malicious code is so narrow, this might result in a lot of false positives. This implies that there must be another level above it for malignity analysis.

There may be programmes that seek more dangerous permission combinations than are actually necessary. In Stopaway (Felt et. al, 2011, p. 627), to establish the amount of vulnerability, the work analyses the code statically to trace API calls and permissions. In a total of 940 results, one-third have over-privileged rights, according to the analysis. API calls made or performed by apps using java reflections, on the other hand, go unnoticed. KIRIN (Enck, et. al, 2009, p. 235), which is a certification tool based on permission analysis. During the installation phase, this is a fairly light-weight technique. Data security criteria have been pre-defined here, as well as a comparison with the permissions requested by the application is done. Due to the failure in following the rules, the application is considered harmful. However, the biggest concern with KIRIN is when it flags some trustworthy programmes as vulnerable, indicating that they may not be safe to use. DroidMat (Wu et. al, 2012, p. 62) is another tool that uses the K-means clustering technique to identify malware and determine whether programmes are benign or potentially susceptible. The system here pulls manifest-based information such as permissions, intent communication, and API call tracking. For analysis, the tool employs the KNN algorithm. DroidMat, on the other hand, can't detect dynamically loaded malicious activities because it runs in a static mode.

C. Dalvik Static Bytecode Analysis

Applications developed in Android are converted to bytecode and further translated into Dalvik bytecode. Here, the feature based behavior of the application is analyzed based on the bytecode. The analysis occurs at the instruction level. This is very time consuming and also high storage is required which might not prove to be efficient for Android.

Scandal (Kim et. al, 2012, p. 1) developed by Jimyung Kim, the tool creates analysis of Dalvik Bytecode and then detects privacy leakage. It detected 11 out of 90 Android applications malicious. It analyzes all possible paths from source to remote server by branch based approach, method invocation for detection and approach involving jump instructions tracking. It does not support reflections calls so they must be manually written. Soot (Arnatovich et. al, 2018, p. 12382), developed by McGill University's Sable Group, was a key tool in this approach. Soot can convert Android apps into a variety of intermediate formats, including Baf, jimple, Shimple, and Grimp. Dexpler is an upgraded version of Soot. Fuchs in ScanDroid (Fuchs et. al, 2009), proposed static analysis during installation and mapping of data flow-based analysis. Its nobility is decided by taking into account data flow analysis at run-time, behavioral factors, and permissions. However, Android Runtime (ART) is Dalvik's successor, and it utilizes the same bytecode and .dex files and not .odex files, with the aim towards performance gains that are transparent to end users, largely replacing Dalvik.

D. Hybrid Static Analysis

The sophisticated malware can even hide behind static class code as well as in metadata, which to a degree can be analyzed by combining the static approaches. Some research has been carried out to augment with the hybrid static analysis by examining both the AndroidManifest.xml file and the disassembled classes.dex code. To describe the pattern of Android malicious samples, Sato et al. (Sato et. al, 2013, p. 23) analyzed a variety of characteristics (including permissions, intent filters, process names, and the number of redefined permissions). Arp et al. (2014, p. 23) presented Drebin, a lightweight approach for detecting Android malware samples directly on the device, in 2014. To classify Android apps, Drebin retrieved four types of feature sets from AndroidManifest.xml and four more from disassembled classes. dex files. FlowDroid (Arzt et. al, 2014, p. 259) is another unique and accurate static taint analysis tool. FlowDroid, unlike earlier efforts, modeled Android's lifecycle or callback methods to lower the false alarm rate.

In all, the static malware analysis technique can be used to analyze the application code but not the dynamic execution associated with it. Also, it analyzes all the flow paths of the application but it may overestimate code execution paths. Static malware analysis can detect the basic malware concerns performed on code but may go wrong in case of composite malwares. Also, a large number of mobile applications are being developed and have various malicious content, increasing every day. In such circumstances, it overburdens the static analysis technique and it may not be possible to deploy such mechanisms for numerous complex malwares.

DYNAMIC ANDROID MALWARE ANALYSIS

This approach examines the application during execution so that they can even detect malwares with obfuscating approaches. Dynamic analysis provides run-time analysis and thus detects dynamic loading concerns. It runs over single path at a time. There are various dynamic analysis approaches used commercially to realize the execution of dynamically loaded code.

A. Anomaly Based Detection

It is also known as Behavioral Malware Detection and relates to the behavior-based examination of apps. Anomaly-based detection may be divided into two stages: training phase and detection phase. The detector tries to learn from its evident actions throughout the training phase. The ability to identify zero-day assaults is a key advantage of anomaly-based detection.

CrowDroid (Burguera et. al, 2011, p. 15) is one such dynamic analysis tool. The application's information are gathered using a trace-based method. The crowdsourcing software creates a logging content file and sends it to a distant server, where a cluster formation technique is used. The findings amassed in the database are referred to as generated. If there are significant system calls, it may label a safe innocent application as malware. Andromaly (Shabtai et. al, 2012, p. 161) is another detection technique based on the behavioral analysis of an application in use. Changes in specific metrics are constantly monitored, and machine learning is utilized to enhance the system's status and detection capabilities. This allows for the identification of benign or susceptible applications. Zhao et. al (2011, p. 158), in AntiMalDroid, uses the SVM method to perform a dynamic study of the application in order to follow its performance. First, an analysis is performed to check if the programme is benign or malicious, after which it is placed in a learning module and a signature is created. The signature is the foundation and is utilized every time an application is checked for vulnerabilities. However, there are false positives when even a safe application shows uncertain behaviors such as battery drainage and may be termed as malicious.

B. Taint Analysis

The tainting is dynamic in this case, and it enables for a system-wide information flow tracking mechanism. This might entail the monitoring and tainting of all system resources. Here the tool performs system analysis in which the flow paths are discovered from source to sinks. These paths are then observed to find the leakage of based on the way information are passed within variables. Each variable through which sensitive information is passed is marked as tainted thereby tracing the paths.

Taintdroid (Enck et. al, 2014, p. 1) is a tracking tool that enables system-wide information flow tracking. It can track numerous sources of sensitive data at the same time, such as the camera, GPS, and microphone, and detect data leaks in third-party developer apps. It classifies sensitive data, maintains track of it, and notifies the app when contaminated data is transferred from the device to the sink. TaintDroid has certain flaws, such as its inability to properly counter specific leaks. XManDroid (Bugiel et. al, 2011) examines application communication links to ensure that they adhere to a set of system policies. XManDroid can prevent privilege escalation attacks such as collusion attacks (e.g., Soundcomber) that take use of the Android core application's hidden channels. Furthermore, a new notion for storing XManDroid's judgments has been included, which can be immediately integrated into Android's standard permission system.

C. Emulation Based Detection

In emulation based technique, system can detect the behavior of malware as well as the sequence of malware. This technique is used to minimize the time of detection. It is also used to detect both the polymorphic malware and even the metamorphic malware. DroidScope () is based on virtual machine introspection and was created by Yan. DroidScope has greater rights than malware programmes since it monitors the whole operating system while remaining outside of the execution context. It also keeps an eye on the Dalvik semantics, allowing privilege escalation attacks on the kernel to be identified. It's based on Quick EMUlator (QEMU). This method was used to detect DroidDream and DroidKungFu.

In all, dynamic malware analysis technique the malwares can be detected for dynamic content loading and execution. Also, behavioral analysis can discover the malicious content missed out during static analysis. However, it does not cover all the paths as in static analysis and thereby executes only one specific path at a time limiting the area of detection.

HYBRID ANDROID MALWARE ANALYSIS

Hybrid malware analysis refers to the combination of both static and dynamic malware analysis techniques. Here, it analyzes not only source code, but also the behavior of the program or application. It combines static and dynamic analysis to acquire information about malware. Security researchers may benefit from both static and dynamic assessments by adopting hybrid analysis (Roundy et. al, 2010, p. 317). As a result, the capacity to accurately detect harmful applications has improved. Both analyses have their own set of benefits and drawbacks. In comparison to dynamic analysis, static analysis is less expensive, faster, and safer. On the other hand Dynamic analysisis dependable and can outperform obfuscation tactics. It may also distinguish between malware variants and undiscovered malware families. However, time and resource expensive. Hybrid analysis thereby comes in where both the approaches are combined accordingly.

Ma et al. (2016, p. 361) developed Ensemble, a method for reducing false positives in malware detection by combining static and dynamic classifiers into a single classifier. To boost accuracy and eliminate false positives, the technique employs a number of characteristics, including static import functions and dynamic call functions. Santos et al. (2013, p. 271) introduced OPEM, a programme that uses a combination of opcode frequency from static analysis and system calls, operations, and raised exceptions from dynamic analysis to discover unknown harmful files. A hybrid analytic approach called AASandbox (Android Application Sandbox) was introduced in 2010 to automatically detect Android malware samples (Bläsing, 2010, p. 55). AASandbox disassembled the classes.dex bytecode into intermediate Smali code in the static analysis section, then pre-checked the code for dangerous code segments. The potential Android applications were run on the emulator for behavior assessment in the dynamic analysis section.

Rehman et al. in their study carry out Malware detection on the Android operating system with a hybrid technique to enhance accuracy in identifying malware and compensate the deficiencies of static and dynamic analysis approaches (Surendran, 2020). This study looks into how to create a malware detection system that can identify a variety of infections. They also cover how to identify malware before the installation process, assess hybrid mobile apps for malware dangers, and notify Android users to malware after the download process, among other things (Hadiprakoso et. al, 2020, p. 252). The first step is to extract the string from the android app and investigate the manifest file. Next, separate the keyword string from the android manifest in the following step. The third step is to use keywords and strings as input characteristics to distinguish malware and genuine apps. The final step is to distinguish between malicious and genuine apps. The findings of this study reveal that, when compared to other algorithms, the SVM is the best suited method for binary, with the suggested model increasing accuracy by up to 85.51 percent.

Android has a significant effect on the current smartphone market. Malware's threat and impact are also growing at an exponential rate. However, many approaches still are undetectable by anti-malware software. Certain aspects like covert channels or unforeseen malwares are so very vulnerable and are also not a part of effective anti-malware solutions. Anti-malware software should employ both a static and dynamic strategy to combat such rapidly evolving threats. Anti-malwares should not only trace the current state of malware, but also should keep updated with the ever-expanding vulnerabilities by putting in hybrid analysis techniques. As a result, the hybrid analysis approach adapts to code obfuscation while simultaneously expanding code coverage. Hybrid analysis is resource intensive and takes longer to generate analysis results (Tam et. al, 2017, p. 1). As a result, hybrid analysis' applicability in a practical scenario is restricted.

Machine Learning Based Android Malware Analysis Techniques

Presently, static and dynamic approaches are utilized for Android malware analysis mainly such as anomaly detection, monitoring events on operating systems such as system calls, Opcodes, API calls, input/output requests, resource utilization and resource locks to determine the suspicious patterns. It is needed that techniques for malware analysis being developed with minimal or no human intervention. Several techniques utilize behavioral analysis and partly machine learning techniques and there also occurs noticeable shifts characteristically. TaintDroid, for example, is a malware analysis system based on irregularities in an app's data usage pattern (Enck et. al, 2014, p. 1). The authors devised a way to detect fraudulent applications by monitoring irregularities in Android Dalvik op-code frequencies (Canfora et. al, 2015, p. 27). Machine learning was used in several techniques to categorize malware based on its character traits. For example, utilizing inter-process interactions and SVM, the authors in (Dash et. al, 2016, p. 252) focused on run-time behavior and categorized Android malware into malware families. In (Alam et. al, 2013, p. 663), a random forest-based technique was used to identify Android malware using a collection of 42 vectors that included battery, CPU, and memory utilization, as well as network activity.

This section brings together Android malware analysis techniques using machine learning. Figure depicts the basic workflow for static android malware analysis system utilizing the machine learning constructs. To identify harmful apps from non-malicious apps, machine learning algorithms can learn typical combinations of malware services, APIs, and system calls. The analysis begins with an apk file taken as an input. The Dalvik Executable file (dex file) must be extracted from the Android application package (APK file). The Dalvik Executable file is then converted into a Java archive in the second phase. Following that, all class files are extracted from the Java archive, and the .class files are decompiled and converted to .java files. Then, for behavioral analysis, combine all Java source code files from the same app into one huge source file for further processing and training using machine learning methods.

A general machine learning approach collects the raw Android applications and associates them with annotations and features to have characterized applications. Android application training set is then to be devised. This would also be termed as labeling in certain scenarios based on analytical results. The feature engineering may be semantic analysis of AndroidManifest.xml, disassembled classes.dex features, smali opcode features, and dynamic behavioral or additional features. Thereafter the machine learning or classification approach to be applied is determined and the training datasets are applied with ML models for malware detection training. Hence, they can predict, evaluate and explain the test data set results (Martin et. al, 2016, p. 817).

Figure 3. Workflow of android file decompiling and machine learning-based detection methodology (Martin et. al, 2016, p. 817)

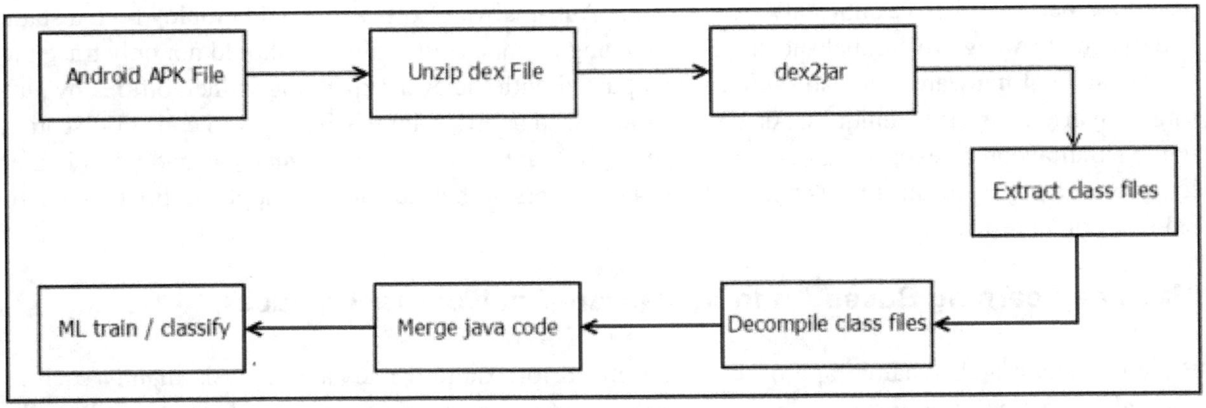

Data Collection for Android Applications

The existing or raw Android apps may be obtained from a variety of sources, including the official Google Play store and third-party app marketplaces (Milosevic ET. AL, 2017, P. 266). It is important to create customized crawlers for different Android application stores to automatically scan, acquire, and download applications into repository for large-scale and up-to-date access to Android apps. API related information may be retrieved from Android APK files, making Android platform analytics very simple. The data can be utilized to evaluate energy use, identify malware, examine graphical aspects, and discover licensing breaches. Difficulties may be encountered while crawling through these websites or repositories. Crawlers are designed to automatically search their data, identify free Android apps, and download them into repository. It's often hard to predict how many applications will be accessible in a specific market beforehand. As a result, specific crawlers are needed with the following goals:

(a) Collect as many apps as feasible,
(b) Have the least influence on market infrastructure.

However, because each market requires a manual review of the website in order to identify and filter out pages with different URLs but identical contents, such as lists that may be sorted according to various criteria, this might raise the expense of building such crawlers. This may involve overhead as it's impossible to tell if two files from different markets are the same unless they're both downloaded and compared, de duplication is local to one market, which means that one file from one market is downloaded exactly once, regardless of whether it's been downloaded from another market (Roy et. al, 2015, p. 81).

The applications may be downloaded from multiple sources. Google Play is most popular and easy to use application. Google Play allows one to search through Android's official market where users may view its applications using a browsing interface. Apps, on the other hand, cannot be downloaded via a web browser. Instead, Google offers an Android app2 that communicates with Google Play servers using a special protocol. However, without a valid Google account, no software may be downloaded from Google Play. Anzhi is another application dataset. It is one of the greatest alternative market and is based in China (Hanna et. al, 2012, p. 62). Benefitting to Chinese Android users, it is a platform for storing and distributing Chinese-language applications. Another Chinese market, AppChina, has implemented severe scraping measures such as a 1Mb/s bandwidth restriction and a multi-hour ban if more than one connection to the service is utilized at the same time. F-Droid (Zeng et. al, 2019, p. 3394) is a collection of free and open-source Android apps that users may download and install on their smartphones and tablets. Many of the programmes on F-Droid are modified versions of apps that their developers have previously launched on other platforms. Some other third party applications are Amazon Appstore, APKMirror, APKPure, APKUpdater, Aptoide, Humble Bundle, QooApp, Samsung Galaxy Apps, Yalp Store are also available with specific applications.

Extract and Study the Android Applications

Once the application is available from the repository, it is necessary to determine the parameters and behavior of the application and label it based on its characteristics. Feature extraction is the process of converting a huge, ambiguous set of inputs into a set of features (Ranveer & Hiray, 2015). Advanced detection is based on extracting features from the malware under investigation (Qu & Hughes, 2013, p. 13). Plaintext strings discovered in disassembled files, malware size, n-gram byte sequences, system resource information such as the collection of Dynamic Link Library (DLLs), and so on are all possible features. These characteristics are supplied as inputs using a machine learning technique. The basic information of any application can be obtained from the following:

a. **Android Package (APK) Files:** Android applications are designed and distributed in the form of APK files. These files contain the source code and hence can be used to know the hash values associated, APK size, market place, certificate and other such details to know its structure.
b. **Manifest:** AndroidManifest.xml is the configuration file of any Android application and consists of parameters like version, application ID, API level, permission list, feature list etc. to determine the global configuration through these artifacts (Arp et. al, 2014, p. 23).
c. **Dalvik Executable Format (DEX):** It stores the actual code within an application and every application has a file classes.dex. This is where one can get the metadata as Dex size, Dex date, native code, crypto code, dynamic code, reflection, class list etc. The classes.dex bytecode stores substantial semantic knowledge about an application's important API calls and data access (Arp et. al, 2014, p. 23). Furthermore, the classes.dex bytecode may be disassembled and analyzed quickly to represent Android apps. Control flow graph (Yang et. al, 2014, p. 163), API dependency graph (Zhang et. al, 2014, p. 1105), code property graph (Yamaguchi et. al, 2014, p. 590), and inter-component call graph are examples of low-level and high-level graph characteristics that may be derived from classes.dex.

d. **Release metadata:** Along with the application metadata, one also needs to determine the metadata associated with the release of the application such as category artifacts, installs, author, updates available, contact information etc. to get the background of the developer and application development. Security exerts need to take care as static analyzers yield false negatives. Providing generality, maintaining backward compatibility, accessing inaccessible APIs may also be legitimate reasons apart from hiding malicious code leading to false negatives. At the end of the analysis, it should yield a control flow graph and data flow graph for further processing.

e. **Intermediate Opcode based analysis:** Smali code is an interpreted intermediary language between Java and the Dalvik Virtual Machine (VM). All of the Smali programmes adhere to a set of grammatical rules. The classes.dex file may be broken down into Smali format files. Each Smali file represents a single class, which includes all of the class's methods, each of which provides human-readable Dalvik instructions (Kang et. al, 2016, p. 1). Each instruction may be broken down into a single opcode and a number of operands. This can be used to extract and analyze n-grams features.

Table 1. Comparison of feature selection algorithms

Evaluation Function	Generality	Calculation Overhead	Classification Accuracy	Applicable Feature Types	Dependence on Machine Learning Algorithm	Feature Selection Algorithm or Evaluation Index
Distance Measure	Strong	Low	Uncertain	Continuous/Discrete	Filter	Absolute value distance, Euclidean distance, Chebyshev distance, Kolmogorov distance. Relief algorithm, Relief-F algorithm.
Information Measure	Strong	Low	Uncertain	Continuous/Discrete	Filter	Information gain (Mutual information). BIF algorithm, MDLM algorithm.
Dependence Measure	Strong	Low	Uncertain	Continuous/Discrete	Filter	Chi-square statistics, T-test, Pearson correlation coefficient, Fisher score. POE1ACC algorithm, PRESET algorithm.
Consistency Measure	Strong	Low	Uncertain	Discrete	Filter	Focus algorithm, LVF algorithm
Classifier Error Rate Measure	Weak	High	High	Continuous/Discrete	Wrapper	SFS algorithm, SBS algorithm, LVW algorithm.

(Liu et. al, 2020)

As shown in the above comparison, various algorithms are applicable for feature selection and can be categorized based on calculation overhead, accuracy, feature types, evaluation index etc. (Liu et. al, 2020) Apart from these, one also needs to analyze the security concerns of the application determining the risk assessment results. Security and its patches become the prime focus when it comes to application analysis and henceforth, study of these aspects hold critical importance and then check the application against multiple parameters. Application may not always appear to be malicious in these types of analysis. Therefore, its behavior analysis would be decision taking component when it comes to distinguishing amongst benign and malicious applications.

Training the Model (Model Learning)

A machine learning training model is a method for providing appropriate training data for a machine learning (ML) algorithm to learn from. It is made up of sample output data as well as the equivalent sets of input data that have an impact on the outcome. The training model is used to process the input data via the algorithm in order to compare the processed output to the sample output. The model is modified based on the results of this association. "Model fitting" is the term for this iterative procedure. The precision of the model is dependent on the correctness of the training or validation dataset. Machine learning models will be trained for Android malware detection and categorization in this stage. So far, malware analysis has used both classical machine learning models (e.g., Support Vector Machine (Zhao et. al, 2011, p. 158), Random Forest (Alam et. al, 2013, p. 663), and K-Nearest Neighbors (Mariconti et. al, 2016)) and deep learning models (e.g., Deep Neural Networks (Yuan et. al, 2014, p. 371) (Yuan et. al, 2016, p. 114), Convolutional Neural Networks (Martinelli et. al, 2007, p. 2372)). In addition, various machine learning methods were used to solve a specific malware detection problem as to automatically learn the weights of different similarity perspectives.

Some machine learning techniques (Narudin et. al, 2016, p. 343) in this direction include Bayesian Reasoning, Bayesian Networks, Naïve Bayes, Clustering, Decision Trees, Nature-inspired, Artificial Immune System, Genetic Algorithms (GA), Swarm Intelligence Neural Networks, Artificial Neural Networks (ANN), Self-Organizing Maps (SOM), Rule based learning, Rule based expert system, Fuzzy Rule based, Association rule discovery, Support Vector Machines (SVM).

Summed Up Analysis

The machine learning algorithm is built on the training data. The algorithm is fed input data by the data scientist, which correlates to an intended output. The model examines the data regularly in order to get a better understanding of the data's behavior, and then modifies itself to fit the model's intended goal. Validation data introduces new data into the model that it hasn't assessed before during training. Validation data serves as the initial test against unknown data, allowing data scientists to assess how well the model predicts new data. Although not all data scientists utilize validation data, it can be useful in optimizing hyper parameters, which impact how the model evaluates data. The trained machine learning model will be used to detect or categorize prospective Android apps at general in this stage. In general, there is no publicly available ground truth data for Android samples. It is usual to split a portion of the labelled Android applications as the testing set to assess the model's efficiency and efficacy in order to evaluate the usefulness of the suggested technique. Since Android malware detection and family attribution are both class-imbalanced classification problems, Accuracy alone is insufficient to assess the models' success. Additional measures, such as Recall, Precision, or F1-score, should be provided. Testing data once the model has been developed confirms that it can make accurate predictions. The testing data should be unlabeled if the training and validation data include labels to track the model's performance metrics. Test data is a last, real-world verification of an unknown dataset to ensure that the machine learning algorithm was properly trained. Static analysis cannot detect malware with high accuracy when the amount of malware and complex functionality of those programs grow. Sophisticated malware authors utilize obfuscation and cryptography methods to hide its qualities thereby making it more difficult to identify malware. Using dynamic analysis approaches, behavioral aspects of malware may be leveraged to improve malware detection rates. Machine learning methods such as Random Forest

(RM), Support Vector Machine (SVM), decision trees, K-Means, K-Nearest Neighbors (K-NN), Decision Tree (DT) and Support Vector Machine (SVM), Hidden Markov Model (HMM) etc. can be used to evaluate the behavior of these malwares(Martinelli et. al, 2007, p. 2372) (Narudin et. al, 2016, p. 343).

As shown in the above figure, various approached have been used in the past and is being used till now for various classifications which depicts the importance and developments in the field through machine learning. Also, with time the number of attempts and results have become advanced depicting the wider scope and applicability of machine learning approaches. Various state-of-art tools today use these approaches and is an open field for development and betterment (Kouliaridis, 2021).

Figure 4. Number of works utilizing each base classification method per year
(Kouliaridis, 2021)

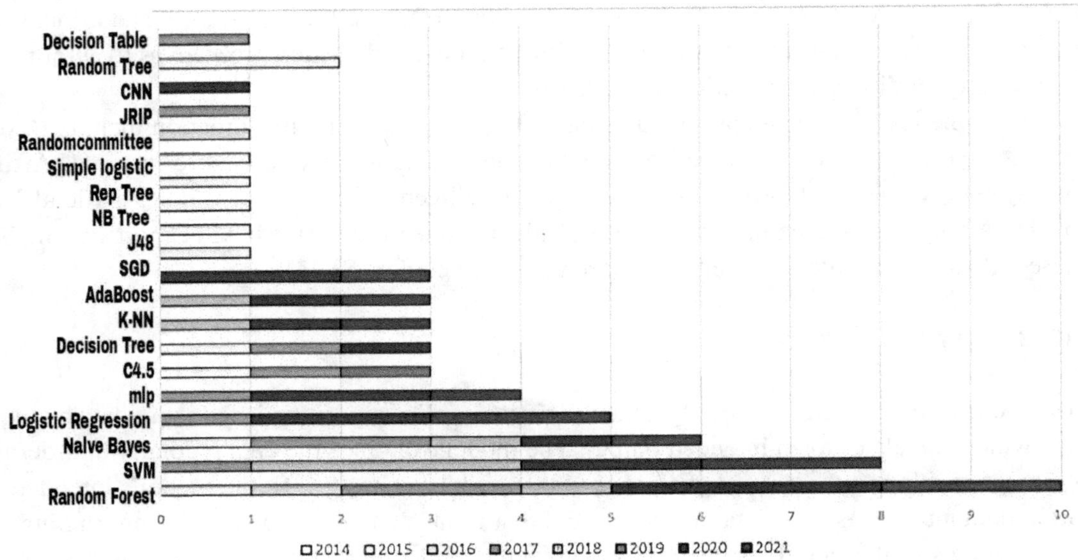

CHALLENGES AND FUTURE RESEARCH SCOPE

As it is already discussed, mobile malware analysis remains a crucial field of research in cyber security due to ever advancing insecurities. It becomes utmost important to continually monitor cyber space for distinct behaviors. Here, there are many types of possible malware attacks as depicted in the previous section. Also, analysis of these attacks is possible with static, dynamic and hybrid techniques. However, there are strategies that may vary and adjust their features in order to evade detection. Malware families are expanding and this demands for constant observation over malware misconducts and executions.

Metamorphism and polymorphism have been shown to be capable of defeating signature-based malware detection. The virus has already mutated into something new by the time detection techniques detect the new signature. During propagation, such malware samples alter their code, such that one instance of the same family has minimal similarity to another variety. They are becoming considerably simpler to construct and spread, especially with the advancement of compiler and binary rewriting capabilities.

Since Android is a primary delivery platform that provides ubiquitous services, malicious actors have used a variety of attack vectors to target Android. Signature-based approaches are insufficient against hidden, encrypted, and altered code because to the enormous growth of unique malware app signature(s) and restricted capabilities inside the Android environment.

Here, for better detection results Natural Language Processing (NLP) and deep learning can be used to get an insight into the execution (Nguyen et. al, 2017, p. 438). Also, algorithms such as genetic algorithm can serve to detect mutating malwares. Malware Sandbox Analysis is another technique for executing malware files in order to determine their behavior; it is a virtual environment in which the file may be executed without affecting the operating system. The malware file is launched while the application analyses its behavior. This is one of the alternatives to signature-based detection, according to (Messmer, 2013).

Individual static or dynamic analysis tools already exist wherein the application is analyzed for malwares. However, a hybrid approach will better conduct the purpose as it would statically go through all possible paths as well as dynamically analyze execution and techniques like reflection. Many detection tools (Nath & Mehtre, 2014, p. 440) (Pattani & Gautam, 2021, p. 1) (Messmer, 2013) have been already working in this direction to detect complex android malwares. However, the detection mechanism is unsatisfactory when considered with respect to the advancing malwares. Also, the mutations in the malwares have made it more difficult for traditional malware analysis methods to perform successful detection. Deep learning would prove to be a better analyzer in such scenario (Yuan et. al, 2016, p. 114) (Costa-jussà et. al, 2017, p. 367). Along with detection it is also necessary to consider the false positives, throughput, accuracy and precision. This opens up an area for detection for combating against such attacks using intelligent techniques.

CONCLUSION

Android is the highest targeted operating system in the mobile environment. This creates a great threat in terms of privacy and security. The work depicts approaches leading to invasion in mobile systems wherein personal or organizational or financial information may be at risk. There are existing security mechanisms involving static, dynamic and hybrid approaches. However, there are many challenges and issues that needs to be addressed in this direction. This offers up a detection and mitigation area in this sector. Rigorous regulations and monitoring are required to mitigate such attacks. In current state-of-the-art tools must also be upgraded to provide protection from such techniques. Machine Learning-based approaches serves as the prime solution to be used to track such behavior and create alerts regarding suspicious activity. It becomes an issue since there may be legitimate apps that work in a similar fashion. Possible research directions are suggested here for malware detection and analysis here.

REFERENCES

Alam, M. S., & Vuong, S. T. (2013, August). Random forest classification for detecting android malware. In IEEE international conference on green computing and communications and IEEE Internet of Things and IEEE cyber, physical and social computing (pp. 663-669). IEEE.

Arnatovich, Y. L., Wang, L., Ngo, N. M., & Soh, C. (2018). A comparison of android reverse engineering tools via program behaviors validation based on intermediate languages transformation. *IEEE Access: Practical Innovations, Open Solutions*, *6*, 12382–12394. doi:10.1109/ACCESS.2018.2808340

Arp, D., Spreitzenbarth, M., Hubner, M., Gascon, H., Rieck, K., & Siemens, C. E. R. T. (2014). Drebin: Effective and explainable detection of android malware in your pocket. In NDSS (Vol. 14, pp. 23-26). Academic Press.

Arzt, S., Rasthofer, S., Fritz, C., Bodden, E., Bartel, A., Klein, J., Le Traon, Y., Octeau, D., & McDaniel, P. (2014). Flowdroid: Precise context, flow, field, object-sensitive and lifecycle-aware taint analysis for android apps. *ACM SIGPLAN Notices*, *49*(6), 259–269. doi:10.1145/2666356.2594299

Beresford, A. R., Rice, A., Skehin, N., & Sohan, R. (2011). Mockdroid: trading privacy for application functionality on smartphones. In *Proceedings of the 12th workshop on mobile computing systems and applications* (pp. 49-54). 10.1145/2184489.2184500

Bläsing, T., Batyuk, L., Schmidt, A. D., Camtepe, S. A., & Albayrak, S. (2010). An android application sandbox system for suspicious software detection. In *5th International Conference on Malicious and Unwanted Software* (pp. 55-62). IEEE.

Bugiel, S., Davi, L., Dmitrienko, A., Fischer, T., & Sadeghi, A. R. (2011). *Xmandroid: A new android evolution to mitigate privilege escalation attacks.* Technische Universität Darmstadt, Technical Report TR-2011-04.

Burguera, I., Zurutuza, U., & Nadjm-Tehrani, S. (2011). Crowdroid: behavior-based malware detection system for android. In *Proceedings of the 1st ACM workshop on Security and privacy in smartphones and mobile devices* (pp. 15-26). 10.1145/2046614.2046619

Canfora, G., Mercaldo, F., & Visaggio, C. A. (2015). Mobile malware detection using op-code frequency histograms. In *12th International Joint Conference on e-Business and Telecommunications (ICETE)* (Vol. 4, pp. 27-38). IEEE.

Ceci. (2021). *Number of available applications in the Google Play Store from December 2009 to July 2021.* Statistica.

Costa-jussà, M. R., Allauzen, A., Barrault, L., Cho, K., & Schwenk, H. (2017). Introduction to the special issue on deep learning approaches for machine translation. *Computer Speech & Language*, *46*, 367–373.

Dash, S. K., Suarez-Tangil, G., Khan, S., Tam, K., Ahmadi, M., Kinder, J., & Cavallaro, L. (2016). Droidscribe: Classifying android malware based on runtime behavior. In 2016 IEEE Security and Privacy Workshops (SPW) (pp. 252-261). IEEE.

Enck, W., Gilbert, P., Han, S., Tendulkar, V., Chun, B. G., Cox, L. P., & Sheth, A. N. (2014). Taintdroid: An information-flow tracking system for realtime privacy monitoring on smartphones. *ACM Transactions on Computer Systems*, *32*(2), 1–29. doi:10.1145/2619091

Enck, W., Ongtang, M., & McDaniel, P. (2009). On lightweight mobile phone application certification. In *Proceedings of the 16th ACM conference on Computer and communications security* (pp. 235-245). ACM.

Faruki, P., Bharmal, A., Laxmi, V., Ganmoor, V., Gaur, M. S., Conti, M., & Rajarajan, M. (2014). Android security: A survey of issues, malware penetration, and defenses. *IEEE Communications Surveys and Tutorials, 17*(2), 998–1022. doi:10.1109/COMST.2014.2386139

Faruki, P., Ganmoor, V., Laxmi, V., Gaur, M. S., & Bharmal, A. (2013). AndroSimilar: robust statistical feature signature for Android malware detection. In *Proceedings of the 6th International Conference on Security of Information and Networks* (pp. 152-159). 10.1145/2523514.2523539

Felt, A. P., Chin, E., Hanna, S., Song, D., & Wagner, D. (2011). Android permissions demystified. In *Proceedings of the 18th ACM conference on Computer and communications security* (pp. 627-638). ACM.

Fuchs, A. P., Chaudhuri, A., & Foster, J. S. (2009). *Scandroid: Automated security certification of android applications.* Manuscript, Univ. of Maryland, http://www. cs. umd. edu/avik/projects/scandroidascaa

Gordon, M. I., Kim, D., Perkins, J. H., Gilham, L., Nguyen, N., & Rinard, M. C. (2015, February). Information flow analysis of android applications in droidsafe. In NDSS (Vol. 15, No. 201, p. 110). Academic Press.

Hadiprakoso, R. B., Buana, I. K. S., & Pramadi, Y. R. (2020). Android malware detection using hybrid-based analysis & deep neural network. In *3rd International Conference on Information and Communications Technology (ICOIACT)* (pp. 252-256). IEEE.

Hanna, S., Huang, L., Wu, E., Li, S., Chen, C., & Song, D. (2012). Juxtapp: A scalable system for detecting code reuse among android applications. In *International Conference on Detection of Intrusions and Malware, and Vulnerability Assessment* (pp. 62-81). Springer.

Kang, B., Yerima, S. Y., McLaughlin, K., & Sezer, S. (2016). N-opcode analysis for android malware classification and categorization. In *International conference on cyber security and protection of digital services (cyber security)* (pp. 1-7). IEEE.

Kim, J., Yoon, Y., Yi, K., Shin, J., & Center, S. W. R. D. (2012). ScanDal: Static analyzer for detecting privacy leaks in android applications. *MoST, 12*(110), 1.

Kouliaridis, V., & Kambourakis, G. (2021). A comprehensive survey on machine learning techniques for android malware detection. *Information, 12*(5), 185.

Kramer, S., & Bradfield, J. C. (2010). A general definition of malware. *Journal in Computer Virology, 6*(2), 105-114.

Li, J., Sun, L., Yan, Q., Li, Z., Srisa-An, W., & Ye, H. (2018). Significant permission identification for machine-learning-based android malware detection. *IEEE Transactions on Industrial Informatics, 14*(7), 3216–3225. doi:10.1109/TII.2017.2789219

Liu, K., Xu, S., Xu, G., Zhang, M., Sun, D., & Liu, H. (2020). A review of android malware detection approaches based on machine learning. *IEEE Access: Practical Innovations, Open Solutions*, 8.

Lu, X., Liu, X., Li, H., Xie, T., Mei, Q., Hao, D., & Feng, F. (2016, May). PRADA: Prioritizing android devices for apps by mining large-scale usage data. In *IEEE/ACM 38th International Conference on Software Engineering (ICSE)* (pp. 3-13). IEEE.

Ma, X., Biao, Q., Yang, W., & Jiang, J. (2016). Using multi-features to reduce false positive in malware classification. In *IEEE Information Technology, Networking, Electronic and Automation Control Conference* (pp. 361-365). IEEE.

Mariconti, E., Onwuzurike, L., Andriotis, P., De Cristofaro, E., Ross, G., & Stringhini, G. (2016). *Mamadroid: Detecting android malware by building markov chains of behavioral models.* arXiv preprint arXiv:1612.04433.

Martin, W., Sarro, F., Jia, Y., Zhang, Y., & Harman, M. (2016). A survey of app store analysis for software engineering. *IEEE Transactions on Software Engineering*, *43*(9), 817–847.

Martinelli, F., Marulli, F., & Mercaldo, F. (2017). Evaluating convolutional neural network for effective mobile malware detection. *Procedia Computer Science*, *112*, 2372–2381.

McGraw, G., & Morrisett, G. (2000). Attacking malicious code: A report to the infosec research council. *IEEE Software*, *17*(5), 33–41. doi:10.1109/52.877857

Messmer, E. (2013, Mar.). Malware-detecting 'sandboxing' technology no silver bullet. *Networkworld.*

Milosevic, N., Dehghantanha, A., & Choo, K. K. R. (2017). Machine learning aided Android malware classification. *Computers & Electrical Engineering*, *61*, 266–274.

Narudin, F. A., Feizollah, A., Anuar, N. B., & Gani, A. (2016). Evaluation of machine learning classifiers for mobile malware detection. *Soft Computing*, *20*(1), 343–357.

Nath, H. V., & Mehtre, B. M. (2014). Static malware analysis using machine learning methods. In *International Conference on Security in Computer Networks and Distributed Systems* (pp. 440-450). Springer. 10.1007/978-3-642-54525-2_39

Nguyen, T. D., Nguyen, A. T., Phan, H. D., & Nguyen, T. N. (2017, May). Exploring API embedding for API usages and applications. In *2017 IEEE/ACM 39th International Conference on Software Engineering (ICSE)* (pp. 438-449). IEEE.

Pattani, K., & Gautam, S. (2021). SonicEvasion: a stealthy ultrasound based invasion using covert communication in smart phones and its security. *International Journal of Information Technology*, 1-11.

Qu, Y., & Hughes, K. (2013, December). Detecting metamorphic malware by using behavior-based aggregated signature. In *World Congress on Internet Security (WorldCIS-2013)* (pp. 13-18). IEEE.

Ranveer, S., & Hiray, S. (2015). Comparative analysis of feature extraction methods of malware detection. *International Journal of Computers and Applications*, *120*(5).

Roundy, K. A., & Miller, B. P. (2010). Hybrid analysis and control of malware. In *International Workshop on Recent Advances in Intrusion Detection* (pp. 317-338). Springer.

Roy, S., DeLoach, J., Li, Y., Herndon, N., Caragea, D., Ou, X., . . . Guevara, N. (2015). Experimental study with real-world data for android app security analysis using machine learning. In *Proceedings of the 31st Annual Computer Security Applications Conference* (pp. 81-90). Academic Press.

Santos, I., Devesa, J., Brezo, F., Nieves, J., & Bringas, P. G. (2013). Opem: A static-dynamic approach for machine-learning-based malware detection. In *International joint conference CISIS'12-ICEUTE´ 12-SOCO´ 12 special sessions* (pp. 271-280). Springer.

Sato, R., Chiba, D., & Goto, S. (2013). Detecting android malware by analyzing manifest files. In *Proceedings of the Asia-Pacific Advanced Network*, (Vol. 36, pp. 23-31). 10.7125/APAN.36.4

Schlegel, R., Zhang, K., Zhou, X. Y., Intwala, M., Kapadia, A., & Wang, X. (2011). Soundcomber: A Stealthy and Context-Aware Sound Trojan for Smartphones. In NDSS (Vol. 11, pp. 17-33). Academic Press.

Shabtai, A., Kanonov, U., Elovici, Y., Glezer, C., & Weiss, Y. (2012). "Andromaly": A behavioral malware detection framework for android devices. *Journal of Intelligent Information Systems*, *38*(1), 161–190. doi:10.100710844-010-0148-x

Statistica. (2020). *Development of new Android malware worldwide from June 2016 to March 2020(in millions)*. Author.

Surendran, R., Thomas, T., & Emmanuel, S. (2020). A TAN based hybrid model for android malware detection. *Journal of Information Security and Applications*, *54*, 102483.

Tam, K., Feizollah, A., Anuar, N. B., Salleh, R., & Cavallaro, L. (2017). The evolution of android malware and android analysis techniques. *ACM Computing Surveys*, *49*(4), 1–41.

Wong, M. Y., & Lie, D. (2016). IntelliDroid: A Targeted Input Generator for the Dynamic Analysis of Android Malware. In NDSS (Vol. 16, pp. 21-24). doi:10.14722/ndss.2016.23118

Wu, D. J., Mao, C. H., Wei, T. E., Lee, H. M., & Wu, K. P. (2012). Droidmat: Android malware detection through manifest and api calls tracing. In *2012 Seventh Asia Joint Conference on Information Security* (pp. 62-69). 10.1109/AsiaJCIS.2012.18

Yamaguchi, F., Golde, N., Arp, D., & Rieck, K. (2014). Modeling and discovering vulnerabilities with code property graphs. In *IEEE Symposium on Security and Privacy* (pp. 590-604). IEEE.

Yan, L. K., & Yin, H. (2012). Droidscope: Seamlessly reconstructing the {OS} and dalvik semantic views for dynamic android malware analysis. In *21st USENIX Security Symposium (USENIX Security 12)* (pp. 569-584). USENIX.

Yang, C., Xu, Z., Gu, G., Yegneswaran, V., & Porras, P. (2014). Droidminer: Automated mining and characterization of fine-grained malicious behaviors in android applications. In *European symposium on research in computer security* (pp. 163-182). Springer.

Yen, Y. S., & Sun, H. M. (2019). An Android mutation malware detection based on deep learning using visualization of importance from codes. *Microelectronics and Reliability*, *93*, 109–114. doi:10.1016/j.microrel.2019.01.007

Yuan, Z., Lu, Y., Wang, Z., & Xue, Y. (2014). Droid-sec: deep learning in android malware detection. In *Proceedings of the ACM conference on SIGCOMM* (pp. 371-372). ACM.

Yuan, Z., Lu, Y., & Xue, Y. (2016). Droiddetector: Android malware characterization and detection using deep learning. *Tsinghua Science and Technology*, *21*(1), 114–123.

Zeng, Y., Chen, J., Shang, W., & Chen, T. H. P. (2019). Studying the characteristics of logging practices in mobile apps: A case study on f-droid. *Empirical Software Engineering*, *24*(6), 3394–3434.

Zhang, M., Duan, Y., Yin, H., & Zhao, Z. (2014). Semantics-aware android malware classification using weighted contextual api dependency graphs. In *Proceedings of the ACM SIGSAC conference on computer and communications security* (pp. 1105-1116). ACM.

Zhao, M., Ge, F., Zhang, T., & Yuan, Z. (2011). AntiMalDroid: An efficient SVM-based malware detection framework for android. In *International conference on information computing and applications* (pp. 158-166). Springer. 10.1007/978-3-642-27503-6_22

Chapter 12
Entropy-Based Feature Selection for Network Intrusion Detection Systems

Sellappan Devaraju
https://orcid.org/0000-0003-3116-4772
VIT Bhopal University, Bhopal, India

Sundaram Jawahar
https://orcid.org/0000-0002-8101-8725
Christ (Deemed), Ghaziabad Campus, India

Srinivasan Ramakrishnan
https://orcid.org/0000-0002-8224-4812
Dr. Mahalingam College of Engineering and Technology, Pollachi, India

Dheresh Soni
VIT Bhopal University, Bhopal, India

Alagappan Somasundaram
Sri Krishna Arts and Science College, Coimbatore, India

ABSTRACT

A network intrusion detection system (NIDS) has a significant role in an industry or organization to protect their data. NIDS should be more reliable to manage huge traffic over the networks to detect the emerging attacks. In this chapter, novel entropy-based feature selection is proposed to select the important features of intrusion detection system. Feature selection reduces the computational time and improves detection rates. In entropy, within-class entropies and between-class entropies are computed for the various classes of intrusion in the KDD dataset. Based on computed entropy values, features are ranked and selected. Radial basis neural network (RBNN) is employed as a classifier. Performances of the proposed entropy-based feature selection algorithm are evaluated using the 10% dataset for training and two other datasets for testing. The proposed system shows significant improvement in the detection rate, reduces the false positive rate (FPR), and also reduces the computational time.

DOI: 10.4018/978-1-6684-3991-3.ch012

INTRODUCTION

The Network Intrusion Detection (NIDS) System is a dependable and secure system that monitors for network vulnerabilities. The flaw will take advantage of a flaw in information assurance. On a daily basis, it is critical to lessen vulnerability from numerous enterprises. The internet is widely utilised for a variety of purposes, including commerce, education, games, entertainment, and other related activities. As a result, any organization's Network Intrusion Detection System (NIDS) is critical in protecting its data from misbehaviors. Despite the fact that every firm uses firewalls and other security measures to protect data, many intruders remain undetected. As a result, information must be better protected. Signature-based and anomaly-based IDS are the two most common types of NIDS. (i)A signature-based NIDS detects an intrusion by comparing it to previously detected intrusions. In the log files, there are signatures. (ii) The anomaly-based NIDS monitors system action and categorises it as either attack or normal (Gupta, 2010). Network-based IDS and host-based IDS are two types of IDS. When the system can converse with each other via the networks, and the network-based IDS identifies misbehaviour. (ii) If there is any misbehaviour, the host-based IDS monitors and analyses the single computer system (Devaraju, 2013; Nie, 2009).

The intrusion is detected using a signature-based or misuse-based intrusion detection system that evaluates previous signatures in log files (Ashara, 2012; Devaraju, 2019; Mansour, 2010; Suseela, 2005). Signature-based assaults rely on the knowledge gathered from previous strikes. Attack signatures, which are sets of rules that uniquely identify attacks, represent this information. Because they have superior accuracy and lower false positive rates, knowledge-based techniques are relatively straightforward for the administrator to sustain the attacks. When users detect an intrusion and compare it to the signatures log files, signature-based assaults are portrayed as known attacks. The log file contains a list of known assaults that have been detected on a computer system or network. Furthermore, the signature-based attack lacks the potential to to detect all types of attacks, particularly new attacks and those involving privilege misuse (Devaraju, 2019; Gang, 2010; Nor, 2008).

Unknown attacks are intrusion detection based on anomalies; these attacks are detected by the network and distinguished from conventional attacks. They can detect attempts to exploit novel and unexpected attacks, which gives them an advantage over signature-based attacks. However, anomaly-based techniques have their own set of drawbacks, including a high false positive rate due to a lack of training data and anomalous behaviour. Signature-based techniques are ideally suited for intrusion detection for these reasons (Devaraju, 2019; Shih-Wei, 2012; Mei, 2011).

Network-based, host-based, and application protocol-based intrusion detection systems are the three types (Jawahar, 2020; Arman, 2009; Yousef, 2014; Pablo, 2012). Misuse or anomaly-based attacks are employed in network-based assaults. The interconnectedness of computer systems is used to detect network-based assaults. For intrusion detection, audit patterns are formed over networks, however extracting the essential information from the audit patterns of the networks is difficult. When two computers are linked together, the attack is sent from one to the other via networking equipment. The assault detection rate and false positive rate are also influenced by large volumes of audit patterns (Jawahar, 2020; Gaik-Yee, 2013; Selvakani, 2011).

Host-based breaches are easy to detect and prevent because they originate from a specific computer system. When some external devices are connected to the computer system, incursions occur. Floppy discs, compact discs, pen drives, and other similar devices are used. The audit patterns generated by the computer system, which comprises system log files and error log files, have been evaluated by the host-based system. Rather than network-based techniques for detecting more reliable attacks, log files collected from a single computer system include more particular information (Gupta, 2010; Devaraju, 2013; Nie, 2009). The host-based solutions, on the other hand, are difficult to manage in large systems.

Intrusion detection systems based on apps are solely concerned with a single application and detect assaults directed at it. The application-based system will check the application access logs or the system calls generated by the processes to detect intrusions (Chi-Ho, 2007; Mohammad, 2011; Pablo, 2013). In this thesis, the proposed approaches are focused on signature-based and network-based intrusion detection systems.

The KDD dataset is a well-known and benchmarking dataset for intrusion detection systems (Gupta, 2010; Devaraju, 2013; Nie, 2009). KDD has millions of instances or records, each of which has 41 characteristics (Nor, 2008; Arman, 2009; Yousef, 2014). Some issues, 41 characteristics, are insufficient to categorise the attacks. To eliminate immaterial features, an entropy-based feature selection is proposed in this study. Support vector machine, principal component analysis, and genetic feature selection methods are some of the most widely used feature selection methods. On the other hand, these methodologies do not allow for the identification of significant features while still delivering low within-class entropy and high between-class entropy values. For feature selection, this technique is known as the between-class and within-class entropy. Feature selection must be focused primarily on research applications for large datasets containing millions of records and tens of variables. For any huge volume of dataset in research applications, feature selection is critical. It's possible that a dataset contains redundant data. Additional characteristics can lengthen computation time while lowering detection rates (Bin, 2014).

LITERATURE REVIEW

Different variety of techniques have been proposed for selection of features to detect the intrusion namely Conditional random, PCA, Entropy, LDA, ICA, Information gain etc. In this view, different techniques are used to detect IDS are discussed.

Conditional random fields, fuzzy logic, neural networks, statistical approaches, data mining, and other techniques are investigated for feature selection (Gupta, 2010; Minjie, 2012; Devaraju, 2011). Conditional random fields are utilised to illustrate attack detection accuracy using the KDD data set. For different layers, automatic feature selection is used (Gupta, 2010). The reduction algorithm is used to pick features for intrusion detection, however it is more sophisticated than other methods (Minjie, 2012). For feature reduction, the genetic algorithm is utilised, which employs encoding as a metric (Hua, 2009), which is difficult to encode the variables every time. For intrusion detection, data mining and clustering techniques are applied, and a reduction procedure is used to cancel the redundant attribute set (Nadiammai, 2014). The sub attributes utilisation method is used to decrease features in order to improve the detection rate in cases when a high number of variables is difficult to process (Shingo, 2011).

(i) Principle Component Analysis: In data analysis and compression, Principle Component Analysis (PCA) is a technique for reducing dimensionality (Gaik-Yee, 2013; Selvakani, 2011; Chi-Ho, 2007). PCA identifies pattern similarities and differences (Shih-Wei, 2012; Mei, 2011; Jawahar, 2020).

The dataset can be represented as a matrix Xnm if each datum contains N features, such as x11,x12,x13,x14... x1N, x21, x22,x23,x24....x2N.

The following formula used to compute the average observation:

$$\mu = \frac{1}{n}\sum_{i=1}^{n} X_i \tag{1}$$

Compression and different dimensionality are two drawbacks of principal component analysis.

(ii) The best transformation matrix is Linear Discriminant Analysis (LDA). It can be used to categorise people into groups (Suseela, 2005; Devaraju, 2019; Gang, 2010; Nor, 2008). The data must have proper class labels, and the optimization technique must be formally formulated (Ashara, 2012; Devaraju, 2019; Mansour, 2010).

The data must have appropriate class labels for the analysis to work. Formulate the optimization technique mathematically,

$$\overline{x}_j = \frac{1}{N_j}\sum_{i=1}^{N_j} x_i \tag{2}$$

Nj signifies the total number of training tokens, while N specifies the number of training tokens in class j. J is, of course, the number of courses. One of the limitations of linear discriminant analysis is that it has a matrix for analysing suitable class label.

(iii) Independent Component Analysis: The redundant technique is used to evaluate the classifier's performance or accuracy. It's to figure out what information isn't useful. It's used to lower the data's dimensionality and classifier may handle enormous amounts of information (Nor, 2008; Devaraju, 2019; Shih-Wei, 2012).

A method for determining a single W, y = Wx, is known as independent component analysis. Finding an acceptable W is the most typical learning approach.

$$W = \eta(I - \Phi(y)yT)W \tag{3}$$

Where is the output vector y's nonlinear function. The limits of independent component analysis are dimensionally different.

(iv) Before and after observing features, the Information Gain (IG) approach is used to calculate the estimated reduction in entropy of classes. The characteristics are chosen based on the computed information gain value; if the values differ significantly, the selected features are considered more essential (Devaraju, 2019; Mansour, 2010; Suseela, 2005). The formula for calculating the amount of information gained

$$\text{InfGain}(S,F) = \text{Entropy}(S) - \sum_{v \in V(F)} \frac{|S_v|}{|S|} . \text{Entropy}(S_v) \tag{4}$$

(v) For feature selection depending on the degree of reliance, relative decision entropy is used. Positive region is a region that can only provide information on the degree of reliance and other associated information that rough set methods can disregard. Positive region data can be used to create a boundary regionIn this way, information from the positive region and information from the boundary region are integrated (Wei, 2015; Shi, 2011; Ramakrishnan, 2016; Ashara, 2012). The disadvantages are that they only provide a degree of dependency for the positive region.

(vi) Maximum Entropy is used to determine the likelihood of belonging to a given set of items. The MaxEnt approach calculates the probability of all possible outcomes for each set of all potential outcomes. It is dependent on the collection of features that are beneficial in making accurate forecasts. To obtain the weights of all features and subsequently normalise them, conditional probability is utilised. The feature selection is done based on the weights (Devaraju, 2019; Mansour, 2010; Suseela, 2005; Devaraju, 2019). The constraint is that the weights of all features must be determined before the feature selection can be made.

Information loss, processing a large number of variables, and increased complexity are all disadvantages of the current system, which are all investigated and compared. These issues are addressed by the proposed entropy-based feature selection method. The number of characteristics is thought to be reduced by the values of between-class and within-class entropy. The between-class entropy is calculated for all features in each class, whereas the within-class entropy is calculated for each feature in each class. Features are chosen when the between-class entropy is high and the within-class entropy is low. After the feature is picked, the detection rate and computational time are measured using a radial basis neural network. In terms of improving performance and lowering the false positive rate, the proposed entropy-based technique is more effective.

The following is the format of the paper: The KDD Dataset is explained in Section 2 depicts the literature review, Section 3, followed by the entropy-based feature selection in Section 4, the radial basis neural network in Section 5, the experimental findings and discussion in Section 6, and the study effort in Section 7.

KDD DATASET DESCRIPTION

The KDD dataset is the benchmarking dataset used to evaluate the intrusion detection system. The dataset is divided into three parts: "10 percent dataset," "Corrected dataset," and "Fulldataset." This dataset contains a million records, each of which has 41 attributes that can be classified as normal or assault. "10 percent KDD" is utilised for training data, whereas "Corrected KDD" and "FullKDD" are used for testing data (Minjie, 2012; http://kdd.ics.uci.edu/databases/kddcup99/kddcup99.html).

The normal value is 97278, the DoS value is 391458, the probe value is 4107, the R2L value is 1126, the U2R value is 52, and the total value is 494021. The normal value is 60593, the DoS value is 229853, the probe value is 4166, the R2L value is 16347, the U2R value is 70, and the total value is 311029. The normal value is 972781, the DoS value is 3883370, the probe value is 41102, the R2L value is 1126, the U2R value is 52, and the total value is 4898431. Some of the attacks in the KDD dataset include guess passwd, buffer overflow, back, ftp write, ipsweep, land, imap, multihop, loadmodule, neptune, nmap, phf, perl, pod, rootkit, portsweep, satan, smurf, spy, warezclient, teardrop, and warezmaster. DoS, Probe, R2L, and U2R are the four types of attacks (Nie, 2009). The KDD Cup 1999 dataset serves as the benchmarking dataset for the Intrusion Detection System. The different sorts of assaults are DoS, Probe, R2L, and U2R.

Denial of Service (DoS) attacks are a sort of cyber-attack that targets machines that are connected to the Internet. DoS attacks take use of weaknesses in TCP/IP implementation. DoS attacks are designed to cause system disruption by flooding them with unauthorised requests. When identifying DoS attacks, traffic level characteristics like "percent of connections with same destination host and same service" and packet level characteristics like "source bytes" and "percentage of packets with defects" are crucial. It makes no difference whether a user is "logged in" or not when detecting DoS attacks like syn flood (http://kdd.ics.uci.edu/databases/kddcup99/kddcup99.html) (Devaraju, 2019; Devaraju, 2020).

Probing: Probing is a type of attack in which a hacker examines a networking device or machine for a flaw that can be exploited by taking advantage of the system's cooperation. Basic connection statistics like "length of connection" and "source bytes" are vital when identifying probes, however aspects like "number of file creations" and "number of files read" are not intended to give information for recognising probes.

Remote-to-Local (R2L) attacks: Detecting host and network level features is more challenging in R2L attacks. Network-level features like "duration of connection" and "service requested," as well as host-level features like "number of failed login attempts," are used to detect R2L attacks. Because the attacker does not have access to the victim's workstation, he or she tries to get unauthorised access from a remote system, such as guessing passwords.

U2R (User-to-Root) attacks: The U2R attacks are the most difficult to capture the semantic aspects of any stage. These attacks are directed towards apps that rely on content for functionality. As a result, U2R attack aspects such as "number of file creations" and "number of shell prompts executed" are picked over elements such as "protocol" and "source bytes." Buffer overflow attacks, for example, give unauthorised access to super user (root) privileges on a local system.

ENTROPY BASED FEATURE SELECTION

Feature selection, a variety of algorithms are examined. For feature selection, the maximum entropy and Shannon entropy approaches are often utilised. Only past data or tested information is taken into account when calculating maximum entropy, and the one with the highest entropy is chosen. The random variable for entropy is uncertainty or randomness of their measures. The entropy model is used to assess a modelling technique's probability estimation and to establish a probability distribution that meets all of the limitations given by the training data. Shannon's contribution to information theory is the broadest and most fundamental definition of the entropy measure. Because Shannon entropy-based approaches take a logarithmic approach to discovering the link between multiple characteristics while considering both within-class and inter-class entropy, they are often employed for feature selection. Using Shannon entropy, these entropy values are determined in order to identify the right strong features. The proposed work, on the other hand, considers both between and within entropy values when choosing relevant features. Between and within class entropy are offered as extensions of Shannon's notion to measure entropy levels in this study. Between-class entropy is a combination of class measurements, while within-class entropy is a class-by-class measurement. These entropy-based feature selection methods are offered to improve their efficiency and lower the rate of false positives. The entropy formula with normalised data is used to compute the within and between class entropy.

Dataset contains a million records and each with 41 variables (7 and 34 variables for discrete and continuous respectively) with values ranging from 0 to 6291668. All 41 features aren't required while developing an NIDS. It's possible that the dataset contains duplicate data (Adel, 2015; Bin, 2014). Additional characteristics can lengthen computation time, reduce intrusion detection system accuracy, and raise the likelihood of false positives. By classifying the dataset, the characteristics with high magnitude have a strong influence on the lower magnitude features. For feature normalisation, only 34 variables (continuous) are used. To boost speed and reduce computing time, feature selection is increasingly crucial. The entropy-based strategy for feature selection is consider to increase the detection rate while reducing computational time. The steps for selecting features are as follows:

Step 1: Before feature selection, feature scaling is a method to standardize the different ranges of feature values. Normalization of data is an important process toscale down [0,1] is done through the following formula:

$$\mathrm{Res}_m[n] = \frac{(\mathrm{Res}_m[n] - \mathrm{Res}_{c\_min}[n])}{(\mathrm{Res}_{c\_max}[n] - \mathrm{Res}_{c\_min}[n])} \tag{5}$$

Where,

$\mathrm{Res}_m[n]$ is the value of mth data instance from n^{th} attribute.
$\mathrm{Res}_{c\_min}[n]$ is the minimum value of attribute j among the dataset Res,
$\mathrm{Res}_{c\_max}[n]$ is the maximum value among all the data instances.

Step 2: Different assaults are classified by NIDS into four categories such as Denial of Service, Probe, U2R and R2L. Feature selection process, four assault classes were examined. There are 41 variables in the dataset, with 7 and 34 variables such as discrete and continuous variables respectively. Only continuous features are used to calculate entropy values. All 34 continuous features have their within and between class entropy of their values are determined. For training, 10% KDD Dataset is used. The entropy formula is used to calculate the entropy values.

$$\mathrm{Entropy} = -\sum_{i=1}^{k} P(value_i).\log_2(P(value_i)) \tag{6}$$

$P(value_i)$ signifies the ith value of probability, and denotes k as number of records in KDD dataset.

Step 3: The entropy values are determined, and then the data is grouped according to the entropy values. The groupings are divided into four categories: "very low," "low," "high," and "very high." The least value is quite low, while the maximum value is very high. After computing the entropy values for 34 features, the classes were divided into four groups depending on their entropy values (low, very low, very high or high). The value of entropy for "very low" range from (0.00 - 0.15), "low" from (0.16 - 0.50), "high" from (0.51 - 0.85) and "very high" from (0.51 - 0.85) respectively (0.86 - 1.00). The ranking has been assigned for feature selection once the variables have been grouped. Table 1 displays the computed entropy values:

Figure 1. Architecture for proposed entropy based feature selection

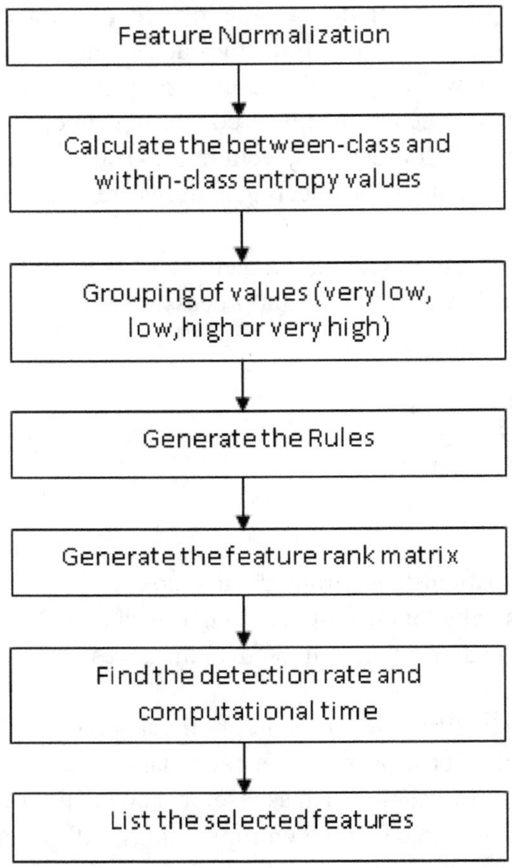

Step 4: After categorization, choose features depending on the ranking, which was determined by the following criteria:

RANK 1: Between-entropy is very high and within-class entropy is very low, then the feature is selected.

RANK 2: Between-entropy is very high and within-class entropy is low, then the feature is selected.

RANK 3: Between-entropy is high and within-class entropy is very low, then the feature is selected.

Three groups of ranking have been generated based on the entropy values. Ranking is considered for all the four classes.

Table 1. Values of between and within class entropy

Features	Within-Class Entropy				Between-Class Entropy
	DoS	Probe	R2L	U2R	
kdd_0	0.01	0.0213	0.0313	0.0087	0.0714
kdd_1	0.4676	0	0.0006	0.0093	0.6885
kdd_2	0.0135	0.0057	0.0014	0.0074	0.0279
kdd_3	0.1035	0	0	0	0.1035
kdd_4	0	0	0.0004	0	0.0004
kdd_5	0.6309	0	0.0322	0.0086	0.6716
kdd_6	0	0	0.0173	0	0.0173
kdd_7	0	0	0.0008	0.0072	0.008
kdd_8	0	0	0	0	0
kdd_9	0	0	0	0	0
kdd_10	0	0	0.0008	0.0028	0.0036
kdd_11	0	0	0.0016	0.0066	0.0082
kdd_12	0	0	0.0007	0.0011	0.0018
kdd_13	0	0	0.0029	0	0.0029
kdd_14	0	0	0	0	0
kdd_15	0.7979	0.1343	0.1323	0.0009	0.8546
kdd_16	0.114	0.4099	0.0127	0.0015	0.4327
kdd_17	0.049	0.3421	0.0024	0	0.3936
kdd_18	0.0158	0.0037	0.0036	0	0.0231
kdd_19	0.1029	0.1789	0.0011	0.0001	0.283
kdd_20	0.2321	0.0041	0.0021	0	0.2383
kdd_21	0.5769	0.1162	0.0004	0.0012	0.6948
kdd_22	0.5805	0.079	0.0003	0.001	0.6609
kdd_23	0.2103	0.0006	0.0045	0	0.2154
kdd_24	0.3223	0.0694	0.0805	0.0017	0.474
kdd_25	0.241	0.2306	0.2702	0.0042	0.746
kdd_26	0.5941	0.0563	0.1233	0.0014	0.7751
kdd_27	0.3478	0.2342	0.0503	0.0016	0.6339
kdd_28	0.4558	0.1137	0.0349	0.0061	0.6105
kdd_29	0.0592	0.3631	0.1743	0.0026	0.5992
kdd_30	0.1054	0.349	0.0368	0	0.4912
kdd_31	0.0347	0.0038	0.0237	0	0.0622
kdd_32	0.4313	0.2828	0.0158	0.0016	0.7316
kdd_33	0.2972	0.0057	0.0038	0.0021	0.3088
Average	0.199815	0.088365	0.031265	0.002285	0.318647

Step 5: Ranked the featres using rank matrix for four classes based on the ranking. The DoS class was cut to 22 features, the Probe class to 21 features, the R2L class to 14 features, and the U2R class to 18 features based on the ranking. The feature rank matrix is shown in Table 2:

Table 2. Feature rank matrix

	DoS	Probe	R2L	U2R	Average	Cumulative
RANK 1	6	5	4	6	5	5
RANK 2	11	9	5	8	8	13
RANK 3	5	7	5	4	5	18

Step 6: A feature rank matrix is used to determine the amount of features for each class. A radial-basis neural network is utilised to determine the detection rate and computational time using the feature rank matrix. If the detection rate is high and the calculation time is short, this characteristic is chosen. Only a few features are used to compute the detection rate and computational time. Table 3 shows the detection rate and computation time:

Table 3. Detection rate and computational time for four classes

	Rank-1	Rank-1 & Rank-2	Rank-1, Rank-2, & Rank-3
Detection Rate and Computational Time for DoS Class			
	(6 features)	(17 features)	(22 features)
Detection Rate	98.46	99.74	99.59
Computational Time (seconds)	103	109	115
Detection Rate and Computational Time for Probe Class			
	(5 features)	(14 features)	(21 features)
Detection Rate	85.23	88.17	88.95
Computational Time (seconds)	14	15	15
Detection Rate and Computational Time for R2L Class			
	(4 features)	(9 features)	(14 features)
Detection Rate	61.48	63.76	64.48
Computational Time (seconds)	9	8	7
Detection Rate and Computational Time for U2R Class			
	(6 features)	(14 features)	(17 features)
Detection Rate	71.93	90.38	81.41
Computational Time (seconds)	<1	<1	<1

The link between the several classes in the proposed system is represented by a scatter plot. A scatter plot is used to demonstrate the discrimination capacity of the suggested feature selection approach. A scatter figure for the variables KDD 2vs KDD 16 and KDD 24vs KDD 28 for the classes DOS, Probe, R2L, U2R, and Normal is displayed using the amended dataset. Six sample records are picked at random for the purpose of determining the relationships. The scatter plot in Figure 2 displays each class. The selected characteristics show a positive relationship. Figure 2 shows a scatter plot for numerous features:

Figure 2. Scatter plot for different features

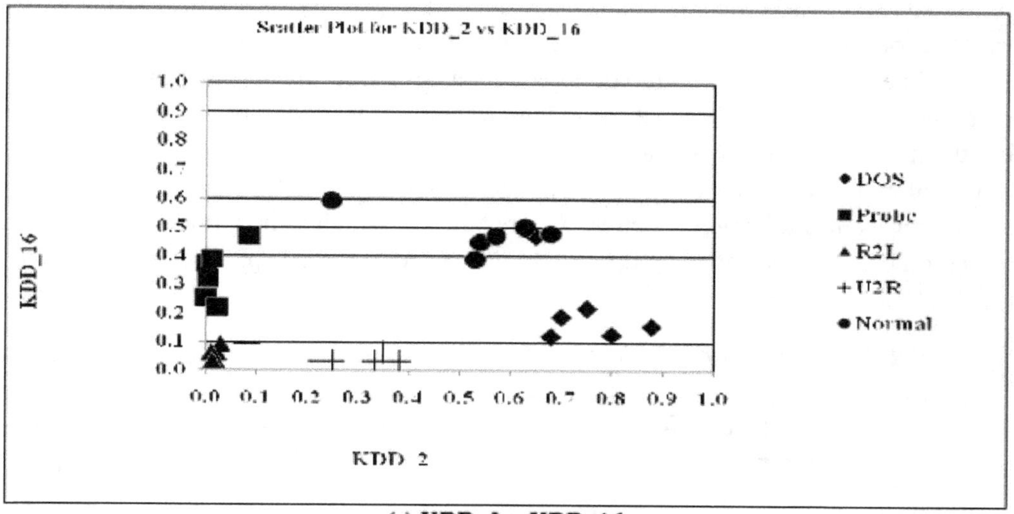

(a) KDD_2vs KDD_16

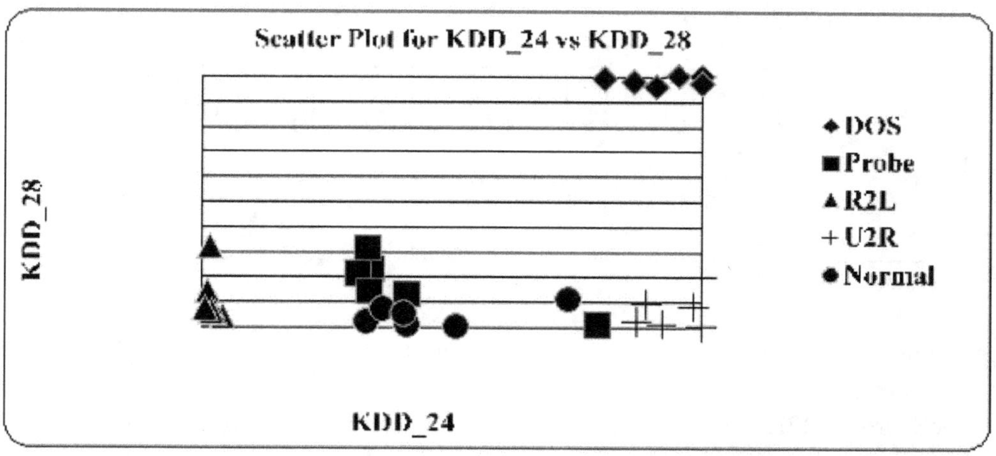

(b) KDD_24vs KDD_28

Features are chosen depends on the rate of detection and the amount of time it takes to compute them. Three discrete variables (protocol, service, and flag) and various continuous variables are included in the proposed system's classification classes, which are mentioned below:

- DoS: protocol, service, flag, kdd_0, kdd_1, kdd_2, kdd_3, kdd_5, kdd_7, kdd_8, kdd_11, kdd_12, kdd_13, kdd_15, kdd_16, kdd_20 and kdd_21.
- Probe: protocol, service, flag, kdd_0, kdd_1, kdd_2, kdd_4, kdd_5, kdd_6, kdd_9, kdd_10, kdd_11, kdd_13, kdd_14, kdd_15, kdd_16, kdd_17, kdd_18, kdd_19, kdd_20 and kdd_21.
- R2L: protocol, service, flag, kdd_0, kdd_1, kdd_2, kdd_3, kdd_4, kdd_5, kdd_6, kdd_11, kdd_13, kdd_14 and kdd_15.
- U2R: protocol, service, flag, kdd_0,kdd_1,kdd_2,kdd_15,kdd_16,kdd_21, kdd_24,kdd_25,kdd_26,kdd_28 and kdd_31.

The proposed feature selection method's discrimination capability is demonstrated using a box plot. The distribution of data into distinct ranges between the variables is represented by the box plot and whisker plots. The x-axis represents the number of classes, while the y-axis represents the variance. To compare the differences between before and after feature selection, a box chart is employed. When comparing the medians of before and after feature selection, before feature selection has a higher median. The box plot and whisker chart are well suited for the suggested feature selection strategy since the variation after feature selection is better. As a result, for reducing the number of features, a feature selection method based on entropy is proposed. Figure 3 shows the box plot and whisker chart.

Figure 3. Box plot & whisker chart for classes

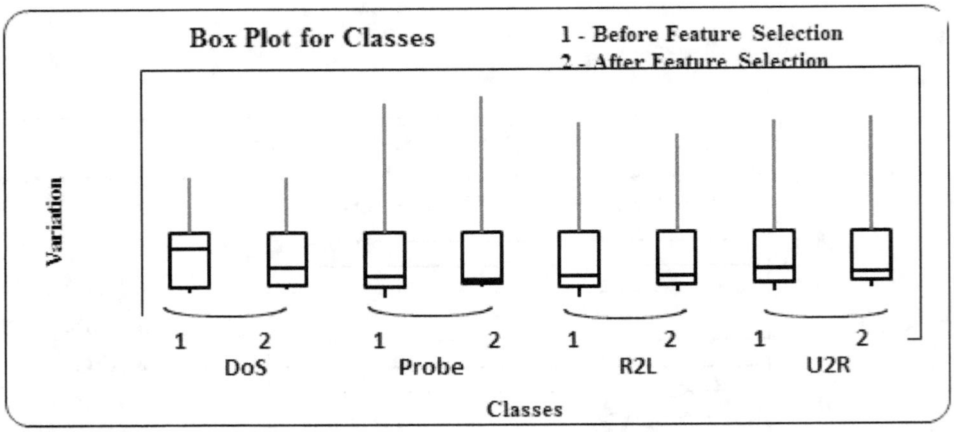

Entropy values are estimated before and after the feature selection procedure to highlight the efficiency of the suggested feature selection approach. Based on these entropy values, the within-class entropy for the classes DoS, probe, R2L, and U2R is high before feature selection and low after feature selection. The between-class entropy is low before feature selection and high after feature selection. The rulesets are generated using the various classes. The rulesets are used to ensure that the radial basis neural network is proper. As a result, the specified features and rulesets are excellent for evaluating proposed systems. Figure 4 depicts the average of between-class and within-class entropy.

Figure 4. Average of between-class and within-class entropy

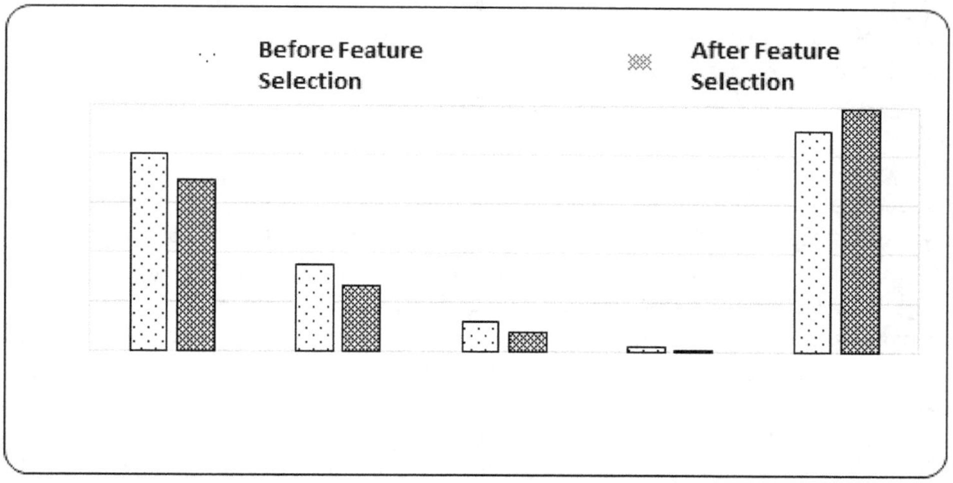

The strategies are used to identify the key characteristics of each class. The training and testing datasets with selected features are employed using the radial basis function. All 41 feature values in the Dataset, records of normal and snmpgetattack are identical. The number of erroneous correlations may rise as a result of this problem. The following are examples of normal and snmpgetattack records.

i. 0,udp,private,SF,105,146,0,0,0,0,0,0,0,0,0,0,0,0,0,0,0,0,2,2,0.00,0.00,0.00,0.00,1.00,0.00,0.00,255,253,0.99,0.01,0.00, 0.00,0.00,0.0 0,0.00,0.00,normal.
ii. 0,udp,private,SF,105,146,0,0,0,0,0,0,0,0,0,0,0,0,0,0,0,0,2,2,0.00,0.00,0.00,0.00,1.00,0.00,0.00,255,253,0.99,0.01,0.00, 0.00,0.00,0.0 0,0.00,0.00,snmpgetattack.
iii. 0,udp,private,SF,105,146,0,0,0,0,0,0,0,0,0,0,0,0,0,0,0,0,2,2,0.00,0.00,0.00,0.00,1.00,0.00,0.00,255,254,1.00,0.01,0.01, 0.00,0.00,0.0 0,0.00,0.00,normal.
iv. 0,udp,private,SF,105,146,0,0,0,0,0,0,0,0,0,0,0,0,0,0,0,0,2,2,0.00,0.00,0.00,0.00,1.00,0.00,0.00,255,254,1.00,0.01,0.01, 0.00,0.00,0.0 0,0.00,0.00,snmpgetattack.
v. 0,udp,private,SF,105,146,0,0,0,0,0,0,0,0,0,0,0,0,0,0,0,0,2,2,0.00,0.00,0.00,0.00,1.00,0.00,0.00,255,255,1.00,0.00,0.01, 0.00,0.00,0.0 0,0.00,0.00,normal.
vi. 0,udp,private,SF,105,146,0,0,0,0,0,0,0,0,0,0,0,0,0,0,0,0,2,2,0.00,0.00,0.00,0.00,1.00,0.00,0.00,255,255,1.00,0.00,0.01, 0.00,0.00,0.0 0,0.00,0.00,snmpgetattack.

The rulesets are measured using a radial basis neural network, KDD dataset is applied. The processing system includes an Intel-(R) Core-(TM)2 Duo - CPU - E7500 @ speed 2.93GHz, RAM - 2.00GB and a 32-bit operating system. Java Development Kit (JDK) is employed using KDD dataset to increase the detection rate (Nie, 2009). The rulesets for each class were produced based on the feature selection. The rulesets were used in a radial basis neural network to improvising the detection rates and reduced less rates of false positive. Rules for each class are listed below.

Rule 1: if protocol=tcp and kdd_0=0 and (service=http or finger or smtp) and (flag=s0 or sf) and kdd_5<=3 and kdd_7<=1 and kdd_8<=1andkdd_11=1and kdd_16=1.00 and kdd_20>=0.50 and kdd_21>=0.01 then DoS.

Rule 2: if protocol=udp and kdd_0=0 and service=private and flag=sf and kdd_1=28 and kdd_2=0 and kdd_3<=3 and kdd_12>=1 and kdd_13>=1 and kdd_15>=1 then DoS.

Rule 3: if kdd_1=icmp and kdd_0=0 and service=ecr_i and flag=sf and (kdd_1>=520 or kdd_1<=1480) and kdd_2=0then DoS.

Rule 4: if protocol =tcp and kdd_0=0 and (service=private or other) and (flag=rej or s0 or rstr) and kdd_15>=1 and kdd_16>=0.50 and kdd_17=1 and kdd_18>=0.97 and kdd_20=1.00 and kdd_21>=0.50then Probe.

Rule 5: if protocol =udp and kdd_0=0 and (service=other or domain_u or private) and flag=sf and (kdd_1<=1 or kdd_2<=1) and kdd_4<=2 and kdd_5<=2 and kdd_6<=1 and kdd_11=0 and kdd_13=0 kdd_14=0 and kdd_15>=20 and kdd_16<=2 and kdd_20<=22 and kdd_21>=0.01then Probe.

Rule 6: if protocol =icmp and kdd_0=0 and (service=ecr_i or eco_i or urp_i) and flag=sf and (kdd_1=0 or kdd_2=0) and kdd_9=0 and kdd_10<=1 and kdd_15<=2 and kdd_16<=50 and kdd_17=1.00 and kdd_18>=0.01 and kdd_19>=0.01then Probe.

Rule 7: if protocol =tcp and kdd_0 <=20 and (service=pop_3 or ftp_data or ftp) and (flag=sf or rsto) and (kdd_1>=20 or kdd_1<=28) and (kdd_2>=90 or kdd_2<=250) and kdd_3<=2 and kdd_4<=2 and kdd_11=1 and kdd_13=1 and kdd_15<=8then R2L.

Rule 8: if protocol =udp and kdd_0=0 and service=other and flag=sf and (kdd_1>=23 or kdd_1<=516) and kdd_2=0 and kdd_5=1 and kdd_6=1and kdd_14<=3 andkdd_15<=6 then R2L.

Rule 9: if protocol =icmp and kdd_0=0 and service=urp_i and flag=sf and kdd_1=552 and kdd_2=0 and kdd_4=1 and kdd_5=1 and kdd_13<=2 and kdd_15=1 and then R2L.

Rule 10: if protocol =tcp and kdd_0=0 and (service=ftp_data or telnet or ftp) and flag=sf and ((kdd_1=0 and kdd_2 !=0) or(kdd_1!=0 and kdd_2 =0)) and kdd_15<=2 and kdd_16<=3 and kdd_21=1.00 and kdd_24<=255 and kdd_25<=52 and kdd_26>=0.01 and kdd_28>=0.01and kdd_31<=0.05then U2R.

RADIAL BASIS NEURAL NETWORK (RBNN)

Radial-Basis Neural Network (RBNN), which is a three-layer feed forward network with input layers (kdd x), a hidden layer () that contains the RBF function, and an output layer (Normal, DoS, Probe, R2L, and U2R) that is a linear summation of hidden layers. Only selected features of each class are used as inputs. The Gaussian transfer function is applied to the hidden layer neuron, these output is inversely proportional from the neuron's centre distance. Weight of link (w) between the hidden and output layers is changed (Devaraju, 2013).

Only a few features are considered in this work for demonstration purposes. The purpose of employing a class-wise model is to lower the rate of false positives while increasing the detection rate. To detect intrusive events using a class-wise model, a large amount of time is necessary, which is achieved by making the class-wise model autonomous and blocking an assault. In the RBNN framework, each class is trained separately and then deployed in a sequential order. This dataset has four different classifications, including Denial of Service, Probe, U2R and R2L. Each class is having different feature sets on its own. To save computational time, the suggested system is a layered classifier. The class-wise model process checks the DoS class to see if it's a DoS attack, then declares it as such; otherwise, it checks the probe class to see if it's a probe attack, then declares it as such; otherwise, it tests the next class, and so on. The class-wise model and layered classifier are effective in lowering false positive rates, increasing detection rates, and requiring less processing effort.

For each neuron, the weight is calculated using the distance.

$$Weight() = RBF(distance) \tag{7}$$

Radial-basis Neural Network function is as follows:

$$f(x) = \sum_{i=1}^{no\_record} w_i \phi_i(x) \tag{8}$$

The 'f' model is a linear combination of a collection of 'no record' fixed functions, often known as basis functions. A linear combination of basis vectors is another name for it.

Class-Wise Layered Classifier

To detect the type of classes, the radial basis neural network is also employed to supplement the multi-layer classifier (Gupta, et al. 2010). The four classes used are DoS, Probe, R2L, and U2R. Attacks are detected if layer-1 is discovered; if not, the next layer-2 is measured, and the process is continued until the process is complete. A layered method is presented to boost detection rate while reducing processing time. The suggested layered classifier for each class is shown in Figure 5.

Figure 5. Proposed class-wise layered classifier

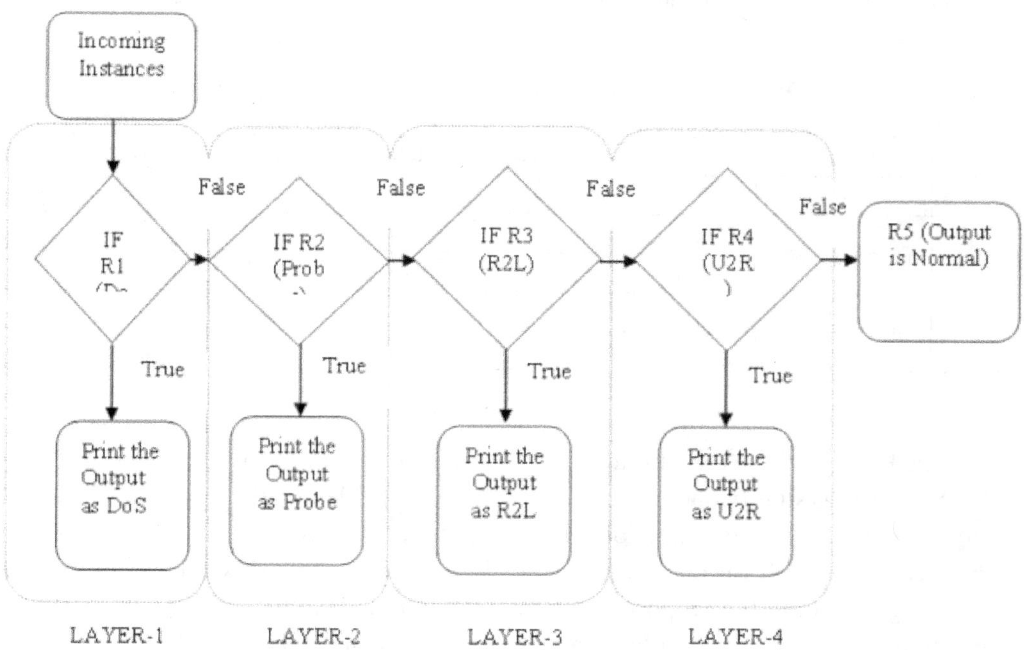

RESULTS AND DISCUSSION

Using a confusion matrix, two separate metrics, Attack Detection-Rate (ADR) and False Positive-Rate (FPR), features are applied to measure the performance (Devaraju, 2013,2014,15,23,40,41). True Negative reflects the amount of connections that were classified as normal but were not. False Positive refers to the amount of connections that were mistakenly labelled as an attack if it normal actually. In False Negative rates reflects the connections that were mistakenly classed as normal if it attack actually. True Positive reflects the connections that were categorised as attack if not.

Attack Detection-Rate (ADR): The ratio of total detected attacks to the total attacks in the specified dataset is known as the ADR.

$$ADR = \frac{\text{True Positive} + \text{True Negative}}{\text{True Positive} + \text{False Positive} + \text{False Negative} + \text{True Negative}} * 100 \quad (9)$$

False Positive-Rate (FPR): Ratio between the misclassified instances of the normal instances.

$$FPR = \frac{\text{Total Number of Instances Misclassified}}{\text{Total Normal Instances}} * 100 \quad (10)$$

Using a 10% of the KDD- Dataset (Training Data)

Table 4 shows the results of five different classes for training, and their efficiency.

Table 4. For 10% of the KDD dataset, the confusion matrix and training results are shown

	Normal	DoS	Probe	R2L	U2R	ADR %	FPR %
Normal	88963	346	0	2303	52	91.45	2.77
DoS	143	390446	0	0	0	99.74	0.04
Probe	244	1	3653	0	0	88.95	5.94
R2L	253	0	0	726	16	64.48	22.46
U2R	1	0	0	2	47	90.38	1.92

For 10% of the KDD Dataset, Figure 6.a depicts a visual representation of the ADR of training results. The ADR for each class is significant in the proposed approach, although the number of features selected is not. Section 4 shows the feature selection by class. The ADR is depicted in Figure 6.a.

Figure 6.b depicts the FPR of training results for 10% of the KDD Dataset in visual form. The FPR of each class is significant in the proposed system, although the features numbers selected is not. The feature selection by class is covered in section 4. The FPR is depicted in Figure 6.b.

The ADR and FPR of a 10% data set were calculated using the radial basis function. For training, 10% of the dataset is used. Table 4 shows the confusion matrix for five different classes, whereas figures 6.a and 6.b show the percentages of ADR and FPR for four different classes. The use of entropy-based feature selection to evaluate the ADR and FPR is proposed in this study. The ADR and FPR are shown using various features. The ADR of a normal attack is 91.45% and the FPR is 2.77 percent; the ADR of a DoS attack is 99.74 percent and the FPR is 0.04 percent; the ADR of a probe attack is 88.95 percent and the FPR is 5.94 percent; the ADR of an R2L attack is 64.48 percent and the FPR is 22.46 percent; and the ADR of a U2R attack is 90.38 percent and the FPR Since 41 features of normal and snmpgetattack values are the similar, the FPR for R2L is high, and sample is addressed in section 4.

Figure 6. ADR and FPR of training results for 10% data set

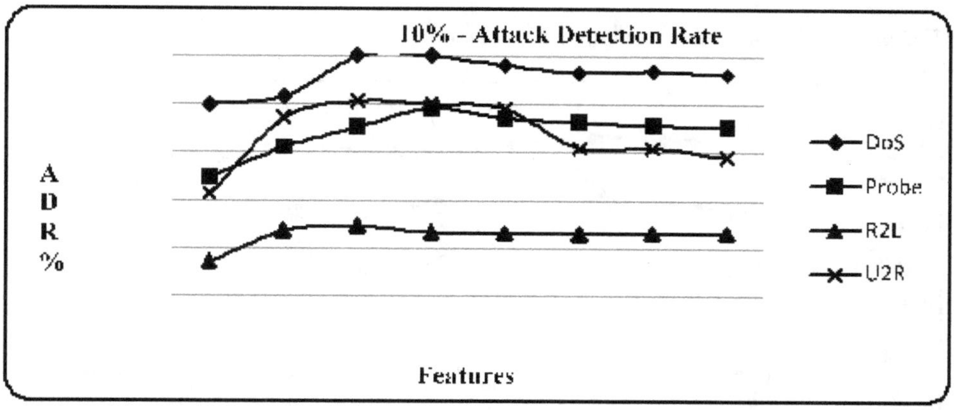

(a) ADR

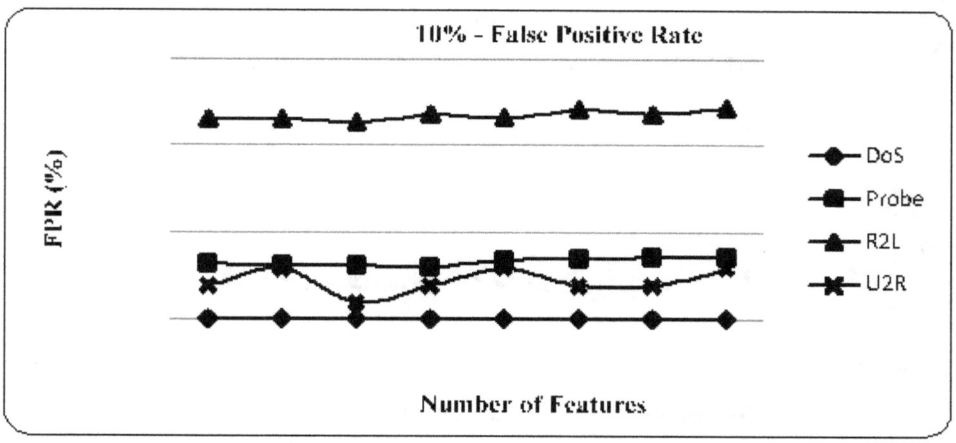

(b) FPR

Using KDD Datasets That Have Been Corrected (Testing Data)

Table 5, represents the results of testing different types of classes, as well as the efficiency.

Figure 7.a depicts the ADR of testing results for the updated KDD Dataset as a graphical representation. The ADR for each class is significant in the proposed approach, although the number of features selected is not. Section 4 shows the feature selection by class. ADR is depicted in Figure 7.a.

Table 5. Testing results and confusion matrix for the corrected KDD dataset

	Normal	DoS	Probe	R2L	U2R	ADR %	FPR %
Normal	58541	258	1012	1023	28	96.61	3.83
DoS	28	229137	0	243	0	99.69	0.01
Probe	209	3	3732	15	0	89.58	5.02
R2L	3018	0	0	9992	13	61.12	18.46
U2R	0	0	0	9	64	91.43	0.00

Figure 7. ADR and FPRof testing results for corrected dataset

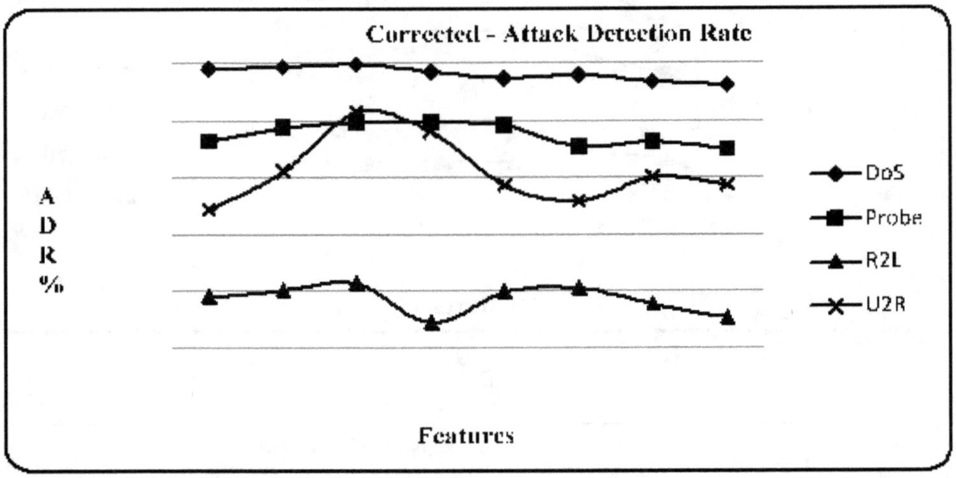

(a) ADR

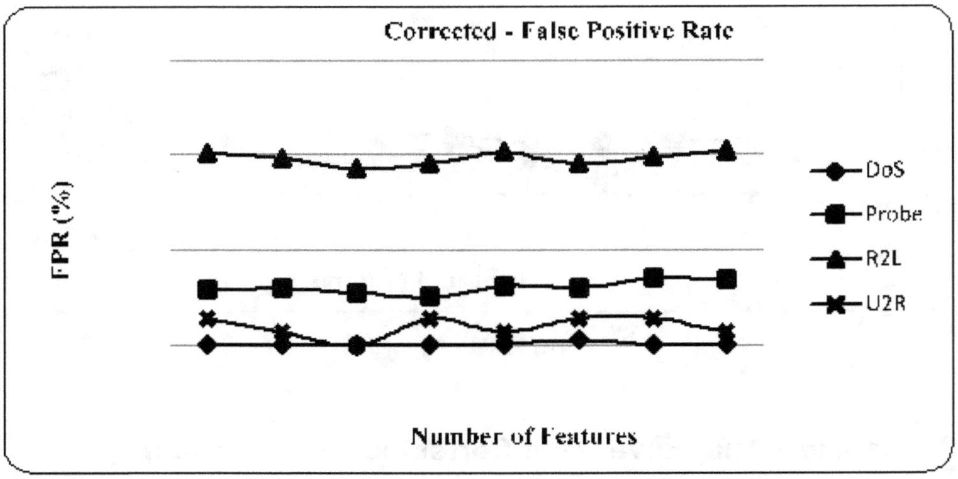

(b) FPR

Figure 7.b shows the representation of pictorial for the FPR of testing results for corrected KDD Dataset. In the proposed system, the FPR for each class is significant, whereas the number of features selected. The class-wise feature selection is shown in section 4. Figure 7.b shows the FPR.

Using the radial basis function, the ADR and FPR of the rectified KDD data set were calculated. For testing used for Corrected dataset. Table 5 represents the confusion matrix for five different classes, while figures 7.a and 7.b illustrate the percentages of ADR and FPR for four different classes. The ADR and FPR are evaluated using selection of entropy-based. The ADR and FPR are shown using various features. The ADR of a normal attack is 96.61 percent, and the FPR is 3.83 percent; the ADR of a DoS attack is 99.69 percent, and the FPR is 0.01 percent; the ADR of a probe attack is 89.58 percent, and the FPR is 5.02 percent; the ADR of an R2L attack is 61.12 percent, and the FPR is 18.46 percent; and the ADR of a U2R attack is 91. Since 41 features of normal and snmpgetattack values are the similar, the FPR for R2L is high, and sample is addressed in section 4.

Using the Complete KDD Dataset (Testing Data)

Table 6 shows the testing results for five different classes and their efficiency.

Table 6. Full KDD dataset confusion matrix & testing results

	Normal	DoS	Probe	R2L	U2R	ADR %	FPR %
Normal	932343	3018	0	21346	321	95.84	2.54
DoS	1087	3879167	0	0	0	99.89	0.03
Probe	2652	15	37478	2	1	91.18	6.45
R2L	238	0	0	694	17	61.63	21.14
U2R	0	0	0	2	47	90.38	0.00

Figure 8.a depicts the ADR of testing findings for the entire KDD dataset as a graphical depiction. The ADR for each class is significant in the proposed approach, although the number of features selected is not. Section 4 shows the feature selection by class. The ADR is depicted in Figure 8.a. Figure 8.b depicts the FPR of testing findings for the entire KDD dataset as a graphical depiction. The FPR for each class is significant in the proposed system, as is the amount of features chosen. The FPR is represented in Figure 8.b.

The ADR and FPR of the entire KDD data set were calculated using the radial basis function. In testing, full dataset is used. In this way, we used two separate datasets to test our hypothesis. Table 6 shows the confusion matrix for five different classes, while figures 8.a and 8.b illustrate the percentages of ADR and FPR for four different classes.

The ADR and FPR are evaluated using features of entropy-based. The ADR and FPR are shown using various features. The ADR of a normal attack is 95.84 percent, and the FPR is 2.54 percent; the ADR of a DoS attack is 99.89 percent, and the FPR is 0.03 percent; the ADR of a probe attack is 91.18 percent, and the FPR is 6.45 percent; the ADR of an R2L attack is 61.63 percent, and the FPR is 21.14 percent; and the ADR of a U2R attack is 90.38 Since 41 features of normal and snmpgetattack values are the similar, the FPR for R2L is high, and sample is addressed in section 4.

Figure 8. ADRand FPRof testing results for full data set

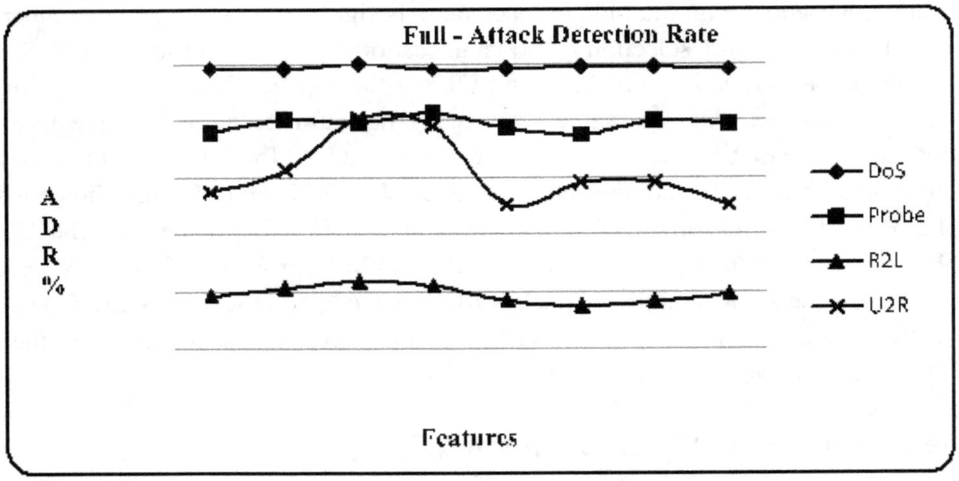

(a) ADR

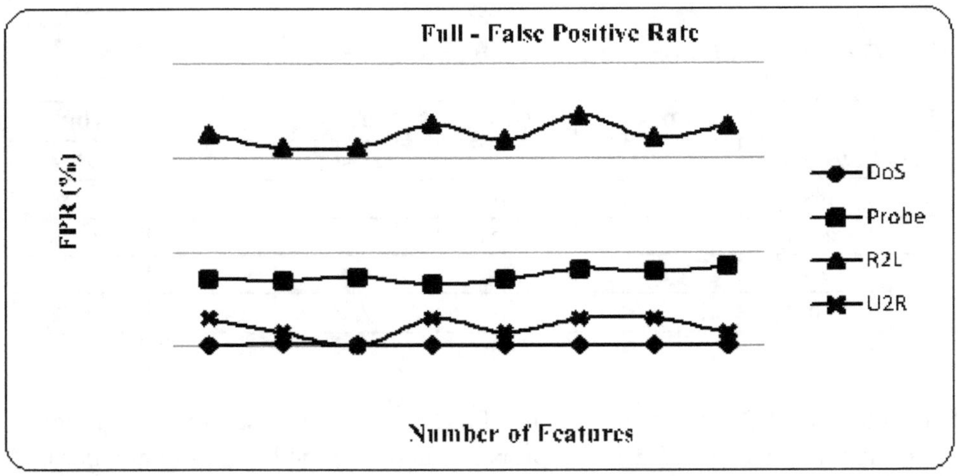

(b) FPR

KDD Dataset Comparison

Because DoS assaults primarily affect networks, entropy-based feature selection and radial basis function can be very successful in detecting DoS attacks and greatly lowering the FPR (Gupta, 2010; Devaraju, 2014)). The repaired dataset and the entire dataset are used to compare the test performance. The repaired dataset is compared to various techniques (table 7) for DoS and U2R attacks that increase ADR, reduce FPR, and reduce Computational Time (CT). The majority of the authors do not evaluate the entire dataset. In this research, the entire dataset is considered for testing, as shown in table 6. ADR alone isn't the sole criterion for evaluating an algorithm's performance. ADR is the major focus of intrusion detection methods. FPR and CT, on the other hand, are crucial parameters. We are attempting to obtain high

ADR and FPR while maintaining a decent CT. Because DoS and U2R attacks primarily affect networks, feature selection outperforms alternative methods when all three parameters of the proposed algorithm are taken into consideration. Proposed strategy is effective in ADR and decreases FPR and CT. Shows the experiments and the comparisons in Table 7. For DoS and U2R attacks, the overall ADR, FPR, and CT are also improved when compared to other algorithms (Revathi, 2015; Wei, 2015; Shi, 2011).

Table 7. Various algorithms are compared in terms of ADR and FPR.

Algorithm		DoS	Probe	R2L	U2R	CT (sec.)
Proposed Approach with feature selection	DR %	**99.69**	89.58	61.12	**91.43**	124
	FPR %	**0.01**	5.02	18.46	**0.00**	
Proposed Approach without feature selection	DR %	98.38	87.27	57.43	80.9	129
	FPR %	0.01	0.55	18.88	0.00	
Layered Conditional Random Fields (Gupta, 2010)	DR %	97.4	98.6	29.6	86.3	156
	FPR %	0.07	0.91	0.35	0.05	
Hybrid fuzzy Weiner method (Revathi, 2015)	DR %	88.92	87.92	89.12	89.1	343
	FPR %	1.34	1.34	1.38	1.29	
Cluster Center and Nearest Neighbor (Wei, 2015)	DR %	99.68	97.61	57.02	3.85	1570
	FPR %	0.09	2.08	14.78	2.64	
Support Vector Machine (Shi, 2011)	DR %	82.85	96.59	78.95	61.5	189
	FPR %	1.05	2.31	11.74	4.69	
SVM with hierarchical clustering (Shi, 2011)	DR %	99.53	97.55	28.81	19.7	193
	FPR %	0.04	1.41	13.54	4.16	

Figure 9. ROC Curves on ADR and FPR

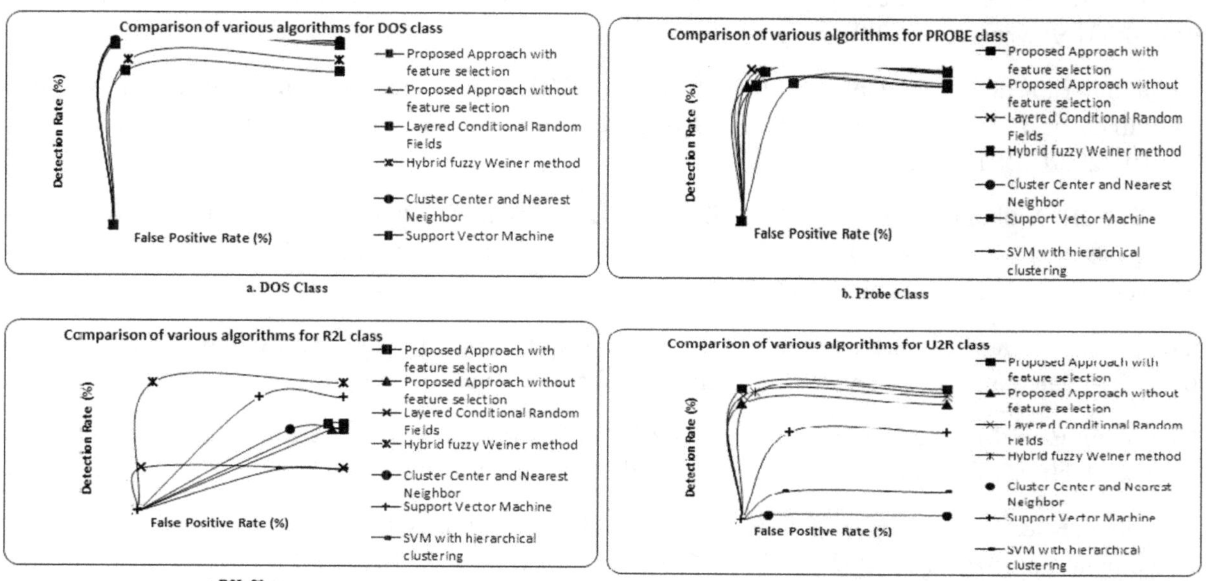

a. DOS Class

b. Probe Class

c. R2L Class

d. U2R Class

Figure 9 The ADR and FPR ROC curves are shown.

The ADR and FPR ROC curves for various classes, including DOS, Probe, R2L, and U2R, are shown in Figure 9. Other algorithms, both with and without feature selection, are compared to the suggested method. The false positive rate is represented by the x-axis in each of these ROC charts, while the detection rate is represented by the y-axis. According to the ROC curves, the suggested model has a good ADR rate and lowers FPR. The DoS class's ADR has improved, whereas the DoS and U2R classes' FPR has declined.

CONCLUSION

This study proposes a feature selection method based on entropy. Because the NIDS system has so many features, choosing the right ones is crucial. By lowering the number of features, entropy-based feature selection is presented for reducing computational time and enhancing detection rate. The radial basis function has been effectively used in a variety of benchmarks. The KDD datasets consist of a 10% training dataset, a corrected dataset, and the entire dataset for testing. When comparing to other methods, the adjusted dataset for testing is taken into account. Since this attack predominantly affected networks, the radial basis function detection rate has improved to 99.69 percent for class DoS and 91.43 percent for U2R. In DoS, the false positive rate has dropped to 0.01 percent, while in U2R, it has dropped to 0.00 percent. The processing time is significant when compared to other algorithms. When compared to existing algorithms, the suggested system's overall performance has improved. The technique to extend this method by personalising the classifiers for future research.

REFERENCES

Abadeh, M. S., Mohamadi, H., & Habibi, J. (2011). Design and analysis of genetic fuzzy systems for intrusion detection in computer networks. *Expert Systems with Applications*, *38*(6), 7067–7075. doi:10.1016/j.eswa.2010.12.006

Abuadlla, Y., Kvascev, G., Gajin, S., & Jovanovic, Z. (2014). Flow-Based Anomaly Intrusion Detection System Using Two Neural Network Stages. *Computer Science and Information Systems*, *11*(2), 601–622. doi:10.2298/CSIS130415035A

Anuar, N. B., Sallehudin, H., Gani, A., & Zakaria, O. (2008). Identifying False Alarm for Network Intrusion Detection System using Hybrid Data Mining and Decision Tree. *Malaysian Journal of Computer Science*, *21*(2), 101–115. doi:10.22452/mjcs.vol21no2.3

Chan, Lee, & Heng. (2013). Discovering fuzzy association rule patterns and increasing sensitivity analysis of XML-related attacks. *Elsevier Jr. of Netw. And Compu. Appl.*, *36*, 829–842.

Cingolani, P. (n.d.). *jFuzzyLogic: Open Source Fuzzy Logic library and FCL language implementation (fcl code).* http://jfuzzylogic.sourceforge.net/html/example_fcl.html

Cingolani, P., & Alcala-Fdez, J. (2012). A Robust and Flexible Fuzzy-Logic Inference System Language Implementation. *WCCI 2012 IEEE World Cong. on Compl. Intelg.*, 1090-1097.

Cingolani, P., & Alcala-Fdez, J. (2013). jFuzzyLogic: A Java Library to Design Fuzzy Logic Controllers According to the Standard for Fuzzy Control Programming. *Intl. Jr. of Compl. Intelg. Sys.*, *6*(1, Supplement 1), 61–75. doi:10.1080/18756891.2013.818190

Cup, K. D. D. (1999). *Intrusion Detection Data*. http://kdd.ics.uci.edu/databases/kddcup99/kddcup99.html

Devaraju S. (2019). Evaluation of Efficiency For Intrusion Detection System Using Gini Index C5 Algorithm. *International Journal of Engineering and Advanced Technology, 8*(6), 2196-2200.

Devaraju, S., & SaravanaPrakash, D. (2019). Developing Efficient Web-Based XML Tool. *International Journal of Recent Technology and Engineering, 8*(3), 8580-8584.

Devaraju, S. & SaravanaPrakash, D. (2019). *Total Benefit Administration for Industry Environment*. TEST Engineering and Management.

Devaraju, S., & Ramakrishnan, S. (2011). Performance Analysis of Intrusion Detection System Using Various Neural Network Classifiers. *IEEE International Conferenceon Recent Trends in Information Technology (ICRTIT 2011),* 3-5.

Devaraju, S., & Ramakrishnan, S. (2013). Performance Comparison of Intrusion Detection System using Various Techniques – A Review. *ICTACT Journal on Communication Technology*, *4*(3), 802–812. doi:10.21917/ijct.2013.0114

Devaraju, S., & Ramakrishnan, S. (2013). Detection of Accuracy for Intrusion Detection System using Neural Network Classifier. *International Journal of Emerging Technology and Advanced Engineering*, *3*(1), 338–345.

Devaraju, S., & Ramakrishnan, S. (2014). Performance Comparison for Intrusion Detection System using Neural Network with KDD Dataset. *ICTACT Journal on Soft Computing*, *4*(3), 743–752. doi:10.21917/ijsc.2014.0106

Devaraju, S., & Ramakrishnan, S. (2015). Detection of Attacks for IDS using Association Rule Mining Algorithm. *Journal of the Institution of Electronics and Telecommunication Engineers*, *61*(6), 624–633. doi:10.1080/03772063.2015.1034197

Devaraju, S., & Ramakrishnan, S. (2019). Association Rule-Mining-Based Intrusion Detection System with Entropy-Based Feature Selection: Intrusion Detection System. In *Handbook of Research on Intelligent Data Processing and Information Security Systems*. IGI Global., doi:10.4018/978-1-7998-1290-6

Devaraju, S., & Ramakrishnan, S. (2020). Fuzzy Rule-Based Layered Classifier and Entropy-Based Feature Selection for Intrusion Detection System. In *Handbook of Research on Cyber Crime and Information Privacy*. IGI Global., doi:10.4018/978-1-7998-5728-0

Eesa, A. S., Orman, Z., & Brifcani, A. M. A. (2015). A novel feature-selection approach based on the cuttlefish optimization algorithm for intrusion detection systems. *Elsevier Expert Systems with Applications*, *42*(5), 2670–2679. doi:10.1016/j.eswa.2014.11.009

Gupta, K. K., Nath, B., & Kotagiri, R. (2010). Layered Approach Using Conditional Random Fields for Intrusion Detection. *IEEE Transactions on Dependable and Secure Computing*, *7*(1), 35–49. doi:10.1109/TDSC.2008.20

Horng, S.-J., Su, M.-Y., Chen, Y.-H., Kao, T.-W., Chen, R.-J., Lai, J.-L., & Perkasa, C. D. (2011). A novel intrusion detection system based on hierarchical clustering and support vector machines. *Elsevier Expert Systems with Applications*, *38*(1), 306–313. doi:10.1016/j.eswa.2010.06.066

Jawahar, S., Harishchander, A., & Devaraju, S. (2020). Closed Sequential Pattern Mining in Biological Data. *International Journal of Life science and Pharma Research, 8*(8), 9-13.

Jawahar, S., Harishchander, A., & Devaraju, S.(2020). Efficiently Mining Closed Sequence Patterns in DNA without Candidate Genertion. *International Journal of Life science and Pharma Research*, (8), 14-18.

Jiang, Gan, Wang, & Wang. (2011). Research of the Intrusion Detection Model Based on Data Mining. *Elsevier Energy Procedia*, *13*, 855–863.

Jiang, H., & Ruan, J. (2009). The Application of Genetic Neural Network in Network Intrusion Detection. *Journal of Computers*, *4*(12), 1223–1230. doi:10.4304/jcp.4.12.1223-1230

Lin, S.-W., Ying, K.-C., Lee, C.-Y., & Lee, Z.-J. (2012). An intelligent algorithm with feature selection and decision rules applied to anomaly intrusion detection. *Elsevier Applied Soft Compu.*, *12*(10), 3285–3290. doi:10.1016/j.asoc.2012.05.004

Lin, W.-C., Ke, S.-W., & Tsai, C.-F. (2015). CANN: An intrusion detection system based on combining cluster centers and nearest neighbors. *Elsevier Knowledge-Based Systems*, *78*, 13–21. doi:10.1016/j.knosys.2015.01.009

Luo, B., & Xia, J. (2014). A novel intrusion detection system based on feature generation with visualization strategy. *Elsevier Expert Systems with Applications*, *41*(9), 4139–4147. doi:10.1016/j.eswa.2013.12.048

Mabu, S., Chen, C., Lu, N., Shimada, K., & Hirasawa, K. (2011). An Intrusion-Detection Model Based on Fuzzy Class-Association-Rule Mining Using Genetic Network Programming. *IEEE Transactions on Systems, Man and Cybernetics. Part C, Applications and Reviews*, *41*(1), 130–139. doi:10.1109/TSMCC.2010.2050685

Mohamed, A. B., Idris, N. B., & Shanmugum, B. (2012). *A Brief Introduction to Intrusion Detection System*. Comm. in Comp. and Info. *Sci.*, *330*, 263–271.

Nadiammai, G. V., & Hemalatha, M. (2014). Effective Approach toward Intrusion Detection System using Data Mining Techniques. *Elsevier Egyptian Informatics Journal*, *15*(1), 37–50. doi:10.1016/j.eij.2013.10.003

Ramakrishnan, S., & Devaraju, S. (2016). *Attack's Feature Selection-Based Network Intrusion Detection System Using Fuzzy Control Language. Springer-International Journal of Fuzzy Systems*. doi:10.100740815-016-0160-6

Sarasamma, S. T., Zhu, Q. A., & Huff, J. (2005). Hierarchical Kohonenen Net for Anomaly Detection in Network Security. *IEEE Tr. on Sys. Man and Cybernetics*, *35*(2), 302–312. doi:10.1109/TSMCB.2005.843274 PMID:15828658

Selvakani,, S., & Kandeeban, R.S. (2011). A Genetic Algorithm Based elucidation for improving Intrusion Detection through condensed feature set by KDD 99 dataset. *Info. And Knowl. Mgnt.*, *1*(1), 1–9.

Sheikhan, & Jadidi, & Farrokhi. (2010). Intrusion detection using reduced-size RNN based on feature grouping. *Neu. Comput&Applic*, *21*(6), 1185–1190.

Sujendran, R., & Arunachalam, M. (2015). Hybrid Fuzzy Adaptive Wiener Filtering with Optimization for Intrusion Detection. *ETRI Journal*, *37*(3), 502–511. doi:10.4218/etrij.15.0114.0275

Tajbakhsh, A., Rahmati, M., & Mirzaei, A. (2009). Intrusion detection using fuzzy association rules. *Elsevier Appl. Soft Compu.*, *9*(2), 462–469. doi:10.1016/j.asoc.2008.06.001

Tsang, C.-H., Kwong, S., & Wang, H. (2007). Genetic-fuzzy rule mining approach and evaluation of feature selection techniques for anomaly intrusion detection. *Elsevier Pattern Recogn*, *40*(9), 2373–2391. doi:10.1016/j.patcog.2006.12.009

Wang, G., Hao, J., Ma, J., & Huang, L. (2010). A new approach to intrusion detection using Artificial Neural Networks and fuzzy clustering. *Elsevier Expert Sys. with Appl.*, *37*(9), 6225–6232. doi:10.1016/j.eswa.2010.02.102

Wang, M., & Zhao, A. (2012). Investigations of Intrusion Detection Based on Data Mining. *Springer Recent Advances in Computer Science and Information EngineeringLecture Notes in Electrical Engineering*, *124*, 275–279. doi:10.1007/978-3-642-25781-0_41

Wei & Di. (2010). A Probability Approach to Anomaly Detection with Twin Support Vector Machines. *Journal of Shanghai Jiaotong Univ. (Sci.)*, *15*(4), 385-391.

Chapter 13
Recognizing User Portraits for Fraudulent Identification on Online Social Networks

Sudha Senthilkumar
School of Computer Science and Engineering, VIT University, Vellore, India

Satha Sivam S.
School of Information Technology and Engineering, VIT University, Vellore, India

Brindha K.
School of Information Technology and Engineering, VIT University, Vellore, India

ABSTRACT

Online social networks (OSNs) are increasingly influencing the way people communicate with each other. Well known sites such as Facebook, LinkedIn, Twitter, and Google+ have millions of users across the globe. With the wide popularity there are lot of security and privacy threats to the users of online social networks (OSN) such as breach of privacy, viral marketing, structural attacks, malware attacks, and profile cloning. Social networks have permitted people to have their own virtual identities which they use to interact with other online users. It is also completely possible and not uncommon for a user to have more than one online profile or even a completely different anonymous online identity. Entity resolution (ER) is the task of matching two different online profiles potentially from social networks. Solving ER has an identification of fake profiles. The solution compares profiles based on similar attributes. The system was tasked with matching two profiles that were in a pool of extremely similar profiles.

INTRODUCTION

People can now have their own virtual identities on social networks, which they can use to connect with other online users. Millions of people use social media sites like Facebook, Twitter, and Google+. Facebook, one of the most popular social networks, recently completed one of the largest initial public

DOI: 10.4018/978-1-6684-3991-3.ch013

offerings in Internet history. These social networks enable real-life users to construct online profiles based on the data they provide. The profiles are online personas that can exist independently of their real-life counterparts. Among these profiles, users communicate directly by posting and sharing content, expressing opinions about one another's posts, etc. The concept of a social network can be described as a graph composed of nodes and vertices, where friends are the nodes and friendships are the vertices. During the registration process, these profiles are created. As the average social network's registration process usually requires the user to type in their information manually, it is very easy and not uncommon to create a profile with inaccurate or fake information. The public information of the profiles from different social networks could be of interest to multiple parties in order to match and correlate data in order to identify a single entity with different profiles. Entity Resolution is the process of matching profiles to create a single entity that represents one real-world entity. We present an alternative form of comparing profiles that takes advantage of other information that is available without using the training phase, while providing a more detailed source of information about people, searching the web for people across social networks, helping businesses know their candidates better before hiring them, improving marketing strategies, detecting fake profiles, etc. The objective is to present an alternative form of comparing profiles that takes advantage of other information that is available without using the training phase. By comparing other types of information if it was publicly available, we went further than just comparing image-based features between profiles to solve ER. A string comparison technique was used for image-based features such as profile images and posted images.

Scope of Proposed System

The proposed system is the first to detect this fraud automatically. We aim to prevent romance scammers from creating fraudulent profiles or engaging with potential victims before they start a romantic relationship. Earlier research found that romantic scam victims score highly on idealized romantic beliefs scales. These beliefs are captured using a combination of structured, unstructured, and deep-learned features.

To solve fake profiles, we designed a method that uses two primary modules. Using a "data hiding" module integrated into the "Comparison distributor" component, we acquire datasets of profiles from social networks, compare the attributes, and discover similarities between them. Through the "Match Selector" component, the "Profile Matching" module works. Based on the previously determined similarities between profiles, the purpose of this second module is to identify potential matches.

LITERATURE SURVEY

Title 1: Design And Evaluation Of A Real-Time Url Spam Filtering Service
Authors: Kurt Thomas, Chris Griery, Justin Ma, Vern Paxsony, Dawn Song
Year: 2011
Description:

Scams, phishing, and malware have all increased as a result of social networks, URL shorteners, and other web services. Although extensive research has been conducted, email-based spam filtering techniques generally fail to protect other web services. URLs are crawled and analyzed in real-time while they are submitted to web services to determine whether they direct to spam. As a result of the diversity of web service spam, we evaluate the viability of Monarch and the fundamental challenges it faces. Social networks, video-sharing sites, blogs, and consumer review pages are among the web services that have proliferated on the Internet. Phishing, malware, and scams have become commonplace since these services have become widespread. Despite extensive research on email spam, many of the solutions do not work for web services. Specifically, recent research indicates that domain and IP blacklists currently used by social network operators are inaccurate. (Thomas, 2011)

Due to Monarch's exclusive focus on URLs, we can reliably detect spam regardless of where a URL appears, or from which account it originates. It is for this reason that spam URL filtering is offered as a service. The properties of email spammers can be inferred from those of Twitter spammers, and vice versa. We compared the similarities and differences between email spam and Twitter spam, including the usage of generic redirectors and public web hosting, and the persistence of spam features over time.

Merits:

- The cloud service allows service providers to filter individual messages posted by users; but it functions in a manner that can be generalized to many types of web services.

Demerits:

- Spam infrastructure IP addresses achieve lower accuracy.

Title 2: Toward Worm Detection in Online Social Networks
Authors: Wei Xu, Fangfang Zhang, Sencun Zhu
Description

In recent years, the proliferation of malware on online social networking (OSN) web-sites has become a major security concern for both the sites and their users. The existing Internet worm detection mechanisms cannot detect these worms because they exhibit unique propagation vectors. Using the propagation characteristics of the OSN worm as well as the topological properties of the network, we propose an early warning OSN worm detection system. The following characteristics of online social networks have made them attractive targets for these worms (hereinafter called OSN worms). Online social networks have the properties of small shortest path lengths and high clustering because they are small-world networks. Meanwhile, it appears that OSN worms are spreading rapidly because users are tightly connected. As well as being scale-free networks, online social networks are also power-law networks, where high-degree nodes tend to form connections. By monitoring only a few hundreds of users, an algorithm based on the topological properties of social graphs is used to keep the OSN websites under surveillance. (Wei, 2010)

Merits:

- By applying a two-level correlation scheme to reduce the noise from normal user accounts, the system can reduce the propagation time between user accounts with a small average shortest path length.
- Identify infected user accounts with a higher level of accuracy through communications.

Demerits:

- Detection systems create decoy friends and add them to a user's friends list as low-interactive honeypots. Users receiving worm evidence can be decoy friends of accounts infected by OSN worms.

Title 3: Spam - The Underground On 140 Characters Or Less
Authors: Chris Grier,Kurt Thomas, Vern Paxson, Michael Zhang
Year: 2010
Description:

We describe spam on Twitter in this work. Our analysis reveals that 8 percent of the 25 million URLs that are posted to our site point to phishing, malware, or scams listed on popular blacklists. In our analysis of spam accounts, we uncover evidence that they originate from previously legitimate accounts that have been compromised and are now puppeteer by spammers. In order to gain insight into the underlying techniques used to attract users, we incorporate spam URLs into campaigns in order to identify trends that differentiate phishing, malware, and spam. It would be possible to significantly curtail Twitter spam by using URL blacklists. With the exception of malware, which is blocked by Google's Safebrowsing API, Twitter does not currently have a spam filtering mechanism, despite an increase in unsolicited messages. Twitter has developed a loose collection of heuristics to quantify spamming activity, such as excessive account creation or requests to befriend other users. We found that 8% of the distinct Twitter links point to spam using over 400 million tweets and 25 million URLs from public Twitter data. According to our analysis of spammer account behavior, only 16% of spam accounts are clearly automated bots, while the rest appear to be compromised accounts being manipulated. We found that Twitter spam is far more effective at convincing users to click on spam URLs than email, with an overall click-through rate of 0.13%.(Chris, 2010)

Merits:

- View a friend's message on Twitter, a single tweet contains the tweet text, the friend's name and icon, the time it was posted, the geo-location data, and the application it was posted with. Users can base their decision on these attributes only if a link is posted

Demerits:

- The limited number of spam URLs posted, an account's owner may have tweeted the URLs unintentionally, unaware that they were spam.

Title 4: Detecting Spammers on Twitter
Authors: Fabricio Benevenuto, Gabriel Magno, Tiago Rodrigues, and Virg´ılio Almeida
Year: 2010
Description

The growth of Twitter users around the world has led to the development of real-time search systems and different types of mining tools that are allowing people to better track the effects of news and events on Twitter. While appealing as a means to spread news and allow users to post their statuses and discuss events, these services open the door to new forms of spam. Users are directed to completely unrelated websites by spammers who use popular words and URLs to post tweets obfuscated by URL shorteners. Our first step was to collect a large dataset of Twitter data that included more than 54 million users, 1.9 billion links, and almost 1.8 billion tweets. We constructed a large labeled collection of users by using tweets related to three famous trending topics from 2009, and classified those users as spammers or non-spammers manually. Our results also highlight the most important attributes for spam detection on Twitter has recently emerged as a popular social system where users share and discuss about everything, including news, jokes, their take about events, and even their mood. (Fabricio, 2010)

Merits:

- A characterization of the users of this labelled collection, bringing to the light several attributes useful to differentiate spammers and non-spammers

Demerits:

- It have different subsets of attributes, our approach is able to detect spammers with high accuracy.

Title 5: Detecting Spammers On Social Networks
Authors: Gianluca Stringhini, Christopher Kruegel, Giovanni Vigna
Year: 2010
Description

Social networking has become a popular way for users to meet and interact online. Users spend a significant amount of time on popular social network platforms (such as Face- book, MySpace, or Twitter), storing and sharing a wealth of personal information. we analyze how spammers who target social networking sites operate. To collect the data about spamming activity, we created a large and diverse set of "honey-profiles" on three large social network- ing sites, and logged the kind of contacts and messages that they received. We then analyzed the collected data and identified anomalous behavior of users who contacted our profiles. Based on the analysis of this behavior, we developed techniques to detect spammers in social networks, and we aggregated their messages in large spam campaigns.Sites such as Facebook, MySpace, and Twitter are consistently among the top 20most-viewed web sites of the Internet. Another important characteristic of social networks is the different levels of user awareness with respect to threats. We believe that these techniques can help social networks to improve their security and detect malicious users. In fact, we develop a tool to detect spammers on Twitter. (Gianluca, 2010)

Merits:

- The information stored in their profiles, attracts spammers and other malicious users

Demerits:

- Twitter limited our machine to execute only 20,000 API calls per hour. Thus, to avoid wasting our limited API calls, we executed Google searches for the most common words in tweets sent by the already detected spammers.

Title 6: Uncovering Social Spammers: Social Honeypots + Machine Learning
Authors: Kyumin Lee, James Caverlee, Steve Webb
Year: 2010
Description

The opportunities for participants to engage, share, and interact. This community value and related services like search and advertising are threatened by spammers, content polluters, and malware disseminators. The conceptual framework and design considerations of the proposed approach, and we present concrete observations from the deployment of social honeypots in MySpace and Twitter. One of the key features of these systems is their reliance on users as primary contributors of content and as annotators and raters of other's content. This reliance on users can lead to many positive effects, including large-scale growth in the size and content in the community, bottom-up discovery of "citizen-experts", serendipitous discovery of new resources beyond the scope of the system designers, and new social-based information search and retrieval algorithms. It develop effective tools for automatically detecting and filtering spammers who target social systems. By focusing on two different communities, we have seen how the general principles of (i) social honeypot deployment, (ii) robust spam profile generation, and (iii) adaptive and ongoing spam detection can effectively harvest spam profiles and support the automatic generation of spam signatures for detecting new and unknown spam. (Kyumin, 2010)

Merits:

- These users post tweets about several things such as online business, marketing and so on. Their posting approach is more sophisticated than duplicate spammers since spam tweets are randomly interspersed with seemingly innocuous legitimate tweets.

Demerits:

- The traditional email spam detection has focused on identifying spam messages which are of relatively low individual value to the spammer.
- This detection is potentially more disruptive to spammers, since these accounts typically represent a more expensive investment by the spammer.

Title 7: Towards Online Spam Filtering in Social Networks
Authors: Hongyu Gao, Yan Chen, Kathy Lee, Diana Palsetia, Alok Choudhary
Year: 2012
Description

Online social networks (OSNs) are extremely popular among Internet users. Unfortunately, in the wrong hands, they are also effective tools for executing spam campaigns. Accordingly, our system adopts a set of novel features that effectively distinguish spam campaigns. It drops messages classified as "spam" before they reach the intended recipients, thus protecting them from various kinds of fraud. We evaluate the system using 187 million wall posts collected from Facebook and 17 million tweets collected from Twitter. Online social networks (OSNs) are extremely popular collaboration and communication tools that have attracted millions of Internet users. Unfortunately, recent evidence shows that they can also be effective mechanisms for spreading attacks. Popular OSNs are increasingly becoming the target of phishing attacks launched from large botnets.URL comparison to incrementally reconstruct spam messages into campaigns, which are then identified by a trained classifier. We evaluate the system on two large datasets composed of over 187 million Facebook wall messages and 17 million tweets, respectively. The experimental results demonstrate that the system achieves high accuracy, low latency and high throughput, which are the crucial properties required for an online system. (Hongyu, 2012)

Merits:

- OSN specific features are those that need social network information to compute. General features are those that could also be used to detect spam outside OSNs.It is used in combination to train the best classifier.

Demerits

- The spam detection work but are not suitable for the OSN environment.

Title 8: Warning Tweet: A Detection System Forsuspicious Urls In Twitter Stream
Authors: Manjeet Chaudhary, A Hingoliwala
Year: 2012

Description

Twitter is a social networking site where users can exchange messages to other users particularly their followers. Usually the messages sent over twitter are known as tweets. Users can sent messages or tweets to users who do not follow thesender . This system find the correlations of URL redirect chains extracted from several tweets. It uses the fact that the malicious users or attackers have limited resources and thus they need to reuse them. URL redirect chains frequently share the same URLs for the attackers or malicious users. Twitter is a social networking Site used to share information between users. Users can send tweets to its followers, to a particular user and also to users who are not the followers of the sender. Twitter tweets can contain only a restricted number of characters thus twitter uses URL shortening services to reduce URL length. Conventional suspicious URL detection systems are ineffective in their protection against conditional redirection servers that distinguish investigators from normal browsers and redirect them to benign pages to cloak malicious landing pages. A new suspicious URL detection system for Twitter that is based on the correlations of URL redirect chains, which are difficult to fabricate. The system can find correlated URL redirect chains using the frequently shared URLs and determine their suspiciousness inalmost real time. (Manjeet, 2012)

Merits:

- The system is highly accurate and can be deployed as a near real-time system to classify large samples of tweets from the Twitter public timeline.

Demerits:

- It is not be successfully fetched, we considered correlations of URL redirect chains extracted from a number of tweets.Because attacker's resources are generally limited and need to be reused, their URL redirect chains usually share the same URLs.

Title 9: Compa: Detecting Compromised Accounts On Social Networks
Authors: Manuel Egele, Gianluca Stringhini, Christopher Kruegel, and Giovanni Vigna
Year: 2013
Description

As social networking sites have risen in popularity, cyber-criminals started to exploit these sites to spread malware and to carry out scams. Previous work has extensively studied the use of fake (Sybil) accounts that attackers set up to distribute spam messages (mostly messages that contain links to scam pages or drive-by download sites). Fake accounts typically exhibit highly anomalous behavior, and hence, are relatively easy to detect. Compromising legitimate accounts is very effective, as attackers can leverage the trust relationships that the account owners have established in the past. Online social networks, such as Facebook and Twitter, have become increasingly popular over the last few years.

People use social networks to stay in touch with family, chat with friends, and share news. The users of a social network build, over time, connections with their friends, colleagues, and, in general, people they consider interesting or trustworthy. We presented a novel approach to detect compromised accounts in social networks. More precisely, we developed statistical models to characterize the behavior of social network users, and we used anomaly detection techniques to identify sudden changes in their behavior. (Manuel, 2013)

Merits:

- The topology of the network that surrounds the spammer. They do not try to distinguish compromised from spam accounts.

Demerits:

- The examining the maliciousness of URLs would fail to detect this XSS worm attack, as the attacker could chose any benign domain. Legitimate account that has been taken over by an attacker.

SYSTEM ARCHITECTURE

In this section, we have provided a clear architecture for our proposed system. Here the system works on two phase one is admin part and the user part. Admin can check the transaction, customer details and cash details. In the user part we stats with the creation of an individual bank account user data will be encrypted for security and before logging in the blocking chain is implemented for pixel comparisons. If the selected pixel matched then used is allowed to login. User can do transaction and while making money transfer an E-coin is generated randomly according to the note and amount is deposited to the account in the form of that e-coin key. (Egele, 2013)

Figure 1. System architecture

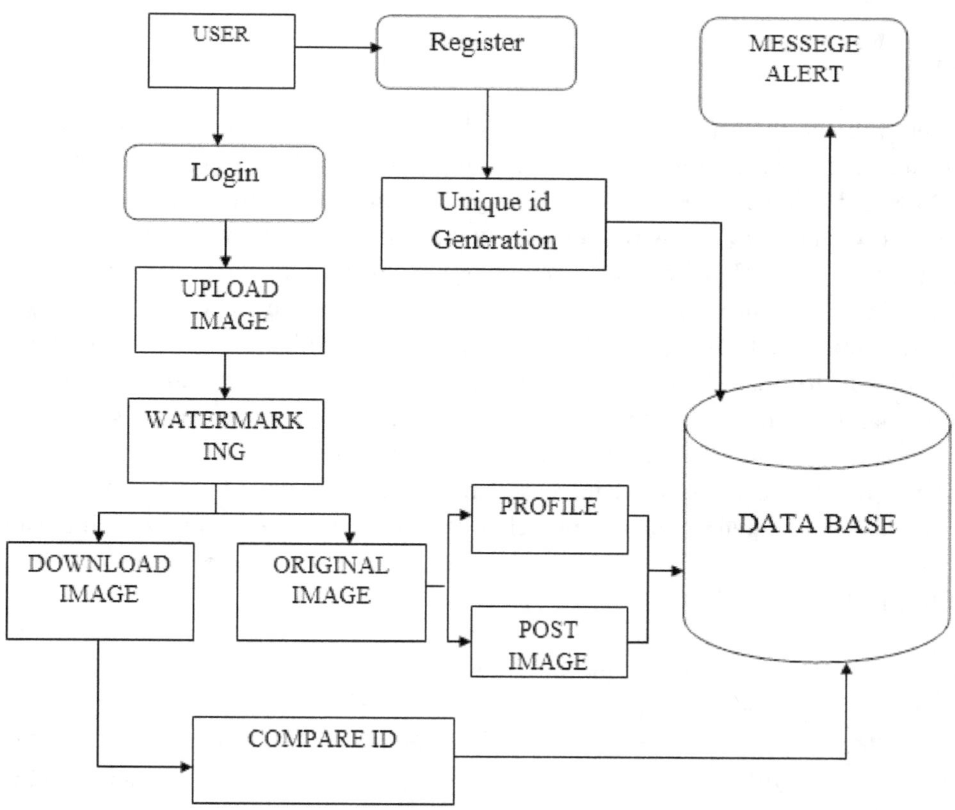

SYSTEM MODULES

Modules

- Login
- Hide data
- Profile Matching
- Altered Profiles

Login

The Login Form module presents site visitors with a form with username and password fields. If the user enters a valid username/password combination they will be granted access to additional resources on your website. Which additional resources they will have access to can be configured separately. Once logged in, the Login Form module presents the user with a Logout button. Logged in users who are inactive for a predetermined period of time will be automatically logged out. The Login Form module will appear in whatever module position it is assigned to in the current template. It is also possible to have a Login Form that will appear in place of regular content when a Menu Item is clicked.

Hide Data

In this module, it consists of a new stenographic algorithm for hiding data in images. Here we have also used a Steganography algorithm. Steganography is the practice of hiding secret message within any media. Most data hiding systems take advantage of human perceptual weaknesses. Steganography is often confused with cryptography because the two are similar in the way that they both are used to protect secret information. Here we have tested few images with different sizes of data to be hidden and concluded that the resulting steno images do not have any noticeable changes. In this module, the concern user who uploads the image will have an id that will be hidden within the image. Once another user who downloads the image cannot see the image as it is hidden. We have also used water mark techniques that will not be visible even for the users. Steganography technique finds its main application in the field of secret communication. The main advantage of this algorithm is to keep the size of the cover image constant while the secret message increased in size. It can be used by intelligence agencies across the world .Hence this new stenographic approach is robust and very efficient for hiding data in images.

Profile Matching

In this module, If the user who uploads the entire image can be viewed by the another user. The another user can downloaded the image but they cannot upload the same image this can be checked by the hidden id. The profile will be checked if the third party who upload the same image, this will be checked by the database. If the profile matches with the another profile, the another user cannot upload the same it consists of a new stenographic algorithm for hiding data in images. Another user can, Use the Image or else can upload the Image internal entry criteria matching system that checks for a primary match based on hard-coded, Already some data inside is there are not check. This profile matching module will check if another user who uploads the image which is in exists with the another user. There by this can avoids the fake user.

Alerted Profiles

If the profiles match, then the concern user will be alerted by the alert message. The user will be notified as their profile image has be tried to upload by another user and the user can block the person or else allow its user wish. User will also be notified with the fake users name, mail id, uploaded image, uploading time and system MAC Address. criteria match fails, no further weighing point match is attempted and the profile is either created newly or rejected based on parameter settings for this interface ID in fake profile. So finally give some Alert Message to the original User.

ALGORITHMNS AND APPROACHES

Algorithm: (Steganography Algorithm)

A new algorithm to hide data inside image using steganography technique. The proposed algorithm uses binary codes and pixels inside an image. The zipped file is used before it is converted to binary codes to maximize the storage of data inside the image. By applying the proposed algorithm, a system called Steganography Imaging System (SIS) is developed. The system is then tested to see the viability of the proposed algorithm. Various sizes of data are stored inside the images and the PSNR (Peak signal-to-noise ratio) is also captured for each of the images tested. Based on the PSNR value of each images, the stego image has a higher PSNR value. Hence this new steganography algorithm is very efficient to hide the data inside the image.

Steganography Imaging System (SIS) is a system that is capable of hiding the data inside the image. The system is using 2 layers of security in order to maintain data privacy. Data security is the practice of keeping data protected from corruption and unauthorized access. The focus behind data security is to ensure privacy while protecting personal or corporate data. Privacy, on the other hand, is the ability of an individual or group to seclude them or information about themselves and thereby reveal them selectively. Data privacy or information privacy is the relationship between collection and dissemination of data, technology, the public expectation of privacy, and the legal issues.

We then tested the algorithm using the PSNR (Peak signal-to-noise ratio). PSNR is a standard measurement used in steganograpy technique in order to test the quality of the stego images. The higher the value of PSNR, the more quality the stego image will have. If the cover image is C of size M × M and the stego image is S of size N × N, then each cover image C and stego image S will have pixel value (x, y) from 0 to M-1 and 0 to N-1 respectively. The PSNR is then calculated as follows:

$$PSNR = 10.\log_{10}\left(\frac{MAX^2}{MSE}\right)$$

where

$$MSE = \frac{1}{MN}\sum_{x=0}^{M-1}\sum_{y=0}^{N-1}\left(C(x,y)-S(x,y)\right)^2 \quad (1)$$

Watermarking

- A watermark is a “secret message” that is embedded into a “cover message”.
- Usually, only the knowledge of a secret key allows us to extract the watermark.
- Has a mathematical property that allows us to argue that its presence is the result of deliberate actions.
- Effectiveness of a watermark is a function of its » Stealth » Resilience » Capacity

SCREENSHOTS

Figures 2 through 9 show screenshots of a proposed system.

Figure 2.

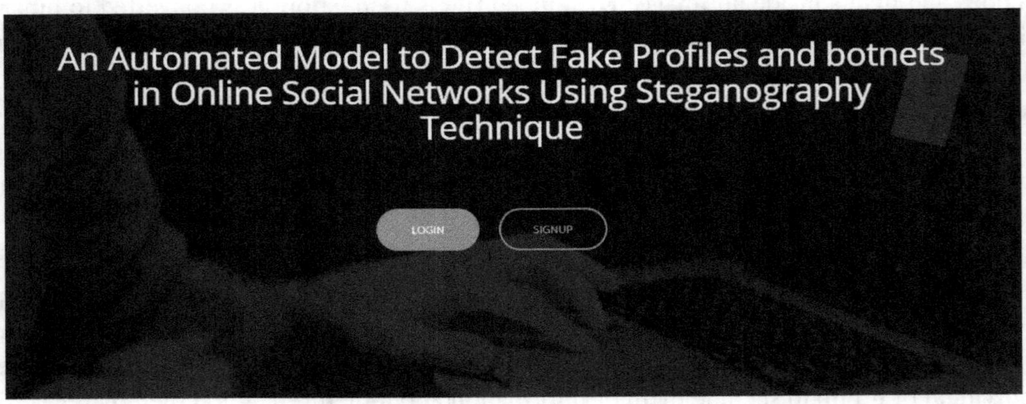

CONCLUSIONS AND FUTURE WORK

We solved Entity Resolution with our system and used it to compare online user profiles from social networks in order to identify matches. Our systems are comparing the two images and identify that fake or not. We are using Steganography Algorithm and that algorithm hides the information inside the image. In this way new images upload in our profile and that image compare to existing user profile. If the image is fake when send notification to original user.

The original user allows the uploading notification that images was uploaded otherwise blocked.

Figure 3.

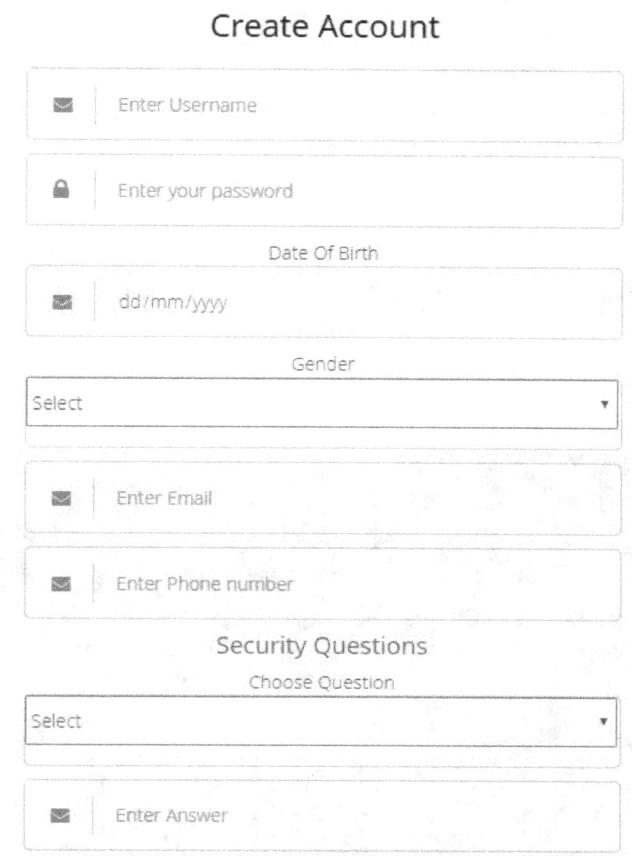

Figure 4.

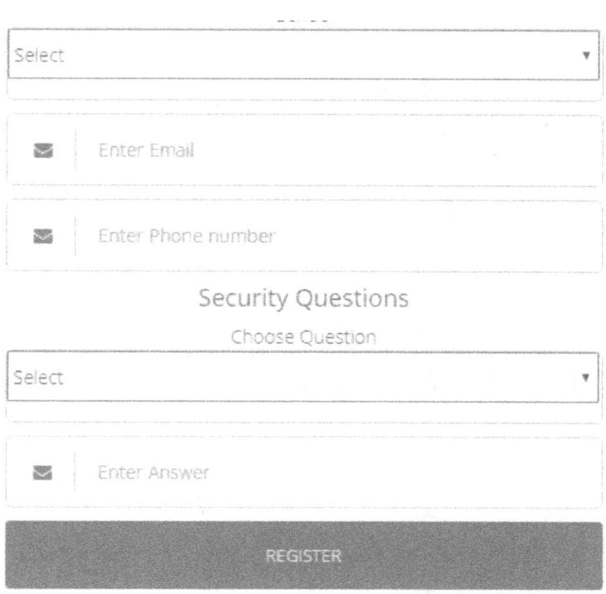

Figure 5.

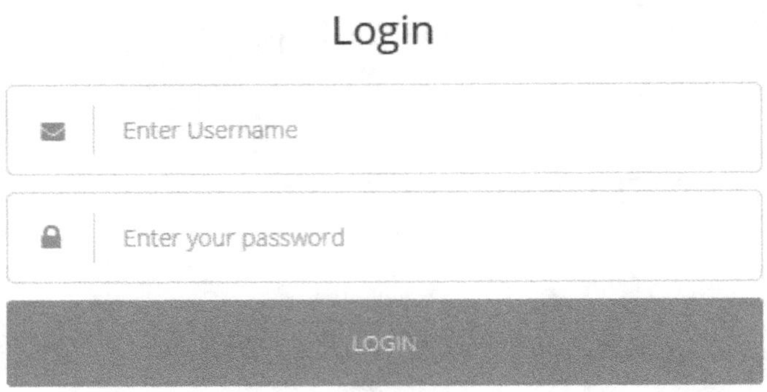

Figure 6.

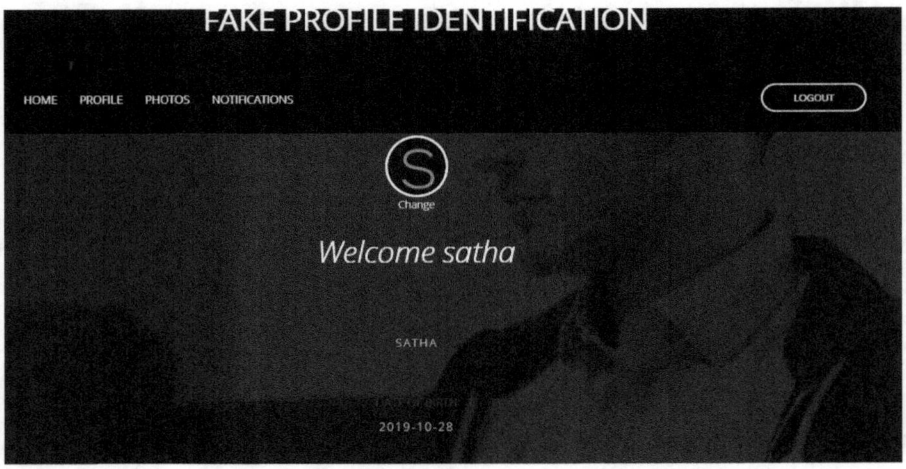

Figure 7.

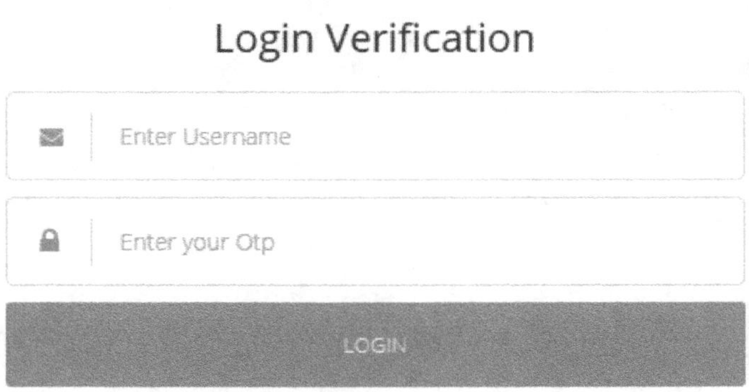

Figure 8.

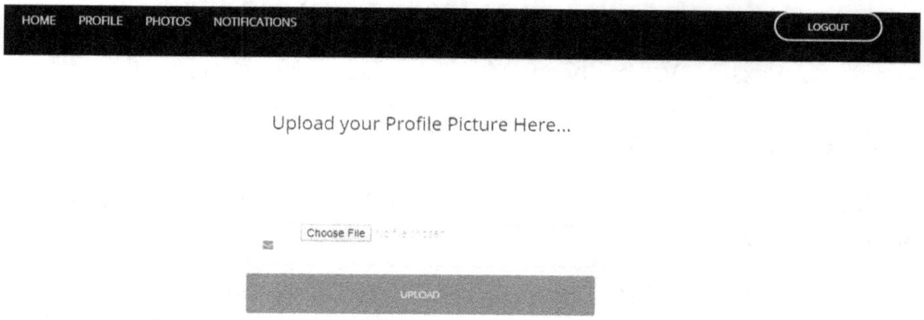

Figure 9.

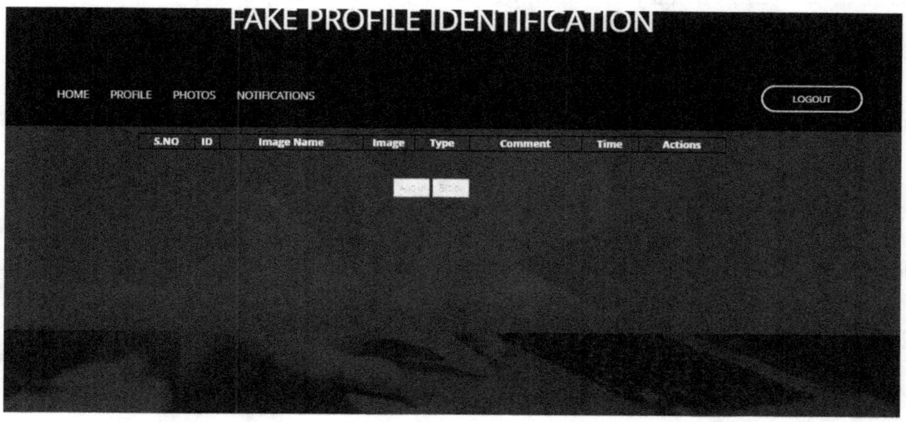

Future Enhancement

Presently multi day's equivalent client make such a significant number of record and have such a significant number of copy account that is make such a large number of issue to another individuals so the future improvement that copy account is have to discover and maintaining a strategic distance from the making of copy clients utilizing for client exercises and help to follow the records and same movement of client.

REFERENCES

Benevenuto, F., Magno, G., Rodrigues, T., & Almeida, V. (2010, July). Detecting spammers on twitter. In Collaboration, electronic messaging, anti-abuse and spam conference (CEAS) (Vol. 6, No. 2010, p. 12). Academic Press.

Egele, M., Stringhini, G., Kruegel, C., & Vigna, G. (2013, February). Compa: Detecting compromised accounts on social networks. NDSS.

Gao, H., Chen, Y., Lee, K., Palsetia, D., & Choudhary, A. N. (2012, February). Towards online spam filtering in social networks. In NDSS (Vol. 12, No. 2012, pp. 1-16). Academic Press.

Grier, C., Thomas, K., Paxson, V., & Zhang, M. (2010, October). @ spam: the underground on 140 characters or less. In *Proceedings of the 17th ACM conference on Computer and communications security* (pp. 27-37). 10.1145/1866307.1866311

Jagatic, T. N., Johnson, N. A., Jakobsson, M., & Menczer, F. (2007). Social phishing. *Communications of the ACM*, *50*(10), 94–100.

Lee, K., Caverlee, J., & Webb, S. (2010, July). Uncovering social spammers: social honeypots+ machine learning. In *Proceedings of the 33rd international ACM SIGIR conference on Research and development in information retrieval* (pp. 435-442). ACM.

Lee, S., & Kim, J. (2012, February). WarningBird: Detecting Suspicious URLs in Twitter Stream. In Ndss (Vol. 12, pp. 1-13). Academic Press.

Stringhini, G., Kruegel, C., & Vigna, G. (2010, December). Detecting spammers on social networks. In *Proceedings of the 26th annual computer security applications conference* (pp. 1-9). Academic Press.

Thomas, K., Grier, C., Ma, J., Paxson, V., & Song, D. (2011, May). *Design and evaluation of a real-time url spam filtering service. In 2011 IEEE symposium on security and privacy*. IEEE.

Xu, W., Zhang, F., & Zhu, S. (2010, December). Toward worm detection in online social networks. In *Proceedings of the 26th annual computer security applications conference* (pp. 11-20). 10.1145/1920261.1920264

Compilation of References

Abadeh, M. S., Mohamadi, H., & Habibi, J. (2011). Design and analysis of genetic fuzzy systems for intrusion detection in computer networks. *Expert Systems with Applications*, *38*(6), 7067–7075. doi:10.1016/j.eswa.2010.12.006

Abawajy, J. (2014). User preference of cyber security awareness delivery methods. *Behaviour & Information Technology*, *33*(3), 237–248.

Abuadlla, Y., Kvascev, G., Gajin, S., & Jovanovic, Z. (2014). Flow-Based Anomaly Intrusion Detection System Using Two Neural Network Stages. *Computer Science and Information Systems*, *11*(2), 601–622. doi:10.2298/CSIS130415035A

Adhikari, M., Das, A., & Mukherjee, A. (2016). Utility Computing and Its Utilization. In *Emerging Research Surrounding Power Consumption and Performance Issues in Utility Computing* (pp. 1–21). IGI Global. doi:10.4018/978-1-4666-8853-7.ch001

Ahanger, T., & Aljumah, A. (2020). Cyber Security Threats, Challenges and Defense mechanisms in Cloud Computing. *IET Communications*, *14*. Advance online publication. doi:10.1049/iet-com.2019.0040

Ahmed, E., & Rehmani, M. H. (2017). *Mobile edge computing: Opportunities, solutions, and challenges*. Elsevier.

Alam, M. S., & Vuong, S. T. (2013, August). Random forest classification for detecting android malware. In IEEE international conference on green computing and communications and IEEE Internet of Things and IEEE cyber, physical and social computing (pp. 663-669). IEEE.

Al-Dalky, R., Abduljaleel, O., Salah, K., Otrok, H., & Al-Qutayri, M. (2014). A Modbus traffic generator for evaluating the security of SCADA systems. *Proceedings of the international symposium on communication systems, networks digital sign*, 809–14. 10.1109/CSNDSP.2014.6923938

Al-Haj, A., & Al-Tameemi, M. (2018). Providing security for NFC-based payment systems using a management authentication server. *2018 4th International Conference on Information Management (ICIM).*

Ali Al-Haj. (2016). *Combining Cryptography and Digital Watermarking for Secured Transmission of Medical Images*. IEEE.

Ali, K. E., Mazen, S. A., & Hassanein, E. E. (2018). A proposed hybrid model for Adopting Cloud Computing in E-government. *Future Computing and Informatics Journal*, *3*(2), 286–295. doi:10.1016/j.fcij.2018.09.001

Al-Juaid, N., & Gutub, A. (2019). Combining rsa and audio steganography on personal computers for enhancing security. *SN Applied Sciences*, *1*(8), 830.

Al-Mutairi, S., & Manimurugan, S. (2017). The clandestine image transmission scheme to prevent from the intruders. *International Journal of Advanced and Applied Sciences*, *4*(2), 52–60.

Almutairi, S., Manimurugan, S., & Aborokbah, M. (2019). A new secure transmission scheme between senders and receivers using HVCHC without any loss. *EURASIP Journal on Wireless Communications and Networking*, *2019*(88), 1–15.

Alshaikh, M. (2020). Developing cybersecurity culture to influence employee behavior: A practice perspective. *Computers & Security, 98*. doi:10.1016/j.cose.2020.102003

Anuar, N. B., Sallehudin, H., Gani, A., & Zakaria, O. (2008). Identifying False Alarm for Network Intrusion Detection System using Hybrid Data Mining and Decision Tree. *Malaysian Journal of Computer Science*, *21*(2), 101–115. doi:10.22452/mjcs.vol21no2.3

Anusree, K., & Binnu, G. S. (2014). *Biometric Privacy using Visual Cryptography Halftoning and Watermarking for Multiple Secrets*. IEEE.

Arnatovich, Y. L., Wang, L., Ngo, N. M., & Soh, C. (2018). A comparison of android reverse engineering tools via program behaviors validation based on intermediate languages transformation. *IEEE Access: Practical Innovations, Open Solutions*, *6*, 12382–12394. doi:10.1109/ACCESS.2018.2808340

Arp, D., Spreitzenbarth, M., Hubner, M., Gascon, H., Rieck, K., & Siemens, C. E. R. T. (2014). Drebin: Effective and explainable detection of android malware in your pocket. In NDSS (Vol. 14, pp. 23-26). Academic Press.

Arzt, S., Rasthofer, S., Fritz, C., Bodden, E., Bartel, A., Klein, J., Le Traon, Y., Octeau, D., & McDaniel, P. (2014). Flowdroid: Precise context, flow, field, object-sensitive and lifecycle-aware taint analysis for android apps. *ACM SIGPLAN Notices*, *49*(6), 259–269. doi:10.1145/2666356.2594299

Attiya, H., & Welch, J. (2004). *Distributed computing: Fundamentals, simulations, and advanced topics* (Vol. 19). John Wiley & Sons. doi:10.1002/0471478210

Axelrod, C. W. (2017). Cybersecurity in the age of autonomous vehicles. intelligent traffic controls and pervasive transportation networks. In Systems, applications and technology conference (LISAT) (pp. 1-6). Academic Press.

Baca, R., Snášel, V., Platos, J., Kratky, M., & El-Qawasmeh, E. (2007). *The fast Fibonacci decompression algorithm.* arXiv preprint arXiv:0712.0811.

Baig, Z. A., Szewczyk, P., Valli, C., Rabadia, P., Hannay, P., Chernyshev, M., ... Peacock, M. (2017). Future challenges for smart cities: Cyber-security and digital forensics. *Digital Investigation*, *22*, 3–13.

Basu, M., & Prasad, B. (2010). Long range variant of Fibonacci universal code. *Journal of Number Theory*, *130*(9), 1925–1931. doi:10.1016/j.jnt.2010.01.013

Belfiore, J., Campbell, D., Capps, S., Cellini, S., Fitzgerald, C., Gundotra, V., … Rudder, E. (2006). *Distributed computing services platform.*

Bellaby, R. W. (2016). Justifying Cyber intelligence. *Journal of Military Ethics*, *15*(4), 299–319. doi:10.1080/15027570.2017.1284463

Benevenuto, F., Magno, G., Rodrigues, T., & Almeida, V. (2010, July). Detecting spammers on twitter. In Collaboration, electronic messaging, anti-abuse and spam conference (CEAS) (Vol. 6, No. 2010, p. 12). Academic Press.

Benson, V. (2017). *The state of global cyber security: Highlights and key findings*. . doi:10.13140/RG.2.2.22825.49761

Beresford, A. R., Rice, A., Skehin, N., & Sohan, R. (2011). Mockdroid: trading privacy for application functionality on smartphones. In *Proceedings of the 12th workshop on mobile computing systems and applications* (pp. 49-54). 10.1145/2184489.2184500

Berman, F., Fox, G., Hey, T., & Hey, A. J. (2003). *Grid computing: Making the global infrastructure a reality* (Vol. 2). John Wiley and Sons. doi:10.1002/0470867167

Bläsing, T., Batyuk, L., Schmidt, A. D., Camtepe, S. A., & Albayrak, S. (2010). An android application sandbox system for suspicious software detection. In *5th International Conference on Malicious and Unwanted Software* (pp. 55-62). IEEE.

Bose, B., Avasarala, B., Tirthapura, S., Chung, Y. Y., & Steiner, D. (2017). Detecting insider threats using radish: A system for real-time anomaly detection in heterogeneous data streams. *IEEE Systems Journal*.

Bouabdellah, M., Kaabouch, N., El Bouanani, F., & Ben-Azza, H. (2018). Network layer attacks and countermeasures in cognitive radio networks: A survey. *J InfSecurAppl*, *38*, 40–49. doi:10.1016/j.jisa.2017.11.010

Branco, T. Jr, de Sá-Soares, F., & Rivero, A. L. (2017). Key issues for the successful adoption of cloud computing. *Procedia Computer Science*, *121*, 115–122. doi:10.1016/j.procs.2017.11.016

Branley-Bell, D., Gómez, Y., Coventry, L., Vila, J., & Briggs, P. (2021). Developing and Validating a Behavioural Model of Cyberinsurance Adoption. *Sustainability*. doi:10.3390/su13179528

Brohi, S. N., Bamiah, M. A., Brohi, M. N., & Kamran, R. (2012). Identifying and analyzing security threats to virtualized cloud computing infrastructures. *2012 International Conference on Cloud Computing Technologies, Applications and Management (ICCCTAM)*, 151–155. 10.1109/ICCCTAM.2012.6488089

Brown, S., Gommers, J., & Serrano, O. (2015, October). From cyber security information sharing to threat management. In *Proceedings of the 2nd ACM workshop on information sharing and collaborative security* (pp. 43-49). ACM.

Buczak, A. L., & Guven, E. (2016). A Survey of Data Mining and Machine Learning Methods for Cyber Security Intrusion Detection. *IEEE Communications Surveys & Tutorials, 18*(2), 1153-1176. Advance online publication. doi:10.1109/COMST.2015.2494502

Bugiel, S., Davi, L., Dmitrienko, A., Fischer, T., & Sadeghi, A. R. (2011). *Xmandroid: A new android evolution to mitigate privilege escalation attacks.* Technische Universität Darmstadt, Technical Report TR-2011-04.

Burguera, I., Zurutuza, U., & Nadjm-Tehrani, S. (2011). Crowdroid: behavior-based malware detection system for android. In *Proceedings of the 1st ACM workshop on Security and privacy in smartphones and mobile devices* (pp. 15-26). 10.1145/2046614.2046619

Buschmann, T., & Bystrykh, L. V. (2013). Levenshtein error-correcting barcodes for multiplexed DNA sequencing. *BMC Bioinformatics*, *14*(1), 1–10. doi:10.1186/1471-2105-14-272 PMID:24021088

Buyya, R. (1999). High performance cluster computing: Architectures and systems (volume 1). *Prentice Hall.*

Buyya, R., & Venugopal, S. (2005). A gentle introduction to grid computing and technologies. *Database*, *2*, R3.

Buyya, R., Yeo, C. S., Venugopal, S., Broberg, J., & Brandic, I. (2009). Cloud computing and emerging IT platforms: Vision, hype, and reality for delivering computing as the 5th utility. *Future Generation Computer Systems*, *25*(6), 599–616. doi:10.1016/j.future.2008.12.001

Canali, C., Rabinovich, M., & Xiao, Z. (2005). Utility computing for Internet applications. In *Web Content Delivery* (pp. 131–151). Springer. doi:10.1007/0-387-27727-7_6

Canfora, G., Mercaldo, F., & Visaggio, C. A. (2015). Mobile malware detection using op-code frequency histograms. In *12th International Joint Conference on e-Business and Telecommunications (ICETE)* (Vol. 4, pp. 27-38). IEEE.

Cao, J., Zhang, Q., & Shi, W. (2018). Challenges and opportunities in edge computing. In *Edge Computing: A Primer* (pp. 59–70). Springer. doi:10.1007/978-3-030-02083-5_5

Carley, K. M., Cervone, G., Agarwal, N., & Liu, H. (2018, July). Social cyber-security. In *International conference on social computing, behavioral-cultural modeling and prediction and behavior representation in modeling and simulation* (pp. 389-394). Springer.

Caron, F. (2018). The Evolving Payments Landscape: Technological Innovation in Payment Systems. *IT Professional, 20*(2), 53–61.

Ceci. (2021). *Number of available applications in the Google Play Store from December 2009 to July 2021.* Statistica.

Chachapara, K., & Bhadlawala, S. (2013). Secure sharing with cryptography in cloud. *2013 Nirma University International Conference on Engineering (NUiCONE).* 10.1109/NUiCONE.2013.6780085

Champakamala, B., Padmini, K., & Radhika, D. (2014). Least significant bit algorithm for image steganography. *International Journal of Advancements in Computing Technology, 3*(4), 34–38.

Chan, Lee, & Heng. (2013). Discovering fuzzy association rule patterns and increasing sensitivity analysis of XML-related attacks. *Elsevier Jr. of Netw. And Compu. Appl., 36*, 829–842.

Choo, Gai, Chiaraviglio, & Yang. (2021). A multidisciplinary approach to Internet of Things (IoT) cybersecurity and risk management. *Computers & Security, 102.*

Choo, K. K. R., Kermani, M. M., Azarderakhsh, R., & Govindarasu, M. (2017). Emerging embedded and cyber physical system security challenges and innovations. *IEEE Transactions on Dependable and Secure Computing, 14*(3), 235–236.

Çimen, C., Akleylek, S., & Akyıldız, E. (2007). *Şifrelerin Matematiği Kriptografi.* ODTÜ Yayıncılık.

Cingolani, P. (n.d.). *jFuzzyLogic: Open Source Fuzzy Logic library and FCL language implementation (fcl code).* http://jfuzzylogic.sourceforge.net/html/example_fcl.html

Cingolani, P., & Alcala-Fdez, J. (2012). A Robust and Flexible Fuzzy-Logic Inference System Language Implementation. *WCCI 2012 IEEE World Cong. on Compl. Intelg.*, 1090-1097.

Cingolani, P., & Alcala-Fdez, J. (2013). jFuzzyLogic: A Java Library to Design Fuzzy Logic Controllers According to the Standard for Fuzzy Control Programming. *Intl. Jr. of Compl. Intelg. Sys., 6*(1, Supplement 1), 61–75. doi:10.1080/18756891.2013.818190

Conti, M., Dehghantanha, A., & Dargahi, T. (2018). Cyber threat Intelligence: Challenges and Opportunities. Cyber Threat Intelligence, 1–6. doi:. doi:10.1007/978-3-319-73951-9_1

Conti, M., Dargahi, T., & Dehghantanha, A. (2018). Cyber threat intelligence: challenges and opportunities. In *Cyber Threat Intelligence* (pp. 1–6). Springer.

Costa-jussà, M. R., Allauzen, A., Barrault, L., Cho, K., & Schwenk, H. (2017). Introduction to the special issue on deep learning approaches for machine translation. *Computer Speech & Language, 46*, 367–373.

Cup, K. D. D. (1999). *Intrusion Detection Data.* http://kdd.ics.uci.edu/databases/kddcup99/kddcup99.html

Cusumano, M. A. (2010). Cloud computing and SaaS as new computing platforms. *Communications of the ACM, 53*(4), 27–29. doi:10.1145/1721654.1721667

Dash, S. K., Suarez-Tangil, G., Khan, S., Tam, K., Ahmadi, M., Kinder, J., & Cavallaro, L. (2016). Droidscribe: Classifying android malware based on runtime behavior. In 2016 IEEE Security and Privacy Workshops (SPW) (pp. 252-261). IEEE.

Das, M., & Sinha, S. (2019). A Variant of the Narayana Coding Scheme. *Control and Cybernetics, 48*(3), 473–484.

Daykin, D. E. (1960). Representation of Natural Numbers as Sums of Generalized Fibonacci Numbers. *Journal of the London Mathematical Society*, *35*(2), 143–160. doi:10.1112/jlms1-35.2.143

Devaraju S. (2019). Evaluation of Efficiency For Intrusion Detection System Using Gini Index C5 Algorithm. *International Journal of Engineering and Advanced Technology, 8*(6), 2196-2200.

Devaraju, S. & SaravanaPrakash, D. (2019). *Total Benefit Administration for Industry Environment*. TEST Engineering and Management.

Devaraju, S., & Ramakrishnan, S. (2011). Performance Analysis of Intrusion Detection System Using Various Neural Network Classifiers. *IEEE International Conferenceon Recent Trends in Information Technology (ICRTIT 2011)*, 3-5.

Devaraju, S., & SaravanaPrakash, D. (2019). Developing Efficient Web-Based XML Tool. *International Journal of Recent Technology and Engineering, 8*(3), 8580-8584.

Devaraju, S., & Ramakrishnan, S. (2013). Detection of Accuracy for Intrusion Detection System using Neural Network Classifier. *International Journal of Emerging Technology and Advanced Engineering*, *3*(1), 338–345.

Devaraju, S., & Ramakrishnan, S. (2013). Performance Comparison of Intrusion Detection System using Various Techniques – A Review. *ICTACT Journal on Communication Technology*, *4*(3), 802–812. doi:10.21917/ijct.2013.0114

Devaraju, S., & Ramakrishnan, S. (2014). Performance Comparison for Intrusion Detection System using Neural Network with KDD Dataset. *ICTACT Journal on Soft Computing*, *4*(3), 743–752. doi:10.21917/ijsc.2014.0106

Devaraju, S., & Ramakrishnan, S. (2015). Detection of Attacks for IDS using Association Rule Mining Algorithm. *Journal of the Institution of Electronics and Telecommunication Engineers*, *61*(6), 624–633. doi:10.1080/03772063.2015.1034197

Devaraju, S., & Ramakrishnan, S. (2019). Association Rule-Mining-Based Intrusion Detection System with Entropy-Based Feature Selection: Intrusion Detection System. In *Handbook of Research on Intelligent Data Processing and Information Security Systems*. IGI Global., doi:10.4018/978-1-7998-1290-6

Devaraju, S., & Ramakrishnan, S. (2020). Fuzzy Rule-Based Layered Classifier and Entropy-Based Feature Selection for Intrusion Detection System. In *Handbook of Research on Cyber Crime and Information Privacy*. IGI Global., doi:10.4018/978-1-7998-5728-0

Dixit, P., & Silakari, S. (2020). Deep Learning Algorithms for Cybersecurity Applications: A Technological and Status Review. *Computer Science Review, 39*. doi:10.1016/j.cosrev.2020.100317

Dooley, J. F. (2013). *A Brief History of Cryptology and Cryptographic Algorithms*. Springer.

Eesa, A. S., Orman, Z., & Brifcani, A. M. A. (2015). A novel feature-selection approach based on the cuttlefish optimization algorithm for intrusion detection systems. *Elsevier Expert Systems with Applications*, *42*(5), 2670–2679. doi:10.1016/j.eswa.2014.11.009

Egele, M., Stringhini, G., Kruegel, C., & Vigna, G. (2013, February). Compa: Detecting compromised accounts on social networks. NDSS.

Elmrabet, Z., Elghazi, H., Sadiki, T., & Elghazi, H. (2016). A new secure network architecture to increase security among virtual machines in cloud computing. *Proceedings of the advances in ubiquitous networking*, 105–16. 10.1007/978-981-287-990-5_9

Enck, W., Gilbert, P., Han, S., Tendulkar, V., Chun, B. G., Cox, L. P., & Sheth, A. N. (2014). Taintdroid: An information-flow tracking system for realtime privacy monitoring on smartphones. *ACM Transactions on Computer Systems*, *32*(2), 1–29. doi:10.1145/2619091

Enck, W., Ongtang, M., & McDaniel, P. (2009). On lightweight mobile phone application certification. In *Proceedings of the 16th ACM conference on Computer and communications security* (pp. 235-245). ACM.

Engebretson, P. (2013). *The basics of hacking and penetration testing: ethical hacking and penetration testing made easy*. Elsevier.

Essaaidi, M., & Dari, Y. (2015). An overview of smart grid cyber-security state of the art study. *Proceedings of the 3rd international renewable and sustainable energy conference*, 1–7.

Evans, D. (2011). The internet of things: How the next evolution of the internet is changing everything. *CISCO White Paper, 1*(2011), 1–11.

Faisal, M. A., Aung, Z., Williams, J. R., & Sanchez, A. (2015). Data-stream-based intrusion detection system for advanced metering infrastructure in smart grid: A feasibility study. *IEEE Systems Journal, 9*(1), 31–44. doi:10.1109/JSYST.2013.2294120

Faruki, P., Bharmal, A., Laxmi, V., Ganmoor, V., Gaur, M. S., Conti, M., & Rajarajan, M. (2014). Android security: A survey of issues, malware penetration, and defenses. *IEEE Communications Surveys and Tutorials, 17*(2), 998–1022. doi:10.1109/COMST.2014.2386139

Faruki, P., Ganmoor, V., Laxmi, V., Gaur, M. S., & Bharmal, A. (2013). AndroSimilar: robust statistical feature signature for Android malware detection. In *Proceedings of the 6th International Conference on Security of Information and Networks* (pp. 152-159). 10.1145/2523514.2523539

Felt, A. P., Chin, E., Hanna, S., Song, D., & Wagner, D. (2011). Android permissions demystified. In *Proceedings of the 18th ACM conference on Computer and communications security* (pp. 627-638). ACM.

Ferrag, M. A., Maglaras, L., Moschoyiannis, S., & Janicke, H. (2019). Deep learning for cyber security intrusion detection: Approaches, datasets, and comparative study. *Journal of Information Security and Applications, 50*. doi:10.1016/j.jisa.2019.102419

Ferreira, P., Le, D. C., & Zincir-Heywood, N. (2019). Exploring Feature Normalization and Temporal Information for Machine Learning Based Insider Threat Detection. *2019 15th International Conference on Network and Service Management (CNSM),* 1-7. 10.23919/CNSM46954.2019.9012708

Fihri, W. F., El Ghazi, H., & Kaabouch, N. (2018). A particle swarm optimization based algorithm for primary user emulation attack detection. IEEE consumer communications and networking conference, 1–6. doi:10.1109/CCWC.2018.8301616

Fihri, W. F., El Ghazi, H., Kaabouch, N., & Abou El Majd, B. (2017). Bayesian decision model with trilateration for primary user emulation attack localization in cognitive radio networks. *Proceedings of the IEEE international symposium on networks, computers, and communications*, 1–6. 10.1109/ISNCC.2017.8071979

Fortino, G., & Palau, C. E. (2012). *Next Generation Content Delivery Infrastructures: Emerging Paradigms and.* Academic Press.

Foster, I. (2003). The grid: Computing without bounds. *Scientific American, 288*(4), 78–85. doi:10.1038cientificamerican0403-78 PMID:12661319

Foster, I., & Kesselman, C. (2003a). *The Grid 2*. Morgan Kauffman.

Foster, I., & Kesselman, C. (2003b). *The Grid 2: Blueprint for a new computing infrastructure*. Elsevier.

Foster, I., Kesselman, C., & Tuecke, S. (2001). The anatomy of the grid: Enabling scalable virtual organizations. *International Journal of High Performance Computing Applications, 15*(3), 200–222. doi:10.1177/109434200101500302

Framework, N. (2012). *Roadmap for smart grid interoperability standards, release 2.0* (Vol. 1108). NIST Special Publication.

Fuchs, A. P., Chaudhuri, A., & Foster, J. S. (2009). *Scandroid: Automated security certification of android applications.* Manuscript, Univ. of Maryland, http://www. cs. umd. edu/avik/projects/scandroidascaa

Gade, N. R., & Reddy, U. (2014). *A Study Of Cyber Security Challenges And Its Emerging Trends On Latest Technologies.* Academic Press.

Gadir, O. M., Subbanna, K., Vayyala, A. R., Shanmugam, H., Bodas, A. P., Tripathy, T. K., … Rao, K. H. (2005). *High-availability cluster virtual server system.* Academic Press.

Gai, K., Qiu, M., Ming, Z., Zhao, H., & Qiu, L. (2017). Spoofing-jamming attack strategy using optimal power distributions in wireless smart grid networks. *IEEE Transactions on Smart Grid, 8*(5), 1–1. doi:10.1109/TSG.2017.2664043

Gao, H., Chen, Y., Lee, K., Palsetia, D., & Choudhary, A. N. (2012, February). Towards online spam filtering in social networks. In NDSS (Vol. 12, No. 2012, pp. 1-16). Academic Press.

Gavai, G., Sricharan, K., Gunning, D., Hanley, J., Singhal, M., & Rolleston, R. (2015). Supervised and unsupervised methods to detect insider threat from enterprise social and online activity data. *Journal of Wireless Mobile Networks, Ubiquitous Computing and Dependable Applications, 6*(4), 2015.

Gayathri, R., & Nagarajan, V. (2015). Secure data hiding using Steganographic technique with Visual Cryptography &Watermarking Scheme. IEEE ICCSP Conference, 118-123.

Ghiasi, M., Dehghani, M., Niknam, T., Kavousi-Fard, A., Siano, P., & Alhelou, H. H. (2021). Cyber-Attack Detection and Cyber-Security Enhancement in Smart DC-Microgrid Based on Blockchain Technology and Hilbert Huang Transform. *IEEE Access: Practical Innovations, Open Solutions, 9*, 29429–29440. https://doi.org/10.1109/ACCESS.2021.3059042

Ghosh, Sayandip, Maity, & Rahaman. (2015). A Novel Dual Purpose Spatial Domain Algorithm for Digital Image Watermarking and Cryptography Using Extended Hamming Code. *IEEE Proceedings of International Conference on Electrical Information and Communication Technology (EICT 2015)*, 167-172.

Gordon, M. I., Kim, D., Perkins, J. H., Gilham, L., Nguyen, N., & Rinard, M. C. (2015, February). Information flow analysis of android applications in droidsafe. In NDSS (Vol. 15, No. 201, p. 110). Academic Press.

Grier, C., Thomas, K., Paxson, V., & Zhang, M. (2010, October). @ spam: the underground on 140 characters or less. In *Proceedings of the 17th ACM conference on Computer and communications security* (pp. 27-37). 10.1145/1866307.1866311

Gudkov, V., & Lepikhova, D. (2018). Fingerprint Model Based on Fingerprint Image Topology and Ridge Count Values. *2018 Global Smart Industry Conference (GloSI).*

Gunavathy, S., & Meena, C. (2019). A Survey: Data Security In Cloud Using Cryptography And Steganography. *International Research Journal of Engineering and Technology, 6*(5), 6792–6797.

Gungor, V. C., Sahin, D., Kocak, T., Ergut, S., Buccella, C., Cecati, C., & Hancke, G. P. (2013). A survey on smart grid potential applications and communication requirements. *IEEE Transactions on Industrial Informatics, 9*(1), 28–42. doi:10.1109/TII.2012.2218253

Gupta, A., & Walia, N. K. (2014). Cryptography Algorithms: A Review. *International Journal of Engineering Development and Research, 2*(2), 1667–1672.

Gupta, B. B. (Ed.). (2018). *Computer and cyber security: principles, algorithm, applications, and perspectives.* CRC Press.

Gupta, K. K., Nath, B., & Kotagiri, R. (2010). Layered Approach Using Conditional Random Fields for Intrusion Detection. *IEEE Transactions on Dependable and Secure Computing*, *7*(1), 35–49. doi:10.1109/TDSC.2008.20

Hadiprakoso, R. B., Buana, I. K. S., & Pramadi, Y. R. (2020). Android malware detection using hybrid-based analysis & deep neural network. In *3rd International Conference on Information and Communications Technology (ICOIACT)* (pp. 252-256). IEEE.

Han, Y., He, W., Shuai, J., & Qing, L. (2014). A Digital Watermarking Algorithm of Colour Image based on Visual Cryptography and Discrete Cosine Transform. *IEEE Ninth International Conference on P2P*, 527-530.

Hanna, S., Huang, L., Wu, E., Li, S., Chen, C., & Song, D. (2012). Juxtapp: A scalable system for detecting code reuse among android applications. In *International Conference on Detection of Intrusions and Malware, and Vulnerability Assessment* (pp. 62-81). Springer.

Hanzal, P., & Homan, J. (2019). Electronic Exchange SAF-T Standard of Data from Organizations to Tax Authorities or Auditors - Situation in the Czech Republic. *2019 9th International Conference on Advanced Computer Information Technologies (ACIT)*, 405-408. 10.1109/ACITT.2019.8780001

Horng, S.-J., Su, M.-Y., Chen, Y.-H., Kao, T.-W., Chen, R.-J., Lai, J.-L., & Perkasa, C. D. (2011). A novel intrusion detection system based on hierarchical clustering and support vector machines. *Elsevier Expert Systems with Applications*, *38*(1), 306–313. doi:10.1016/j.eswa.2010.06.066

Hualin, Z., Qiqi, W., & Yujing, H. (2018). Design Fingerprint Attendance Machine Based on C51 Single-chip Microcomputer. *2018 IEEE International Conference of Safety Produce Informatization (IICSPI).*

Humayun, M., Niazi, M., & Jhanjhi, N. (2020). Cyber Security Threats and Vulnerabilities: A Systematic Mapping Study. *Arabian Journal for Science and Engineering*, *45*, 3171–3189. https://doi.org/10.1007/s13369-019-04319-2

Hu, P., Dhelim, S., Ning, H., & Qiu, T. (2017). Survey on fog computing: Architecture, key technologies, applications and open issues. *Journal of Network and Computer Applications*, *98*, 27–42. doi:10.1016/j.jnca.2017.09.002

Hussain, M., Wahab, A. W. A., Idris, Y. I. B., Ho, A. T., & Jung, K.-H. (2018). Image steganography in spatial domain: A survey. *Signal Processing Image Communication*, *65*, 46–66.

Islamiati, D., Agata, D., & Anom Besari, A. (2019). Design and Implementation of Various Payment System for Product Transaction in Mobile Application. *2019 International Electronics Symposium (IES).*

Jagatic, T. N., Johnson, N. A., Jakobsson, M., & Menczer, F. (2007). Social phishing. *Communications of the ACM*, *50*(10), 94–100.

Javaid, A. (2013). *Cyber security: Challenges ahead.* Available at SSRN 3281086.

Jawahar, S., Harishchander, A., & Devaraju, S. (2020). Closed Sequential Pattern Mining in Biological Data. *International Journal of Life science and Pharma Research, 8*(8), 9-13.

Jawahar, S., Harishchander, A., & Devaraju, S.(2020). Efficiently Mining Closed Sequence Patterns in DNA without Candidate Genertion. *International Journal of Life science and Pharma Research*, (8), 14-18.

Jiang, Gan, Wang, & Wang. (2011). Research of the Intrusion Detection Model Based on Data Mining. *Elsevier Energy Procedia*, *13*, 855–863.

Jiang, H., & Ruan, J. (2009). The Application of Genetic Neural Network in Network Intrusion Detection. *Journal of Computers*, *4*(12), 1223–1230. doi:10.4304/jcp.4.12.1223-1230

Kaabouch, N., & Hu, W. C. (2014). *Software-defined and cognitive radio technologies for dynamic spectrum management* (Vol. 1 and 2). IGI Global.

Kang, B., Yerima, S. Y., McLaughlin, K., & Sezer, S. (2016). N-opcode analysis for android malware classification and categorization. In *International conference on cyber security and protection of digital services (cyber security)* (pp. 1-7). IEEE.

Karakış, R., Güler, İ., Çapraz, İ., & Bilir, E. (2015, December). A novel fuzzy logic-based image steganography method to ensure medical data security. *Computers in Biology and Medicine*, *67*, 172–183.

Kaur, J. P., & Kaur, R. (2014). *Security Issues and Use of Cryptography in Cloud Computing*. Academic Press.

Kavis, M. J. (2014). *Architecting the cloud: Design decisions for cloud computing service models (SaaS, PaaS, and IaaS)*. John Wiley & Sons. doi:10.1002/9781118691779

Kavya Rani, S. R., Soundarya, B. C., Gururaj, H. L., & Janhavi, V. (2021). Comprehensive Analysis of Various Cyber Attacks. *2021 IEEE Mysore Sub Section International Conference (MysuruCon)*, 255-262. doi:10.1109/MysuruCon52639.2021.9641089

Kaw, J. A., Loan, N. A., Parah, S. A., Muhammad, K., Sheikh, J. A., & Bhat, G. M. (2019, April). A reversible and secure patient information hiding system for IoT driven e-health. *International Journal of Information Management*, *45*, 262–275.

Khan, N., Brohi, S., & Zaman, N. (2020). *Ten Deadly Cyber Security Threats Amid COVID-19 Pandemic*. doi:10.36227/techrxiv.12278792

Kim, J., Yoon, Y., Yi, K., Shin, J., & Center, S. W. R. D. (2012). ScanDal: Static analyzer for detecting privacy leaks in android applications. *MoST*, *12*(110), 1.

Kirthi, K., & Kak, S. (2016). *The Narayana Universal Code*. arXiv: 1601.07110.

Kleinrock, L. (2005). A vision for the Internet. *ST Journal of Research*, *2*(1), 4–5.

Klein, S. T., & Ben-Nissan, M. K. (2010). On the Usefulness of Fibonacci cCmpression Codes. *The Computer Journal*, *53*(6), 701–716. doi:10.1093/comjnl/bxp046

Klimburg A, (2012). *National cyber security framework manual*. NATO Cooperative Cyber Defense Center of Excellence.

Knapp, E. D., & Samani, R. (2013). *Applied cyber security and the smart grid: implementing security controls into the modern power infrastructure*. Elsevier, Syngress.

Koo, D., Shin, Y., Yun, J., & Hur, J. (2016). A Hybrid Deduplication for Secure and Efficient Data Outsourcing in Fog Computing. *2016 IEEE International Conference on Cloud Computing Technology and Science (CloudCom)*, 285–293. 10.1109/CloudCom.2016.0054

Koshy, T. (2001). *Fibonacci and Lucas Numbers with Applications*. Wiley. doi:10.1002/9781118033067

Kouliaridis, V., & Kambourakis, G. (2021). A comprehensive survey on machine learning techniques for android malware detection. *Information*, *12*(5), 185.

Kramer, S., & Bradfield, J. C. (2010). A general definition of malware. *Journal in Computer Virology*, *6*(2), 105-114.

Kremer, J. F., & Müller, B. (2012, June). *Cyber security: developing a framework to understand the emerging challenges to states in an interconnected world*. In *British International Studies Association and International Studies Association Joint International Conference*, Edinburgh, UK.

Kshetri, N. (2016). Cybersecurity and development. *Markets Globalization Dev Rev., 1*(2). doi:10.23860/MGDR-2016-01-02-03

Lallie, Shepherd, Nurse, Erola, Epiphaniou, Maple, & Bellekens. (2021). Cyber security in the age of COVID-19: A timeline and analysis of cyber-crime and cyber-attacks during the pandemic. *Computers & Security, 105*. doi:10.1016/j.cose.2021.102248

Laniepce, S., Lacoste, M., Kassi-Lahlou, M., Bignon, F., Lazri, K., & Wailly, A. (2013). Engineering intrusion prevention services for iaas clouds: The way of the hypervisor. *2013 IEEE Seventh International Symposium on Service-Oriented System Engineering*, 25–36. 10.1109/SOSE.2013.27

Le, D. C., & Nur Zincir-Heywood, A. (2019). Machine learning based Insider Threat Modelling and Detection. *2019 IFIP/IEEE Symposium on Integrated Network and Service Management (IM)*, 1-6.

Lee, I. (2020). Internet of Things (IoT) Cybersecurity: Literature Review and IoT Cyber Risk Management. *Future Internet, 12*(9), 157. doi:10.3390/fi12090157

Lee, S., & Kim, J. (2012, February). WarningBird: Detecting Suspicious URLs in Twitter Stream. In Ndss (Vol. 12, pp. 1-13). Academic Press.

Lee, K., Caverlee, J., & Webb, S. (2010, July). Uncovering social spammers: social honeypots+ machine learning. In *Proceedings of the 33rd international ACM SIGIR conference on Research and development in information retrieval* (pp. 435-442). ACM.

Li, Y., & Liu, Q. (2021). A comprehensive review study of cyber-attacks and cyber security. Emerging trends and recent developments. *Energy Reports, 7*. doi:10.1016/j.egyr.2021.08.126

Liang, X., Gao, K., Zheng, X., & Zhao, T. (2013). A study on cyber security of smart grid on public networks. *Proceedings of the IEEE green technologies conference*, 301–8. 10.1109/GreenTech.2013.53

Liban, A., & Hilles, S. (2018). Latent Fingerprint Enhancement Based On Directional Total Variation Model With Lost Minutiae Reconstruction. *2018 International Conference on Smart Computing and Electronic Enterprise (ICSCEE).*

Li, J., Sun, L., Yan, Q., Li, Z., Srisa-An, W., & Ye, H. (2018). Significant permission identification for machine-learning-based android malware detection. *IEEE Transactions on Industrial Informatics, 14*(7), 3216–3225. doi:10.1109/TII.2017.2789219

Lin, S.-W., Ying, K.-C., Lee, C.-Y., & Lee, Z.-J. (2012). An intelligent algorithm with feature selection and decision rules applied to anomaly intrusion detection. *Elsevier Applied Soft Compu., 12*(10), 3285–3290. doi:10.1016/j.asoc.2012.05.004

Lin, W.-C., Ke, S.-W., & Tsai, C.-F. (2015). CANN: An intrusion detection system based on combining cluster centers and nearest neighbors. *Elsevier Knowledge-Based Systems, 78*, 13–21. doi:10.1016/j.knosys.2015.01.009

Liu, J., Xiao, Y., & Gao, J. (2014). Achieving accountability in smart grid. *IEEE Systems Journal, 8*(2), 493–508. doi:10.1109/JSYST.2013.2260697

Liu, K., Xu, S., Xu, G., Zhang, M., Sun, D., & Liu, H. (2020). A review of android malware detection approaches based on machine learning. *IEEE Access: Practical Innovations, Open Solutions*, 8.

Liu, W., Wang, X., & Peng, W. (2020). State of the Art: Secure Mobile Payment. *IEEE Access: Practical Innovations, Open Solutions, 8*, 13898–13914. doi:10.1109/ACCESS.2019.2963480

Lu, X., Liu, X., Li, H., Xie, T., Mei, Q., Hao, D., & Feng, F. (2016, May). PRADA: Prioritizing android devices for apps by mining large-scale usage data. In *IEEE/ACM 38th International Conference on Software Engineering (ICSE)* (pp. 3-13). IEEE.

Luo, B., & Xia, J. (2014). A novel intrusion detection system based on feature generation with visualization strategy. *Elsevier Expert Systems with Applications*, *41*(9), 4139–4147. doi:10.1016/j.eswa.2013.12.048

Luo, J., & Yang, M. (2018). Offline Transferable E-Cash Mechanism. *2018 IEEE Conference on Dependable and Secure Computing (DSC).*

Lu, Y., & Da Xu, L. (2018). Internet of Things (IoT) cybersecurity research: A review of current research topics. *IEEE Internet of Things Journal*, *6*(2), 2103–2115.

Ly, K., Sun, W., & Jin, Y. (2016). Emerging challenges in cyber-physical systems: A balance of performance, correctness, and security. *2016 IEEE Conference on Computer Communications Workshops (INFOCOM WKSHPS)*, 498-502. doi: 10.1109/INFCOMW.2016.7562128

Lyytinen, K., & Yoo, Y. (2002). Ubiquitous computing. *Communications of the ACM*, *45*(12), 63–96.

Ma, X., Biao, Q., Yang, W., & Jiang, J. (2016). Using multi-features to reduce false positive in malware classification. In *IEEE Information Technology, Networking, Electronic and Automation Control Conference* (pp. 361-365). IEEE.

Mabu, S., Chen, C., Lu, N., Shimada, K., & Hirasawa, K. (2011). An Intrusion-Detection Model Based on Fuzzy Class-Association-Rule Mining Using Genetic Network Programming. *IEEE Transactions on Systems, Man and Cybernetics. Part C, Applications and Reviews*, *41*(1), 130–139. doi:10.1109/TSMCC.2010.2050685

Maglaras, L. A., Kim, K. H., Janicke, H., Ferrag, M. A., Rallis, S., Fragkou, P., ... Cruz, T. J. (2018). Cyber security of critical infrastructures. *Ict Express*, *4*(1), 42–45.

Mahbub, M. (2020). Progressive researches on IoT security: An exhaustive analysis from the perspective of protocols, vulnerabilities, and preemptive architectonics. *Journal of Network and Computer Applications*, *168*, 1–32. doi:10.1016/j.jnca.2020.102761

Mahmud, R., Kotagiri, R., & Buyya, R. (2018). Fog computing: A taxonomy, survey and future directions. In *Internet of everything* (pp. 103–130). Springer. doi:10.1007/978-981-10-5861-5_5

Malonia & Kumar. (2016). Digital Image Watermarking using Discrete Wavelet Transform and Arithmetic Progression. *IEEE Students's Conference on Electrical.*

Mandlekar, V. G., Mahale, V., Sancheti, S. S., & Rais, M. S. (2014). Survey on fog computing mitigating data theft attacks in cloud. *Int. J. Innov. Res. Comput. Sci. Technol*, *2*, 13–16.

Manesh, M. R., Mullins, M., Forerster, K., & Kaabouch, N. (2018). A preliminary work toward investigating the impacts of injection attacks on air traffic. IEEE Aerospace Conference, 1–6.

Manesh, M. R., & Kaabouch, N. (2017). *Security threats and countermeasures of MAC layer in cognitive radio networks. J Ad Hoc Netw.*

Manimurugan & Al-Mutari. (2017). A Novel Secret Image Hiding Technique for Secure Transmission. *Journal of Theoretical and Applied Information Technology*, *95* (1), 166-176.

Manimurugan, S., Al-Mutairi, S., Aborokbah, M. M., Chilamkurti, N., Ganesan, S., & Patan, R. (2020). Effective Attack Detection in Internet of Medical Things Smart Environment Using a Deep Belief Neural Network. *IEEE Access: Practical Innovations, Open Solutions*, *8*, 77396–77404.

Mariconti, E., Onwuzurike, L., Andriotis, P., De Cristofaro, E., Ross, G., & Stringhini, G. (2016). *Mamadroid: Detecting android malware by building markov chains of behavioral models.* arXiv preprint arXiv:1612.04433.

Martinelli, F., Marulli, F., & Mercaldo, F. (2017). Evaluating convolutional neural network for effective mobile malware detection. *Procedia Computer Science*, *112*, 2372–2381.

Martin, W., Sarro, F., Jia, Y., Zhang, Y., & Harman, M. (2016). A survey of app store analysis for software engineering. *IEEE Transactions on Software Engineering*, *43*(9), 817–847.

Masys, A. J. (Ed.). (2016). *Exploring the Security Landscape: non-traditional security challenges.* Springer International Publishing.

Mattern, T., Felker, J., Borum, R., & Bamford, G. (2014). Operational Levels of Cyber Intelligence. *International Journal of Intelligence and CounterIntelligence*, *27*(4), 702–719. doi:10.1080/08850607.2014.924811

Mavroeidis, V., & Bromander, S. (2017) Cyber Threat Intelligence Model: An Evaluation of Taxonomies, Sharing Standards, and Ontologies within Cyber Threat Intelligence. *European Intelligence and Security Informatics Conference (EISIC)*. 10.1109/EISIC.2017.20

McGraw, G., & Morrisett, G. (2000). Attacking malicious code: A report to the infosec research council. *IEEE Software*, *17*(5), 33–41. doi:10.1109/52.877857

Mckinnel, D. R., Dargahi, T., Dehghantanha, A., & Choo, K. R. (2019). A systematic literature review and meta-analysis on artificial Intelligence in Vulnerability Analysis and Penetration Testing. *Computers & Electrical Engineering*, *75*, 175–188. doi:10.1016/j.compeleceng.2019.02.022

Messmer, E. (2013, Mar.). Malware-detecting 'sandboxing' technology no silver bullet. *Networkworld.*

Meza, J., Campbell, S., & Bailey, D. (2009). *Mathematical and statistical opportunities in cyber security.* arXiv preprint arXiv:0904.1616.

Miao, Y., Ruan, Z., Pan, L., Wang, Y., Zhang, J., & Xiang, Y. (2018). *Automated big traffic analytics for cyber security.* arXiv preprint arXiv:1804.09023.

Milosevic, N., Dehghantanha, A., & Choo, K. K. R. (2017). Machine learning aided Android malware classification. *Computers & Electrical Engineering*, *61*, 266–274.

Mohamed, Ahmed, & Hassan. (2013). Securing biometric data by combining watermarking and cryptography. *IEEE 2nd International Conference on Advances in Biomedical Engineering*, 179-182.

Mohamed, A. B., Idris, N. B., & Shanmugum, B. (2012). *A Brief Introduction to Intrusion Detection System.* Comm. in Comp. and Info. *Sci.*, *330*, 263–271.

Mohammad, Zeki, Chebil, & Gunawan. (2013). Properties of Digital Image Watermarking. *IEEE 9th International Colloquium on Signal processing and its Applications*, 8-10.

Mohanta, B. K., Jena, D., Satapathy, U., & Patnaik, S. (2020). Survey on IoT Security: Challenges and Solution using Machine Learning, Artificial Intelligence and Blockchain Technology. *Journal of Internet of Things*, *11*, 1–32. doi:10.1016/j.iot.2020.100227

Mondal, R. K., & Sarddar, D. (2015). Utility Computing. *International Journal of Grid and Distributed Computing*, *8*(4), 115–122. doi:10.14257/ijgdc.2015.8.4.11

Moniruzzaman, Hawlader, & Hossain. (2014). Wavelet Based Watermarking Approach of Hiding Patient Information in Medical Image for Medical Image Authentication. *IEEE 17th International Conference on Computer and Information Technology (ICCIT)*, 374-378.

Mouradian, C., Naboulsi, D., Yangui, S., Glitho, R. H., Morrow, M. J., & Polakos, P. A. (2017). A comprehensive survey on fog computing: State-of-the-art and research challenges. *IEEE Communications Surveys and Tutorials*, *20*(1), 416–464. doi:10.1109/COMST.2017.2771153

Mstafa, R. J., Elleithy, K. M., & Abdelfattah, E. (2017). A robust and secure video steganography method in dwt-dct domains based on multiple object tracking and ecc. *IEEE Access: Practical Innovations, Open Solutions*, *5*, 5354–5365.

Mthunzi, S. N., Benkhelifa, E., Bosakowski, T., & Hariri, S. (2019). A bio-inspired approach to cyber security. In *Machine Learning for Computer and Cyber Security* (pp. 75–104). CRC Press.

Mukherjee, M., Shu, L., & Wang, D. (2018). Survey of fog computing: Fundamental, network applications, and research challenges. *IEEE Communications Surveys and Tutorials*, *20*(3), 1826–1857. doi:10.1109/COMST.2018.2814571

Nadiammai, G. V., & Hemalatha, M. (2014). Effective Approach toward Intrusion Detection System using Data Mining Techniques. *Elsevier Egyptian Informatics Journal*, *15*(1), 37–50. doi:10.1016/j.eij.2013.10.003

Nalli, A., & Ozyilmaz, C. (2015). The third order variations on the Fibonacci universal code. *Journal of Number Theory*, *149*, 15–32. doi:10.1016/j.jnt.2014.07.010

Narudin, F. A., Feizollah, A., Anuar, N. B., & Gani, A. (2016). Evaluation of machine learning classifiers for mobile malware detection. *Soft Computing*, *20*(1), 343–357.

Nath, H. V., & Mehtre, B. M. (2014). Static malware analysis using machine learning methods. In *International Conference on Security in Computer Networks and Distributed Systems* (pp. 440-450). Springer. 10.1007/978-3-642-54525-2_39

Nguyen, T. D., Nguyen, A. T., Phan, H. D., & Nguyen, T. N. (2017, May). Exploring API embedding for API usages and applications. In *2017 IEEE/ACM 39th International Conference on Software Engineering (ICSE)* (pp. 438-449). IEEE.

Nie, Q., Xu, X., Feng, B., & Zhang, L. Y. (2018). Defining embedding distortion for intra prediction mode-based video steganography. *Comput. Mater. Continua*, *55*(1), 59–70.

Nifakos, S., Chandramouli, K., Nikolaou, C. K., Papachristou, P., Koch, S., Panaousis, E., & Bonacina, S. (2021, July 28). Influence of Human Factors on Cyber Security within Healthcare Organisations: A Systematic Review. *Sensors (Basel)*, *21*(15), 5119. https://doi.org/10.3390/s21155119

Olaniyan, R., Fadahunsi, O., Maheswaran, M., & Zhani, M. F. (2018). Opportunistic edge computing: Concepts, opportunities and research challenges. *Future Generation Computer Systems*, *89*, 633–645. doi:10.1016/j.future.2018.07.040

Overeinder, B. J., Sloot, P. M., Heederik, R. N., & Hertzberger, L. O. (1996). A dynamic load balancing system for parallel cluster computing. *Future Generation Computer Systems*, *12*(1), 101–115. doi:10.1016/0167-739X(95)00038-T

Pahl, C. (2015). Containerization and the paas cloud. *IEEE Cloud Computing*, *2*(3), 24–31. doi:10.1109/MCC.2015.51

Pal, P., Schantz, R., Rohloff, K., & Loyall, J. (2009, July). Cyber physical systems security challenges and research ideas. In *Workshop on Future Directions in Cyber-physical Systems Security* (pp. 1-5). Academic Press.

Parah, S. A., Ahad, F., Sheikh, J. A., Loan, N. A., & Bhat, G. M. (2017). A new reversible and high capacity data hiding technique for e-healthcare applications. *Multimedia Tools and Applications*, *76*(3), 3943–3975.

Parkinson, S., Ward, P., Wilson, K., & Miller, J. (2017). Cyber threats facing autonomous and connected vehicles: Future challenges. *IEEE Transactions on Intelligent Transportation Systems*, *18*(11), 2898–2915.

Pattani, K., & Gautam, S. (2021). SonicEvasion: a stealthy ultrasound based invasion using covert communication in smart phones and its security. *International Journal of Information Technology*, 1-11.

Pelosi, Poudel, Lamichhane, Lam, Kessler, & Mac-Monagle. (2018). *Positive identification of lsb image steganography using cover image comparisons.* Academic Press.

Polverari, J. (2018). *Why less is more when it comes to cybersecurity 2018.* Available: https://www.forbes.com/sites/forbestechcouncil/2018/06/01/whyless-is-more-when-it-comes-to-cybersecurity/

Powell, J. L. II. (2016). *Utilizing Cyber Threat Intelligence to Enhance Cybersecurity*. ProQuest LLC.

Prabhjot & Kaur. (2017). A Review of Information Security using Cryptography Technique. *International Journal of Advanced Research in Computer Science, 8*, 323-326.

Pradeep. (2019). An Efficient Framework for Sharing a File in a Secure Manner Using Asymmetric Key Distribution Management in Cloud Environment. *Journal of Computer Networks and Communications,* 1-8.

Pradhan, A., Sahu, A. K., Swain, G., & Sekhar, K. R. (2016). Performance evaluation parameters of image steganography techniques. *2016 International Conference on Research Advances in Integrated Navigation Systems (RAINS)*, 1-8.

Preuveneers, D., & Joosen, W. (2021). Sharing Machine Learning Models as Indicators of Compromise for Cyber Threat Intelligence. *J. Cybersecur. Priv.*, *1*(1), 140–163. doi:10.3390/jcp1010008

Quayyum, F., Cruzes, D. S., & Jaccheri, L. (2021). Cybersecurity awareness for children: A systematic literature review. *International Journal of Child-Computer Interaction, 30*. doi:10.1016/j.ijcci.2021.100343

Qu, Y., & Hughes, K. (2013, December). Detecting metamorphic malware by using behavior-based aggregated signature. In *World Congress on Internet Security (WorldCIS-2013)* (pp. 13-18). IEEE.

Rabah, K. (2004, March). Steganography—The art of hiding data. *Information Technology Journal*, *3*(3), 245–269.

Rajasekharaiah, K., Dule, C., & Sudarshan, E. (2020). Cyber Security Challenges and its Emerging Trends on Latest Technologies. *IOP Conference Series. Materials Science and Engineering*, *981*, 022062. doi:10.1088/1757-899X/981/2/022062

Rajawat, M., & Tomar, D. S. (2015). A Secure Watermarking and Tampering detection technique on RGB Image using 2 Level DWT. *IEEE Fifth International Conference on Communication Systems and Network Technologies*, 638-642.

Ramakrishnan, S., & Devaraju, S. (2016). *Attack's Feature Selection-Based Network Intrusion Detection System Using Fuzzy Control Language*. *Springer-International Journal of Fuzzy Systems.* doi:10.100740815-016-0160-6

Ramsdale, A., Shiaeles, S., & Kolokotronis, N. (2020). A Comparative Analysis of Cyber-Threat Intelligence Sources, Formats and Languages. *Electronics (Basel)*, *9*(5), 824. doi:10.3390/electronics9050824

Rani & Arora. (2015). Image security system using encryption and steganography. *IJIRSET, 4*(6).

Ranveer, S., & Hiray, S. (2015). Comparative analysis of feature extraction methods of malware detection. *International Journal of Computers and Applications*, *120*(5).

Rashid, T., Agrafiotis, I., & Nurse, J. R. (2016). A new take on detecting insider threats. *Intl. Workshop on Managing Insider Security Threats.*

Rawat & Bajracharya. (2015a). Detection of false data injection attacks in smart grid communication systems. *IEEE Signal Process Lett, 22*(10), 1652–6.

Rawat, D. B., & Bajracharya, C. (2015b). Cyber security for smart grid systems: status, challenges and perspectives. *Proceedings of the SoutheastCon*, 1–6. 10.1109/SECON.2015.7132891

Reyes, H., & Kaabouch, N. (2013). Jamming and lost link detection in wireless networks with fuzzy logic. *International Journal of Scientific and Engineering Research, 4*(2), 1–7.

Rodofile, N. R., Radke, K., & Foo, E. (2016). DNP3 network scanning and reconnaissance for critical infrastructure. *Proceedings of the Australasian Computer science week multiconference*, 39:1–39:10. 10.1145/2843043.2843350

Roundy, K. A., & Miller, B. P. (2010). Hybrid analysis and control of malware. In *International Workshop on Recent Advances in Intrusion Detection* (pp. 317-338). Springer.

Roy, S., DeLoach, J., Li, Y., Herndon, N., Caragea, D., Ou, X., . . . Guevara, N. (2015). Experimental study with real-world data for android app security analysis using machine learning. In *Proceedings of the 31st Annual Computer Security Applications Conference* (pp. 81-90). Academic Press.

Runder, M. (2015). Cyber-Threats to Critical National Infrastructure - An Intelligence Challenge. *International Journal of Intelligence and Counter Intelligence.*

Sadkhan, S. B. (2004). Cryptography: current status and future trends. *International Conference on Information and Communication Technologies: From Theory to Applications.*

Sahu & Swain. (2019). A novel n-Rightmost bit replacement image steganography technique. *3D Res., 10*(1), 2.

Sahu, A. K., & Swain, G. (2019, September). An optimal information hiding approach based on pixel value differencing and modulus function. *Wireless Personal Communications, 108*(1), 159–174.

Sajić, M., Bundalo, D., Bundalo, Z., Stojanović, R., & Sajić, L. (2018). Design of Digital Modular Bank Safety Deposit Box Using Modern Information and Communication Technologies. In *7th Mediterranean Conference on Embedded Computing MECO 2018. 10-14 June 2018* (pp. 107-112). 10.1109/MECO.2018.8406014

Saleh, Aly, & Omara. (2016). Data security using cryptography and steganography techniques. *International Journal of Advanced Computer Science and Applications, 7*(6), 390-397.

Sanjay, K., & Ambar, D. (2016). A Novel Spatial Domain Technique for Digital Image Watermarking Using Block Entropy. *IEEE Fifth International Conference on Recent Trends in Information Technology.*

Santos, I., Devesa, J., Brezo, F., Nieves, J., & Bringas, P. G. (2013). Opem: A static-dynamic approach for machine-learning-based malware detection. In *International joint conference CISIS'12-ICEUTE´ 12-SOCO´ 12 special sessions* (pp. 271-280). Springer.

Sarasamma, S. T., Zhu, Q. A., & Huff, J. (2005). Hierarchical Kohonenen Net for Anomaly Detection in Network Security. *IEEE Tr. on Sys. Man and Cybernetics, 35*(2), 302–312. doi:10.1109/TSMCB.2005.843274 PMID:15828658

Sarker, I. H., Kayes, A. S. M., & Badsha, S. (2020). Cybersecurity data science: An overview from a machine learning perspective. *Journal of Big Data, 7*, 41. https://doi.org/10.1186/s40537-020-00318-5

Sarker, I. H., Kayes, A. S. M., Badsha, S., Alqahtani, H., Watters, P., & Ng, A. (2020). Cybersecurity data science: An overview from machine learning perspective. *Journal of Big Data*, *7*(1), 1–29. doi:10.118640537-020-00318-5

Sato, R., Chiba, D., & Goto, S. (2013). Detecting android malware by analyzing manifest files. In *Proceedings of the Asia-Pacific Advanced Network*, (Vol. 36, pp. 23-31). 10.7125/APAN.36.4

Schlegel, R., Zhang, K., Zhou, X. Y., Intwala, M., Kapadia, A., & Wang, X. (2011). Soundcomber: A Stealthy and Context-Aware Sound Trojan for Smartphones. In NDSS (Vol. 11, pp. 17-33). Academic Press.

Selvakani,, S., & Kandeeban, R.S. (2011). A Genetic Algorithm Based elucidation for improving Intrusion Detection through condensed feature set by KDD 99 dataset. *Info. And Knowl. Mgnt.*, *1*(1), 1–9.

Shabtai, A., Kanonov, U., Elovici, Y., Glezer, C., & Weiss, Y. (2012). "Andromaly": A behavioral malware detection framework for android devices. *Journal of Intelligent Information Systems*, *38*(1), 161–190. doi:10.100710844-010-0148-x

Shanmuganathan, M., Almutairi, S., Aborokbah, M. M., Ganesan, S., & Ramachandran, V. (2020, August). Review of advanced computational approaches on multiple sclerosis segmentation and classification. *IET Signal Processing*, *14*(6), 333–341.

Shapsough, S., Qatan, F., Aburukba, R., Aloul, F., & Al Ali, A. (2015). Smart grid cyber security: challenges and solutions. *Proceedings of the international conference on smart grid and clean energy technologies*, 170–5. 10.1109/ICSGCE.2015.7454291

Shea, D. (2016, May). Preventing Cyber Intelligence Failures by Analyzing Intelligence Failures and Intelligence Reforms. *ProQuest.*

Sheikhan, & Jadidi, & Farrokhi. (2010). Intrusion detection using reduced-size RNN based on feature grouping. *Neu. Comput&Applic*, *21*(6), 1185–1190.

Sherazi, H. H. R., Iqbal, R., Ahmad, F., Khan, Z. A., & Chaudhary, M. H. (2019). *DDoS attack detection A key enabler for sustainable communication in internet of vehicles.* Elsevier.

Shin, M., Cornelius, C., Peebles, D., Kapadia, A., Kotz, D. K., & Triandopoulos, N. (2011). AnonySense: A system for anonymous opportunistic sensing. *Pervasive and Mobile Computing*, *7*(1), 16–30.

Singh, Hasan & Kumar. (2015). Enhance security for image encryption and decryption by appling hybrid techniques using MATLAB. *IJIRCCE, 3*(7).

Sobb, T., Turnbull, B., & Moustafa, N. (2020). Supply Chain 4.0: A Survey of Cyber Security Challenges, Solutions and Future Directions. *Electronics (Basel)*, *9*(11), 1864. https://doi.org/10.3390/electronics9111864

Soleymani, S. A., Abdullah, A. H., Zareei, M., Anisi, M. H., Vargas-Rosales, C., Khan, M. K., & Goudarzi, S. (2017). A secure trust model based on fuzzy logic in vehicular ad hoc networks with fog computing. *IEEE Access: Practical Innovations, Open Solutions*, *5*, 15619–15629. doi:10.1109/ACCESS.2017.2733225

Statistica. (2020). *Development of new Android malware worldwide from June 2016 to March 2020(in millions)*. Author.

Stinson, D. R. (2002). *Cryptography Theory and Practice*. Chapman & Hall / CRC.

Stringhini, G., Kruegel, C., & Vigna, G. (2010, December). Detecting spammers on social networks. In *Proceedings of the 26th annual computer security applications conference* (pp. 1-9). Academic Press.

Subramanian, N., & Jeyaraj, A. (2018). Recent security challenges in cloud computing. *Computers & Electrical Engineering*, *71*, 28–42. doi:10.1016/j.compeleceng.2018.06.006

Sujendran, R., & Arunachalam, M. (2015). Hybrid Fuzzy Adaptive Wiener Filtering with Optimization for Intrusion Detection. *ETRI Journal*, *37*(3), 502–511. doi:10.4218/etrij.15.0114.0275

Sun, Y., & Havidz, S. (2019). Factors Impacting the Intention to Use M-Payment. *2019 International Conference on Information Management and Technology (ICIMTech)*. 10.1109/ICIMTech.2019.8843758

Surendran, R., Thomas, T., & Emmanuel, S. (2020). A TAN based hybrid model for android malware detection. *Journal of Information Security and Applications*, *54*, 102483.

Suwald, T., & Rottschafer, T. (2019). Capacitive Fingerprint Sensor for Contactless Payment Cards. *2019 26th IEEE International Conference on Electronics, Circuits and Systems (ICECS)*.

Tajbakhsh, A., Rahmati, M., & Mirzaei, A. (2009). Intrusion detection using fuzzy association rules. *Elsevier Appl. Soft Compu.*, *9*(2), 462–469. doi:10.1016/j.asoc.2008.06.001

Tam, K., Feizollah, A., Anuar, N. B., Salleh, R., & Cavallaro, L. (2017). The evolution of android malware and android analysis techniques. *ACM Computing Surveys*, *49*(4), 1–41.

Thakur, K., Hayajneh, T., & Tseng, J. (2019). Cyber security in social media: Challenges and the way forward. *IT Professional*, *21*(2), 41–49.

Thomas, J. H. (2007). *Variant of Fibonacci Universal Code*. arXiv: cs/0701085v2.

Thomas, K., Grier, C., Ma, J., Paxson, V., & Song, D. (2011, May). *Design and evaluation of a real-time url spam filtering service. In 2011 IEEE symposium on security and privacy*. IEEE.

Thuraisingham, B. (2020, May). The Role of Artificial Intelligence and Cyber Security for Social Media. In *2020 IEEE International Parallel and Distributed Processing Symposium Workshops (IPDPSW)* (pp. 1-3). IEEE.

Tonge, A. (2013). Cyber security: Challenges for society- literature review. *IOSR Journal of Computer Engineering.*, *12*, 67–75. doi:10.9790/0661-1226775

Tounekti, O., Ruiz-Martinez, A., & Skarmeta Gomez, A. (2020). Users Supporting Multiple (Mobile) Electronic Payment Systems in Online Purchases: An Empirical Study of Their Payment Transaction Preferences. *IEEE Access: Practical Innovations, Open Solutions*, *8*, 735–766. doi:10.1109/ACCESS.2019.2961785

Tounsi, W., & Rais, H. (2018). A survey on technical threat intelligence in the age of sophisticated Cyber Attacks. *Computers & Security*, *72*, 212–233. doi:10.1016/j.cose.2017.09.001

Tsang, C.-H., Kwong, S., & Wang, H. (2007). Genetic-fuzzy rule mining approach and evaluation of feature selection techniques for anomaly intrusion detection. *Elsevier Pattern Recogn*, *40*(9), 2373–2391. doi:10.1016/j.patcog.2006.12.009

Tu, W., Pop, F., Jia, W., Wu, J., & Iacono, M. (2019). *High-Performance Computing in Edge Computing Networks*. Elsevier. doi:10.1016/j.jpdc.2018.10.014

Tyagi, V. (2012). Image steganography using least significant bit with cryptography. Journal of Global Research in Computer Science, 3(3), 53-55.

Ujjwal, K. C., Garg, S., Hilton, J., Aryal, J., & Forbes-Smith, N. (2019). Cloud Computing in natural hazard modeling systems: Current research trends and future directions. *International Journal of Disaster Risk Reduction*.

Upadhyay, N. (2018). Fogology: What is (not) Fog Computing? *Procedia Computer Science*, *139*, 199–203. doi:10.1016/j.procs.2018.10.243

van Steen, M., & Tanenbaum, A. S. (2016). A brief introduction to distributed systems. *Computing*, *98*(10), 967–1009. doi:10.100700607-016-0508-7

Varghese, B., & Buyya, R. (2018). Next generation cloud computing: New trends and research directions. *Future Generation Computer Systems*, *79*, 849–861. doi:10.1016/j.future.2017.09.020

Varghese, B., Wang, N., Barbhuiya, S., Kilpatrick, P., & Nikolopoulos, D. S. (2016). Challenges and opportunities in edge computing. *2016 IEEE International Conference on Smart Cloud (SmartCloud)*, 20–26. 10.1109/SmartCloud.2016.18

Varol, N., Aydogan, F., & Varol, A. (2017). Cyber Attacks Targetting Android Cellphones. *The 5th International Symposium on Digital Forensics and Security (ISDFS 2017)*.

Wang, G., Hao, J., Ma, J., & Huang, L. (2010). A new approach to intrusion detection using Artificial Neural Networks and fuzzy clustering. *Elsevier Expert Sys. with Appl.*, *37*(9), 6225–6232. doi:10.1016/j.eswa.2010.02.102

Wang, M., & Zhao, A. (2012). Investigations of Intrusion Detection Based on Data Mining. *Springer Recent Advances in Computer Science and Information EngineeringLecture Notes in Electrical Engineering*, *124*, 275–279. doi:10.1007/978-3-642-25781-0_41

Wang, W., & Lu, Z. (2013). Cyber security in the smart grid: Survey and challenges. *Computer Networks*, *57*(5), 1344–1371. doi:10.1016/j.comnet.2012.12.017

Wei & Di. (2010). A Probability Approach to Anomaly Detection with Twin Support Vector Machines. *Journal of Shanghai Jiaotong Univ. (Sci.)*, *15*(4), 385-391.

Weiss, A. (2007). Computing in the clouds. *Networker (Washington, D.C.)*, *11*(4), 16–25.

Weng, Li, Chi, & Mu. (2018). *Convolutional video steganography with temporal residual modeling.* arXiv preprint.

Werstein, P., Situ, H., & Huang, Z. (2006). Load balancing in a cluster computer. *2006 Seventh International Conference on Parallel and Distributed Computing, Applications and Technologies (PDCAT'06)*, 569–577. 10.1109/PDCAT.2006.77

Wong, M. Y., & Lie, D. (2016). IntelliDroid: A Targeted Input Generator for the Dynamic Analysis of Android Malware. In NDSS (Vol. 16, pp. 21-24). doi:10.14722/ndss.2016.23118

Wu, D. J., Mao, C. H., Wei, T. E., Lee, H. M., & Wu, K. P. (2012). Droidmat: Android malware detection through manifest and api calls tracing. In *2012 Seventh Asia Joint Conference on Information Security* (pp. 62-69). 10.1109/AsiaJCIS.2012.18

Wu, W., Kang, R., & Li, Z. (2015). Risk assessment method for cyber security of cyber physical systems. *1st International Conference on Reliability Systems Engineering*, Beijing, China.

Ximenes, A., Sukaridhoto, S., Sudarsono, A., Ulil Albaab, M., Basri, H., Hidayat Yani, M., Chang Choon, C., & Islam, E. (2019). Implementation QR Code Biometric Authentication for Online Payment. *2019 International Electronics Symposium (IES)*.

Xiong, W., & Lagerström, R. (2019). Threat modeling – A systematic literature review. *Computers & Security*, *84*, 53–69. doi:10.1016/j.cose.2019.03.010

Xu, W., Zhang, F., & Zhu, S. (2010, December). Toward worm detection in online social networks. In *Proceedings of the 26th annual computer security applications conference* (pp. 11-20). 10.1145/1920261.1920264

Yamaguchi, F., Golde, N., Arp, D., & Rieck, K. (2014). Modeling and discovering vulnerabilities with code property graphs. In *IEEE Symposium on Security and Privacy* (pp. 590-604). IEEE.

Yan, L. K., & Yin, H. (2012). Droidscope: Seamlessly reconstructing the {OS} and dalvik semantic views for dynamic android malware analysis. In *21st USENIX Security Symposium (USENIX Security 12)* (pp. 569-584). USENIX.

Yang, C., Xu, Z., Gu, G., Yegneswaran, V., & Porras, P. (2014). Droidminer: Automated mining and characterization of fine-grained malicious behaviors in android applications. In *European symposium on research in computer security* (pp. 163-182). Springer.

Yankov, M. P., Olsen, M. A., Stegmann, M. B., Christensen, S. S., & Forchhammer, S. (2020). Fingerprint Entropy and Identification Capacity Estimation Based on Pixel-Level Generative Modelling. *IEEE Transactions on Information Forensics and Security, 15*, 56–65. doi:10.1109/TIFS.2019.2916406

Yaqoob, I.,, & Ahmed, E., Habib ur Rehman, M., Ahmed, A. I. A., Al-Garadi, M., Imran, M., & Guizani, M. (2017). The rise of ransomware and emerging security challenges in the Internet of Things. *Computer Networks, 129*. Advance online publication. doi:10.1016/j.comnet.2017.09.003

Yen, Y. S., & Sun, H. M. (2019). An Android mutation malware detection based on deep learning using visualization of importance from codes. *Microelectronics and Reliability, 93*, 109–114. doi:10.1016/j.microrel.2019.01.007

Yi, P., Zhu, T., Zhang, Q., Wu, Y., & Li, J. (2014). A denial of service attack in advanced metering infrastructure network. *Proceedings of the IEEE international conference on communications,* 1029–34. 10.1109/ICC.2014.6883456

Yousefpour, A., Fung, C., Nguyen, T., Kadiyala, K., Jalali, F., Niakanlahiji, A., … Jue, J. P. (n.d.). *All One Needs to Know about Fog Computing and Related Edge Computing Paradigms*. Academic Press.

Yuan, Z., Lu, Y., Wang, Z., & Xue, Y. (2014). Droid-sec: deep learning in android malware detection. In *Proceedings of the ACM conference on SIGCOMM* (pp. 371-372). ACM.

Yuan, Z., Lu, Y., & Xue, Y. (2016). Droiddetector: Android malware characterization and detection using deep learning. *Tsinghua Science and Technology, 21*(1), 114–123.

Zeckendorf, E. (1972). Representations des nombres naturels par une somme de nombres de fibonacci on de nombres de lucas. *Bulletin de la Société Royale des Sciences de Liège*, 179–182.

Zeng, Y., Chen, J., Shang, W., & Chen, T. H. P. (2019). Studying the characteristics of logging practices in mobile apps: A case study on f-droid. *Empirical Software Engineering, 24*(6), 3394–3434.

Zhang, M., Duan, Y., Yin, H., & Zhao, Z. (2014). Semantics-aware android malware classification using weighted contextual api dependency graphs. In *Proceedings of the ACM SIGSAC conference on computer and communications security* (pp. 1105-1116). ACM.

Zhang, P., Zhou, M., & Fortino, G. (2018). Security and trust issues in Fog computing: A survey. *Future Generation Computer Systems, 88*, 16–27. doi:10.1016/j.future.2018.05.008

Zhang, Q., Cheng, L., & Boutaba, R. (2010). Cloud computing: State-of-the-art and research challenges. *Journal of Internet Services and Applications, 1*(1), 7–18. doi:10.100713174-010-0007-6

Zhang, W., & Kang, M. (2019). Factors Affecting the Use of Facial-Recognition Payment: An Example of Chinese Consumers. *IEEE Access: Practical Innovations, Open Solutions, 7*, 154360–154374.

Zhang, Y., Qin, C., Zhang, W., Liu, F., & Luo, X. (2018). On the fault tolerant performance for a class of robust image steganography. *Signal Processing, 146*, 99–111.

Zhang, Z., Gong, S., Dimitrovski, A. D., & Li, H. (2013). Time synchronization attack in smart grid: Impact and analysis. *IEEE Transactions on Smart Grid, 4*(1), 87–98. doi:10.1109/TSG.2012.2227342

Zhao, M., Ge, F., Zhang, T., & Yuan, Z. (2011). AntiMalDroid: An efficient SVM-based malware detection framework for android. In *International conference on information computing and applications* (pp. 158-166). Springer. 10.1007/978-3-642-27503-6_22

Zissis, D., & Lekkas, D. (2012). Addressing cloud computing security issues. *Future Generation Computer Systems*, *28*(3), 583–592. doi:10.1016/j.future.2010.12.006

Zolotukhin, O., & Kudryavtseva, M. (2018). Authentication Method in Contactless Payment Systems. 2018 International Scientific-Practical Conference Problems of Infocommunications. *Science and Technology (PIC S&T).*

Zou, B., Choobchian, P., & Rozenberg, J. (2021). (2021). Cyber resilience of autonomous mobility systems: Cyber-attacks and resilience-enhancing strategies. *Journal of Transportation Security*, *14*, 137–155. https://doi.org/10.1007/s12198-021-00230-w

Zubair, Szewczyk, Valli, & Rabadia. (2017). Future challenges for smart cities: Cyber-security and digital forensics. *Digital Investigation, 22*, 3-13. doi:10.1016/j.diin.2017.06.015

Zwilling, M., Klien, G., Lesjak, D., Wiechetek, Ł., Cetin, F., & Basim, H. N. (2022). Cyber Security Awareness, Knowledge and Behavior: A Comparative Study. *Journal of Computer Information Systems*, *62*(1), 82–97. doi:10.1080/08874417.2020.1712269

About the Contributors

Jena Om Prakash is currently working as an Assistant Professor in the Department of Computer Science, Ravenshaw University, Cuttack, Odisha, India. He has 11 years of teaching and research experience in under graduate and post graduate level. He has published several technical papers in international journals/conferences/edited book chapters of reputed publications. He is a member of IEEE, ACM, IETA, IAAC, IRED, IAENG and WACAMLDS. His current research interest includes Database, Pattern Recognition, Cryptography, Network Security, Artificial Intelligence, Machine Learning, Soft Computing, Natural Language Processing, Data Science, Compiler Design, Data Analytics and Machine Automation. He has many edited books, published by Wiley, CRC press, Bentham, IEEE, IGI Global, DeGryuter Publication in to his credit and also author of two text book under Kalyani Publisher. He also serves as reviewer committee member, International Advisory Board Member, TPC member and editor of many international journals.

H. L. Gururaj is currently working as an Associate Professor in the Department of Computer Science and Engineering. He has 10 years of teaching and research experience. He is a professional member of ACM. Dr. Gururaj appointed as ACM Distinguish Speaker (2018-2021) by the ACM US council. He received YOUNG SCIENTIST AWARD from ITS-SERB, Department of Science & Technology, Government of India in December 2016. He worked as Special Editor of EAI publisher and Guest Editor of Multimedia Springer Journal. He is an Editorial Board member of various International Journals. He has published more than 75 research papers in various reputed conferences and journals. He worked as a reviewer for various journals and conferences. He also received the Best paper awards at various National and International Conferences. He has authored one Book on Network Simulators and one Edited Book on Blockchain by Springer Publishers. He has honored as Session Chair, Keynote Speaker, TPC member, Advisory committee member at National and International Seminars, Workshops, and Conferences across Globe. His research interests include BlockChain Technology and Cyber Security.

* * *

Kathirvel A. acquired B.E.(CSE), M.E. (CSE) from Crescent Engineering College affiliated to University of Madras and Ph. D (CSE.) from Anna University. He is currently working as Professor, Dept of Computer Science and Engineering, Karunya Institute of Technology and Sciences, Coimbatore. He is a studious researcher by himself, completed 18 sponsored research projects worth of Rs 103 lakhs and published more than 110 articles in journals and conferences. 4 research scholars have completed Ph. D and 3 under progress under his guidance. He is working as scientific and editorial board member of many journals. He has reviewed dozens of papers in many journals. He has author of 13 books. His research interests are protocol development for wireless ad hoc networks, security in ad hoc network, data communication and networks, mobile computing, wireless networks, WSN and DTN. He is a Life member of the ISTE (India), IACSIT (Singapore), Life Member IAENG (Hong Kong), Member ICST (Europe), IAES, etc. He has given a number of guest lecturers/expert talks and seminars, workshops and symposiums.

Somasundaram A. is working as Assistant Professor, Department of Computer Applications, Sri Krishna Arts and Science College, Coimbatore, India. He did his UG program; B.Sc., (Computer Science) at Sree Saraswathi Thyagaraja College, India, and completed his Masters's degree M.C.A at Dr. Mahalingam College of Engineering & Technology, Pollachi, India, He qualified himself in SET and NET examinations. He is currently is pursuing a Ph.D. at Chikkanna Government Arts College affiliated to Bharathiyar University, Coimbatore. His research interests include Cloud Security and Distributed computing.

Amrita received her Ph.D. in Computer Science and Engineering from Sharda University, Greater Noida, India and M.Tech in Computer Science from Banasthali Vidyapith, Rajasthan, India. She is currently working as an Associate Professor in the Department of Computer Science & Engineering, School of Engineering & Technology, Sharda University and also associated with Center of Excellence in Cyber Security and Cryptology, Sharda University. She has more than 16 years of experience in academic and research in addition to 5 years of Industry experience. She has published several papers in reputed refereed journals, international conferences and book chapters. She has 9 published patents and one granted copyright. Her research interests include Intrusion Detection System, Cyber Security, Machine Learning, Soft Computing, Data Mining, Hybrid System, and Feature Selection. She has guided and guiding B.Tech, M. Tech students and Ph.D. scholar. She has served as session chair, member of advisory/technical program committee and reviewer for many International / National conferences. Moreover, she has also delivered expert talk as resource person in faculty development programs, member of organizing committee for faculty development programs and hackathon. She is a Senior Member of The Institute of Electrical and Electronics Engineers (IEEE).

Ajay Kumar Balmiki pursued his undergraduate degree in the branch of Computer Science and Engineering from Budge Budge Institute of Technology, and he pursued his masters program in the branch of Computer Science and from Maulana Abul Kalam Azad University of Technology (formerly known as West Bengal University of Technology), West Bengal, India. His field of interest are Artificial Intelligence (AI), Machine Learning (ML), Internet of Things (IoT), Cyber Security and in Cloud Computing. He is having a handful knowledge of Java (Object oriented programming Language) as well as in Ruby on rails. He is holding the experience of guest lecturer of one year in the subjects like Java and in Computer Network and Security.

Çagla Çelemoğlu is a Dr. research assistant in the Department of Mathematics,Ondokuz Mayis University, Samsun, Turkey. She received her B.Sc and M. Sc in Mathematics from Uludağ University and Karabük University from Turkey, respectively. Her research interests cryptography, coding theory, algebra and number theory. She has papers in magazines with index and conference paper publications in this areas.

Sima Das has achieved M.Tech, in the field of Computer Science and Engineering from Maulana Abul Kalam Azad University of Technology (Main-Campus), West Bengal, India. She is currently affianced as assistant professor in the department of Computer Science and Engineering at Camellia Institute of Technology and Management, Hooghly, West Bengal, India, having the research field of Human computer interaction, Brain Computer Interaction, Cognitive Load, Internet of Things, Cyber Security, Artificial Intelligence, Machine Learning, Deep Learning. She has achieved Research Excellence Award from Global Innovation & Excellence Award 2021. She is Reviewer of International Journal Science & Management, Editor at Dr. BGR Publications and associate member of Institute of Engineers, professional member of IEEE.

Prasanna Kumar G. is currently working as Assistant Professor, Department of Information Science and Engineering, NIE Institute of Technology, Mysuru, India. He's doing his research (Ph.D. Degree) in Computer Science and Engineering at Visvesvaraya Technological University, Belagavi. He has 10 years of teaching experience at both UG and PG level. His research interests include Ubiquitous Network, Network Security, Wireless Senor Network, Ad-hoc networks, IoT and Social IoT. He has published research papers in various international journals, International conferences and UGC referred journals. He worked as reviewer for various journals and conferences.

Srinidhi H. R. is currently working as Assistant Professor, Department of Information Science and Engineering, NIE Institute of Technology, Mysuru, India. He has 3 years of teaching experience at UG & PG level . His research interests include machine learning, AI, IoT and Social IoT. He has published few research papers in various conferences and symposium.

Sudeep Jadey is currently working as Assistant Professor, Department of Information Science & Engineering, NIE Institute of Technology, Mysuru, India. He is doing his research(Ph.D. Degree) in Computer Science & Engineering at Visvesvaraya Technological University, Belagavi. He has 16 years of teaching experience at both UG & PG level. His research interests include Blockchain Technology, Internet of things, cybersecurity and Webmining. He has published many research papers in various international journals includes IEEE, Taylor and Francis, Elsevier and Springer Book Chapter, WoS, Scopus. He is a Life member for IETE.

S. Jawahar received the B.Sc degree in Electronics in 2003 from Bharathiar University, Coimbatore, and the M.C.A. degree in Computer Applications in 2006 from the Anna University, Chennai, and the M.Phil. degree in Computer Science in 2013 from Bharathiar University, Coimbatore. He completed Ph.D degree in Computer Science in 2021 from Bharathiar University, Coimbatore. He has 13+ years of teaching experience and 1+ year industry experience. He is working as Assistant Professor, Department of Computer Technology, PSG college of Arts and Science, Coimbatore, Tamil Nadu, India. Dr.S.Jawahar is a Reviewer for various reputed Journals and Conferences. He has published more than 18 papers in international journals and conference proceedings. His area of research includes Data Mining, Bio-Informatics, and Image Processing.

Karnavel K. is working as Associate Professor in the Department of Artificial Intelligence and Data Science, Anand Institute of Higher Technology, Chennai. He had completed B.Tech from IFET College of Engineering and Technology, Affiliate to Anna University, Chennai specialized with Information

Technology in the year 2006. He completed M.Tech in Computer Science and Engineering from Sri Manakula Vinayagar College of Engineering and Technology, Affiliated to Pondicherry University during 2009. He has been awarded Ph.D in Information and Communication Engineering from Anna University, Chennai in 2015. He started his career in the field of teaching since 2009. He is passionate in teaching with 12 years of experience, administering education and classroom procedures by fostering academic development. He has vast experience in mentoring students in the field of academic and research. He is very much interested to organize workshop, seminar and conference for faculty and student fraternity. He is a dedicated teacher and motivator. He teaches with latest pedagogy and real time examples. He encourages student to gain practical experience by attending guest lecture, case study, Industrial visit, soft skills training and Internship program. He is an active member in different Engineering associations. He won several awards for his excellence in teaching.

Raghavendra K. is currently working as a Assistant Professor, Department of Computer Science & Engineering, NIE Institute of Technology, Mysuru, India. He is doing his research (Ph.D. Degree) in Computer Science & Engineering at Visvesvaraya Technological University, Belagavi. He has 10 years of teaching experience at both UG and PG level. His research interests include Computer Networks, Network Security, Wireless Sensor Network, Adhoc Networks, Blockchain and IOT. He has published more than 10 research papers in various international journals includes IEEE, Springer, Elsevier, WoS, Wiley Book Chapter and UGC referred journals.

Neelima Kant, CISM,CEH is currently working as a Lead Consultant in Wipro technologies and Pursuing Ph.D from Sharda University. She is a competent professional with 11 plus years of experience in the security domain and 8 plus years of experience in Information Security Management where she has established, managed and successfully transformed security frameworks to achieve strategic alignment with business goals. With extensive experience in managing comprehensive and diverse Information Security projects, She has successfully carried out security audits and Information Security standard implementations to help organizations realize value in security, creating robust information security posture, achieving security certifications for applicable standards. Her area of interests for research is cyber security, threat Intelligence, Deep Learning , AI and ML.

Anil Kumar K. M. is currently working as Associate Professor, Department of Computer Science & Engineering, JSS Science and Technology University, Mysuru, Karnataka, India. He has 22 years of Teaching and 10 years of Research experience. He has taught courses like Algorithmic approach to Problem-solving, Data Structures, C++, Data communication, Computer Network, Java & J2EE, Cryptography & Network Security, Big Data Analytics. His research interest includes Text mining, Sentiment Analysis, Data mining, Opinion mining, Web Mining, Data Analytics, and Computer Networks. He has received 5 grants from different Government and Private funding agencies for Training and Research. He has Published nearly 45+ Research papers in National and International proceedings. Currently, 5 students are pursuing Ph.D under his guidance.

A. V. Senthil Kumar is working as a Director & Professor in the Department of Research and PG in Computer Applications, Hindusthan College of Arts and Science, Coimbatore since 05/03/2010. He has to his credit 11 Book Chapters, 250 papers in International Journals, 15 papers in National Journals, 25 papers in International Conferences, 5 papers in National Conferences, and edited 8 books (IGI Global,

USA). He is an Editor-in-Chief for International Journal titled "International Journal of Data Mining and Emerging Technologies", "International Journal of Image Processing and Applications", "International Journal of Advances in Knowledge Engineering & Computer Science", " International Journal of Advances in Computers and Information Engineering" and "International Journal of Research and Reviews in Computer Science".. Key Member for India, Machine Intelligence Research Lab (MIR Labs). He is an Editorial Board Member and Reviewer for various International Journals.

Kaushik Mazumdar (Member, IEEE), BTech., MTech., PhD. (Tech.), has achieved his qualification in the field of Radio Physics and Electronics from the University of Calcutta, India. He is currently affianced as Prof. in Department of Electronics Engineering, Indian Institute of Technology, Dhanbad, having the Research field of Mobile Communication, Mobile Edge Computing, Carrier Transport in Semiconductor and Heterostructure Devices, VLSI design, Electronic Low dimensional Systems, Analysis and Design of MOSFET, Embedded Systems and Nano-Devices. Under him, an ample number of Govt. of India funded Research and Development Projects are ongoing as Principal Investigator (PI, no Co-PI in them).

Pooja M. R. is currently working as Professor in the Department of Computer Science & Engineering. She has more than 40 research publications in peer reviewed international journals and international conferences. Her research interests include Machine Learning and Artificial Intelligence, Big Data Analytics and Health informatics. She has received appreciation for her multidisciplinary research with substantial contributions in the field of Health Informatics and Artificial Intelligence. She is nominated as Bentham Ambassador from INDIA in recognition of her research in Medical Informatics.She has been selected as Editorial Board Member for various peer reviewed international journals and also is a member of the Technical Program Committee of various international conferences. She has delivered talks as an invited speaker at various virtual international conferences including COPD-2021 ,CWC-2021, World Conference on Pediatrics and Neonatal Healthcare,Global Conference on Healthcare held in North Macedonia, Turkey, US and UK. She has been a resource person for AICTE sponsored Faculty Development Programmes on both Data Science and Cyber Security.She has been the Speaker on Data Science at Annual International Meet on Women in Data Science -2021.Shehas been a member of various international professional bodies and has been an invited resource person for AICTE sponsored Faculty Development programmes on AI and Machine Learning.She has attended various workshops organized by AICTE for implementation of Outcome Based Education and has been a resource person for the same.

Saravanan N. received his B.E. degree from National Engineering College, Manonmanium Sundaranar University, M.E. and Ph.D degree from CEG, Anna University Chennai. He has got teaching, research and administrative experience of more than 17 years in various engineering colleges. He is currently working as Professor in Computer Science and Engineering at Misrimal Navajee Munoth Jain Engineering College, Chennai. He has worked as Lecturer, Assistant Professor, Associate Professor & Head in various institutions. He has published more than 10 papers in national and international conferences and in international journals. He is a member of the ISTE (India), Life Member IAENG. He has organized a number of guest lecturers/expert talks and seminars, workshops and symposiums. He has also guided more than 3 dozen projects (B.E/B.Tech/M.E/M.Tech/MCA). His research interests are Software Engineering, Software Quality Assurance and Web Services.

Ayşe Nalli is a professor and supervisor of doctoral and master students in the Department of Mathematics, Karabük University, Karabük, Turkey. She received her B.Sc, M. Sc and Ph. D in Mathematics from Selçuk University, Konya, Turkey. Her research interests cryptography, algebra, Diophantine equation and number theory. She has well over 30 journal in magazines with index.

Meera S. is working as a Professor and Head at Department of Computer Science and Engineering in Agni College of Technology, Chennai. She had completed her Bachelor of Engineering in Computer Sciences and Engineering from Vinayaga Mission University, Salem, and Tamil Nadu. She pursued her Masters of Engineering from Jayaram College of Engineering and Technology Affiliated to Anna University, Trichy. She achieved the Degree of Doctor of Philosophy under the Faculty of Information and Communication Engineering from Anna University, Chennai. She Had Published more than 10 Papers in Scopus Indexed,2 SCI journals and 3 Conference Papers. She had authored books titles Innovative Technology & Unstructured Data Analytics Using Machine Learning and 2 Chapters. She had filed 2 Innovation Patent. She is having 23 years of Experience in teaching. Her areas of Interest are Data Mining, Bid Data, Networks, Security Machine Learning, etc.

Girish S. C. is currently working as Assistant Professor, Department of Information Science and Engineering, NIE Institute of Technology, Mysuru, India. He's doing his research(Ph.D. Degree) in Computer Science and Engineering at Visvesvaraya Technological University, Belagavi. He has 8 years of teaching experience at both UG and PG level. His research interests include Machine Learning, AI, Image Processing, and Data Science. He has published 3 research papers in various international journals and 1 book chapter in Springer International Publishing.

Sudip Kumar Sahana, male, is currently working as Asst. Prof. in the Department of Computer Science and Engineering, B.I.T(Mesra), Ranchi, India. He received the B.E. Degree in Computer Technology from Nagpur University, India in 2001, the M.Tech. Degree in Computer Science in 2006 and Ph.D in Engineering in 2013 from the B.I.T (Mesra), Ranchi, India. His major field of study is in Computer Science. His research and teaching interests include Soft Computing, Evolutionary Algorithms, Artificial Intelligence and High Performance Computing. He is author of number of research papers in the field of Computer Science.

Devaraju Sellappan received the B.Sc degree in Chemistry in 1997 from the University of Madras, Chennai, and the M.C.A. degree in Computer Applications in 2001 from the Periyar University, Salem, and the M.Phil. degree in Computer Science in 2004 from Periyar University, Salem and also received M.B.A. degree in Human Resource from Madurai Kamaraj University, Madurai in 2007. He received Ph.D degree in Computer Applications from Anna University, Chennai in 2017. He has 19+ years of teaching experience and 2 years industry experience. He is an Senior Assistant Professor, School of Computing Science and Engineering (SCSE), VIT Bhopal University, Sehore, Madhya Pradesh, India. Dr. S. Devaraju has published 2 patents, 5 Book Chapters and Reviewer for various reputed Journals and Conferences. He has published more than 40+ papers in international journals and conference proceedings. His area of research includes Network Security, Intrusion Detection, Soft Computing, and Wireless Communication.

Dheresh Soni is a Doctorate in Computer Science & Engineering from Mewar University, Chittourgarh, and an M.E. in Computer Science & Engineering from RGPV Bhopal. He has an academic experience of 13+ years and has worked as Assistant Professor in renowned Universities/Colleges like Chandigarh University, Swami Rama Himalayan University, Sagar Group of Institute (Now known as SAGE University), & Patel Group of Institute (Now known as Madhyanchal Professional University). He has published 04 patents of which 01 German patent has been granted (03 Published, 01 Granted). He has published nine articles in peer-reviewed International/National journals with high impact factors and has 10 publications in various International Conferences held in India. Dr. Dheresh Soni has attended a number of Conferences, FDP, Seminars, Workshops, and Short term courses. He has supervised 03 M.Tech students and active member of Computer Societies like CSI, ACM, IAENG, ICST, and IACSIT. He has good skills in Cloud Computing, Security, and IoT. Many students have completed their projects in the field of web development, and IoT under his supervision.

Ramakrishnan Srinivasan received the B.E. degree in Electronics and Communication Engineering in 1998 from the Bharathidasan University, Trichy, and the M.E. degree in Communication Systems in 2000 from the Madurai Kamaraj University, Madurai. He received his PhD degree in Information and Communication Engineering from Anna University, Chennai in 2007. He has 21+ years of teaching experience and 1 year industry experience. He is a Professor and the Head of the Department of Information Technology, Dr. Mahalingam College of Engineering and Technology, Pollachi, India. Dr. Ramakrishnan is an Associate Editor for IEEE Access and he is a Reviewer of 28 International Journals including 7 IEEE Transactions, 5 Elsevier Science Journals, 3 IET Journals, ACM Computing Reviews, Springer Journals, Wiley Journals, etc. He is in the editorial board of 7 International Journals. He is a Guest Editor of special issues in 3 International Journals including Telecommunication Systems Journal of Springer. He has filed one Indian patent. He has published 181 papers in international, national journals and conference proceedings. Dr. S. Ramakrishnan has published 12 books on Image Processing, Pattern Recognition, Soft Computing, Information Security and Wireless Sensor Networks for CRC Press, USA, InTech Publisher, Croatia and Lambert Academic Publishing, Germany. He has also reviewed 3 books for McGraw Hill International Edition and 20 books for ACM Computing Reviews. He was the convenor of IT board in Anna University of Technology- Coimbatore Board of Studies(BoS). He had successfully guided 7 PhD scholars and guiding 2 scholars. His biography has been included in Marquis Whos's Who in the World 2012 & 2016 edition. His areas of research include digital image processing, information security, and soft computing.

Index

M

N

O

P

R

S

T

U

V

Z

www.ingramcontent.com/pod-product-compliance
Lightning Source LLC
LaVergne TN
LVHW080034210926
848481LV00011B/3188